Problem Books in Mathematics

Edited by P. R. Halmos

Problem Books in Mathematics

Series Editor: P.R. Halmos

Unsolved Problems in Intuitive Mathematics, Volume I: Unsolved Problems in Number Theory
by *Richard K. Guy*
1981. xviii, 161 pages. 17 illus.

Theorems and Problems in Functional Analysis
by *A.A. Kirillov* and *A.D. Gvishiani*
1982. ix, 347 pages. 6 illus.

Problems in Analysis
by *Bernard Gelbaum*
1982. vii, 228 pages. 9 illus.

A Problem Seminar
by *Donald J. Newman*
1982. viii, 113 pages.

Problem-Solving Through Problems
by *Loren C. Larson*
1983. xi, 344 pages. 104 illus.

Demography Through Problems
by *N. Keyfitz* and *J.A. Beekman*
1984. viii, 141 pages. 22 illus.

Problem Book for First Year Calculus
by *George W. Bluman*
1984. xvi. 384 pages. 384 illus.

Exercises in Integration
by *Claude George*
1984. x. 550 pages. 6 illus.

Exercises in Number Theory
by *D.P. Parent*
1984. x. 541 pages.

Problems in Geometry
by *Marcel Berger et al.*
1984. viii. 266 pages. 244 illus.

D.P. Parent

A Pseudonym for

D. Barsky F. Bertrandias G. Christol
A. Decomps H. Delange J.-M. Deshouillers
K. Gérardin J. Lagrange J.-L. Nicolas
M. Pathiaux G. Rauzy M. Waldschmidt

Exercises in Number Theory

Springer-Verlag
New York Berlin Heidelberg Tokyo

Series Editor
Paul R. Halmos
Department of Mathematics
Indiana University
Bloomington, IN 47405
U.S.A.

AMS Classification: 00A07, 12-01, 10-01

Library of Congress Cataloging in Publication Data
Parent, D. P.
Exercises in number theory.
(Problem books in mathematics)
Translation of: Exercises de théorie des nombres.
Bibliography: p.
Includes indexes.
1. Numbers, Theory of–Problems, exercises, etc.
I. Title. II. Series.
QA241.P2913 1984 512′.7 84-16056

Title of the original French edition: *Exercices de théorie des nombres,* © BORDAS, Paris, 1978.

Printed and bound by R.R. Donnelley & Sons, Harrisonburg, Virginia.
Printed in the United States of America.

9 8 7 6 5 4 3 2 1

ISBN 0-387-96063-5 Springer-Verlag New York Berlin Heidelberg Tokyo
ISBN 3-540-96063-5 Springer-Verlag Berlin Heidelberg New York Tokyo

Preface

After an eclipse of some 50 years, Number Theory, that is to say the study of the properties of the integers, has regained in France a vitality worthy of its distinguished past. More and more researchers have been attracted by problems which, though it is possible to express in simple statements, whose solutions require all their ingenuity and talent. In so doing, their work enriches the whole of mathematics with new and fertile methods.

To be in a position to tackle these problems, it is necessary to be familiar with many specific aspects of number theory. These are very different from those encountered in analysis or geometry. The necessary know-how can only be acquired by studying and solving numerous problems. Now it is very easy to formulate problems whose solutions, while sometimes obvious, more often go beyond current methods. Moreover, there is no doubt that, even more than in other disciplines, in mathematics one must have exercises available whose solutions are accessible. This is the objective realised by this work. It is the collaborative work of several successful young number theorists. They have drawn these exercises from their own work, from the work of their associated research groups as well as from published work. Without running through all the areas in number theory, which would have been excessive, the exercises given here deal with those directions that now appear most important. The solutions are rarely easy, but this book always gives a method to solve them. The appearance of this volume is gratefully welcomed. To all those who have made the effort to delve into the problems proposed here, it will solidify their attachment to the theory of numbers.

Ch. Pisot

Introduction

This book is the work of a group of mathematicians, including a large number of participants from the Seminaire Delange-Pisot-Poitou. It was written as follows: first many exercises were collected, mostly from university examinations. Then a choice was made, and they were grouped so as to constitute 10 chapters arranged in a random order. Certain important areas in number theory, notably the area of Diophantine Equations, have been left out.

Many books on number theory have appeared recently in France, among them those of Y. Amice, A. Blanchard, W.J. Ellison and M. Mendes-France, G. Rauzy, P. Samuel, J.P. Serre and M. Waldschmidt (see the bibliography). Readers of these books will find here supplementary and illustrative material. Conversely, a neophyte mathematician, who is curious about, say, p-adic anaylsis, or distributions modulo 1, after having been interested by certain attractive exercises, will have a strong motive for learning the theory. A part of our problems are accessible to a reader at the level of good second or third year students of a university. The introduction of each chapter is there to recall the definitions and principal theorems that one will need.

On behalf of the authors, I would like to thank those who have aided us, either by suggesting problems or in making up the solutions: J.P. Borel, P. Cassou-Nogues, H. Cohen, P. Damey, F. Dress, A. Durand, J. Fresnel, E. Helsmoortel, J. Martinet, B. de Mathan, M. Mendes-France, M. Mignotte, J.J. Payan, J. Queyrut, G. Revuz, G. Rhin, M.F. Vigneras. We would equally like to thank the Mathematical Society of France which helped in the publication of the original version of this book.

J.L. Nicolas

While reading the proofs of this introduction, I learned of the death of Professor Charles Pisot. All the contributors of this book dedicate this English translation to his memory.

Contents

CHAPTER 1

Prime Numbers: Arithmetic Functions: Selberg's Sieve

INTRODUCTION

1.1 PRIME NUMBERS

With x a real number and $x \geqslant 0$, we *denote* by $\pi(x)$ the *NUMBER OF PRIME NUMBERS NOT EXCEEDING* x and we set

$$\vartheta(x)= \sum_{p\leqslant x} \log p.$$

The essential result in the theory of prime numbers is the following:

THEOREM: (Prime Number Theorem): *When* $x \to +\infty$

$$\pi(x) \sim \frac{x}{\log x} .$$

This relation is equivalent to:

$$\vartheta(x) \sim x$$

(as will be seen in Exercise 1·3*).*

In fact, results quite a bit more precise than this are known (cf., for example, [E11]).

We will also give estimates for certain quantities related to the primes, notably the sum $\sum_{p\leqslant x}\frac{1}{p}$ and the product $\prod_{p\leqslant x}\left(1-\frac{1}{p}\right)$, for which we have the celebrated *MERTENS' FORMULA*:

$$\prod_{p\leqslant x}\left(1-\frac{1}{p}\right)\sim\frac{e^{-\gamma}}{\log x}\qquad(x\to+\infty),$$

where γ is Euler's constant (cf., Exercises 1·5 and 1·6).

1.2 A FREQUENTLY USEFUL LEMMA

The following Lemma is often useful, in particular for the evaluation of the sum $\sum_{p\leqslant x}\frac{1}{p}$ mentioned above.

LEMMA: *Let E be a set of real numbers such that for every real t, $E\cap(-\infty,t]$ is empty or finite.*

Let g be a real or complex function defined on E, and let

$$G(t)=\sum_{\substack{u\in E\\ u\leqslant t}}g(u)$$

If F is a C^1 function on the closed interval $[a,b]$, then

$$\sum_{\substack{u\in E\\ a<u\leqslant b}}F(u)g(u)=F(b)G(b)-F(a)G(a)-\int_a^b G(t)F'(t)dt.$$

This formula is obtained immediately after noticing that for all $u\in E$ such that $a<u\leqslant b$ we have:

$$F(b)g(u)-F(u)g(u)=\int_a^b g(u)Y(t-u)F'(t)dt,$$

where

$$Y(x)=\begin{cases}0 & \text{for } x<0,\\ 1 & \text{for } x\geqslant 0,\end{cases}$$

and adding.

In fact, this is the formula for integrating by parts the Stieltjes integral $\int_a^b F(t)\mathrm{d}G(t)$, which is equal to the sum $\sum_{\substack{u\in E\\ a<u\leqslant b}} F(u)g(u)$ (cf., [Wid]). It is the best way to remember it.

We have

$$\int_a^b F(t)\mathrm{d}G(t) = F(b)G(b) - F(a)G(a) - \int_a^b G(t)F'(t)\mathrm{d}t.$$

EXAMPLE: Let $E=\mathbb{N}^*$, $g(u)=1$, $[a,b]=[1,x]$, $F(t)=\frac{1}{t}$. We have:

$$\sum_{1<n\leqslant x} \frac{1}{n} = \int_1^x \frac{1}{t}\,\mathrm{d}[t] = \frac{[x]}{x} - 1 + \int_1^x \frac{[t]}{t^2}\,\mathrm{d}t$$

$$= \log x - \frac{x-[x]}{x} - \int_1^x \frac{t-[t]}{t^2}\,\mathrm{d}t,$$

whence

$$\sum_{n\leqslant x} \frac{1}{n} = \log x + 1 - \int_1^{+\infty} \frac{t-[t]}{t^2}\,\mathrm{d}t - \frac{x-[x]}{x} + \int_x^{+\infty} \frac{t-[t]}{t^2}\,\mathrm{d}t$$

$$= \log x + \gamma + \eta(x),$$

where

$$\gamma = 1 - \int_1^{+\infty} \frac{t-[t]}{t^2}\,\mathrm{d}t \quad \text{and} \quad |\eta(x)| \leqslant \frac{1}{x}.$$

1.3 IDEA OF AN ARITHMETIC FUNCTION: THE USUAL ARITHMETIC FUNCTIONS

Any real or complex function defined on $\mathbb{N}^*$ is called an *ARITHMETIC FUNCTION*.

Usually, we consider functions for which $f(n)$ is determined from arithmetic properties of the integer n.

Some classical examples are the following:

$d(n)$ = *NUMBER OF DIVISORS* of n;

$\sigma(n)$ = *SUM OF THE DIVISORS* of n;

$\varphi(n)$ = *number of integers* m *such that* $1 \leqslant m \leqslant n$ *and* $(m,n) = 1$ (*EULER'S* TOTIENT *FUNCTION*);

$\nu(n)$ = *NUMBER OF* PRIME *DIVISORS* of n;

$\Omega(n)$ = *TOTAL NUMBER OF FACTORS* in the decomposition of n into *prime* factors;

($\Omega(1) = 0$ and, if $n = q_1^{\alpha_1} q_2^{\alpha_2} \cdots q^{\alpha_k}$, where $q_1, q_2, \ldots, q_k$ are prime numbers and $\alpha_1, \alpha_2, \ldots, \alpha_k$ are some integers greater than zero, $\Omega(n) = \alpha_1 + \alpha_2 + \cdots + \alpha_k$).

$\lambda(n) = (-1)^{\Omega(n)}$ (*LIOUVILLE'S FUNCTION*)

Other examples are:

$$\mu(n) = \begin{cases} (-1)^{\nu(n)} & \text{if } n \text{ is divisible by the square of no prime number,} \\ 0 & \text{in the opposite case;} \end{cases}$$

the *MÖBIUS FUNCTION*, and *VON MANGOLDT'S FUNCTION* Λ defined by:

$$\Lambda(n) = \begin{cases} \log p \text{ if } n = p^k \text{ with } p \text{ a prime number and } k \geqslant 1, \\ 0 \quad \text{if } n \text{ is not of that form.} \end{cases}$$

We shall *denote* by z the *function which is identically one on the integers*, and *denote* by i the *identity mapping of* $\mathbb{N}^*$ *onto* $\mathbb{N}^*$.

1.4 ADDITIVE FUNCTIONS AND MULTIPLICATIVE FUNCTIONS

An arithmetic function is called an *ADDITIVE FUNCTION* if:

$$f(mn) = f(m) + f(n) \quad \textit{whenever } (m,n) = 1. \tag{1}$$

This clearly implies that $f(1) = 0$.

Furthermore, if $n_1, n_2, \ldots, n_k$ *are pairwise relatively prime*, then:

$$f(n_1 n_2 \cdots n_k) = f(n_1) + f(n_2) + \cdots + f(n_k).$$

The functions ν *and* Ω *are additive.* The function log clearly *is as well.*

A function f is called a *MULTIPLICATIVE FUNCTION* if

$$f(1) = 1$$

and

$$f(mn) = f(m)f(n) \quad \textit{whenever } (m,n) = 1. \tag{2}$$

(The condition $f(1) = 1$, ignored by certain authors, serves only to avoid considering as multiplicative the function that is identically zero).

If f is multiplicative and if $n_1, n_2, \ldots, n_k$ are pairwise relatively prime, then:

$$f(n_1 n_2 \cdots n_k) = f(n_1) f(n_2) \cdots f(n_k).$$

The functions $d, \sigma, \varphi, \lambda, \mu$ *are multiplicative.* It is obvious that z *and* i *are as well.*

PROPOSITION: *An additive or multiplicative function is completely determined by its values on the powers of the primes.*

In fact, if $n = q_1^{\alpha_1} q_2^{\alpha_2} \cdots q_k^{\alpha_k}$, where $q_1, q_2, \ldots, q_k$ are distinct prime numbers and $\alpha_1, \alpha_2, \ldots, \alpha_k$ are integers greater than zero, then:

$$f(n) = f(q_1^{\alpha_1}) + f(q_2^{\alpha_2}) + \cdots + f(q_k^{\alpha_k}) \quad \text{if } f \text{ is additive,}$$

$$f(n) = f(q_1^{\alpha_1}) f(q_2^{\alpha_2}) \cdots f(q_k^{\alpha_k}) \quad \text{if } f \text{ is multiplicative,}$$

which can be written more concisely as:

$$f(n) = \sum_{p^r \| n} f(p^r) \quad \text{or} \quad f(n) = \prod_{p^r \| n} f(p^r).$$

We see, moreover, that *there exists exactly one additive function and one multiplicative function taking given values on the powers of the prime numbers.*

A function f is a *COMPLETELY ADDITIVE FUNCTION* if:

$$f(mn) = f(m) + f(n),$$

for arbitrary m and n, and not only when $(m,n) = 1$

Thus the function Ω is *completely additive.*

A function f is a *COMPLETELY MULTIPLICATIVE FUNCTION* if:

$$f(1) = 1 \quad \text{and} \quad f(mn) = f(m)f(n),$$

for all m and n.

A completely additive or completely multiplicative function is determined by its values on the prime numbers, for,

$$f(p^r) = rf(p) \quad \text{or} \quad f(p^r) = (f(p))^r,$$

respectively.

1.5 ALGEBRAIC STRUCTURE ON THE SET OF ARITHMETIC FUNCTIONS

The set A of arithmetic functions is made into an algebra over $\mathbb{C}$ by defining addition and multiplication by scalars in the usual way, and by taking as the product of two elements the *CONVOLUTION* defined as follows:

$f*g$ is the arithmetic function h given by the formula:

$$h(n) = \sum_{d/n} f(d)g\left(\frac{n}{d}\right) \qquad \left[\text{or } h(n) = \sum_{dd'=n} f(d)g(d')\right].$$

This *convolution is commutative, associative, and distributive with respect to addition.* It possesses an *IDENTITY*, the function *e defined by*:

$$e(1) = 1 \quad \text{and} \quad e(n) = 0 \text{ for } n > 1.$$

Moreover, for all $f,g \in A$ and $\lambda \in \mathbb{C}$,

$$(\lambda f)*g = f*(\lambda g) = \lambda(f*g).$$

A *is* therefore *a unitary commutative algebra.*

PROPOSITION: *The group* G *of invertible elements is the set of arithmetic functions* f *such that* $f(1) \neq 0$.

If $f \in G$ the *inverse* of f will be *denoted* by f^{-*}.

PROPOSITION: *The Möbius function* μ *is the inverse of the function* z.

The relation $\mu*z = e$, which says that $\mu = z^{-*}$, translates as:

$$\sum_{d/n} \mu(d) = \begin{cases} 1 & \text{for } n = 1, \\ 0 & \text{for } n > 1. \end{cases}$$

PROPOSITION: *The von Mangoldt function* Λ *is characterised by:*

$$\Lambda * z = \log .$$

PROPOSITION: *The set* M *of multiplicative functions is a sub-group of* G.

1.6 THE FIRST MÖBIUS INVERSION FORMULA

With $a \in G$, it is clear that if $f, g \in A$,

$$g = a * g \iff g = a^{-*} * g.$$

In particular, by taking $a = z$ we see that *for* $f, g \in A$ *we have the equivalence*:

$$g = z * f \iff f = \mu * g,$$

in other words (as $z * f = f * z$),

$$g(n) = \sum_{d/n} f(d) \quad \text{for all } n \in \mathbb{N}^*$$

$$\iff f(n) = \sum_{d/n} \mu(d) g\left(\frac{n}{d}\right) \quad \text{for all } n \in \mathbb{N}^*.$$

This is usually called the *FIRST MÖBIUS INVERSION FORMULA*.

1.7 ARITHMETIC FUNCTIONS AS OPERATORS ON A VECTOR SPACE

Let X be the vector space of complex functions defined on the interval $[1, +\infty)$.

Given $a \in A$ and $F \in X$ we shall *denote* by $a \top F$ the *function defined on* $[1, +\infty)$ *by*:

$$a \top F \equiv G(x) = \sum_{n \leqslant x} a(n) F\left(\frac{x}{n}\right).$$

Thus *with* a *fixed, the mapping* $F \to a \top F$ *is a linear operator on* X.

It has the following properties:

For all $F \in X$: $e_\top F = F$; (α)

For all $a,b \in A$, for all $F \in X$: $\begin{cases} (a + b)_\top F = a_\top F + b_\top F, \\ \text{and} \\ a_\top(b_\top F) = (a * b)_\top F; \end{cases}$ (β)

For all $a \in A$, for all $\lambda \in \mathbb{C}$, for all $F \in X$:

$$(\lambda a)_\top F = \lambda(a_\top F).$$

1.8 THE SECOND MÖBIUS INVERSION FORMULA

With $a \in A$, if $F,G \in X$, it is clear that by Formulae (α) and (β) we have:

$$G = a_\top F \iff F = a^{-*}{}_\top G.$$

In particular, upon taking $a = z$ we see that *for* $F,G \in X$

$$G = z_\top F \iff F = \mu_\top G;$$

in other words:

$$G(x) = \sum_{n \leqslant x} F\left(\frac{x}{n}\right) \quad \text{for all } x \geqslant 1$$

$$\iff F(x) = \sum_{n \leqslant x} \mu(n) G\left(\frac{x}{n}\right) \quad \text{for all } x \geqslant 1.$$

This is usually called the *SECOND MÖBIUS INVERSION FORMULA*

1.9 SUMMATION FUNCTION OF AN ARITHMETIC FUNCTION

The *SUMMATION FUNCTION of an arithmetic function* is the function A defined on the interval $[1,+\infty)$ by:

$$A(x) = \sum_{n \leqslant x} a(n).$$

We observe that A is nothing but the function $a_\top F$ corresponding to $F(x) = 1$ for all $x \geqslant 1$. Property (β) then gives the following result:

PROPOSITION: *Let $a, b \in A$ and $c = a * b$. Let A, B, C be the summation functions of a, b, c. We have:*

$$C = a_\top B = b_\top A.$$

In other words, for all $x \geqslant 1$:

$$C(x) = \sum_{n \leqslant x} a(n) B\left(\frac{x}{n}\right) = \sum_{n \leqslant x} b(n) A\left(\frac{x}{n}\right). \tag{2}$$

EXAMPLE 1: Let $a = \mu$, $b = z$. Then $c = e$, and for all $x \geqslant 1$:

$$B(x) = [x] \quad \text{and} \quad C(x) = 1.$$

Therefore,

$$\sum_{n \leqslant x} \mu(n) \left[\frac{x}{n}\right] = 1 \text{ for all } x \geqslant 1.$$

The same result would be obtained by applying the Second Möbius Inversion Formula with $F(x) = 1$ for all $x \geqslant 1$.

EXAMPLE 2: It is customary to denote by ψ the summation function of von Mangoldt's function Λ. By taking $a = \Lambda$, $b = z$, one has $c = \log$, and relation (2) gives:

$$\sum_{n \leqslant x} \log n = \sum_{n \leqslant x} \Lambda(n) \left[\frac{x}{n}\right] = \sum_{n \leqslant x} \psi\left(\frac{x}{n}\right).$$

EULER PRODUCTS

The following result about multiplicative functions is fundamental.

THEOREM: *Let g be a multiplicative function. If:*

$$\sum_{p,r} |g(p^r)| < +\infty \tag{3}$$

where p runs over the set of prime numbers and r runs over $\mathbb{N}^*$, *then the series:*

$$\sum_{n=1}^{\infty} g(n)$$

is absolutely convergent, and we have:

$$\sum_{n=1}^{\infty} g(n) = \prod \left(1 + \sum_{r=1} g(p^r)\right), \tag{4}$$

where the infinite product, in which p runs over the sequence of prime numbers, is absolutely convergent.

(It is clear that each of its factors is well defined, for the series $\sum_{r=1}^{\infty} g(p^r)$ is absolutely convergent).

Notice that if the series $\sum_{n=1}^{\infty} g(n)$ is absolutely convergent, relation (3) obviously holds (so that the absolute convergence of this series is equivalent to (3)). Hence *if the series* $\sum_{n=1}^{\infty} g(n)$ *is absolutely convergent, relation* (4) *holds.*

1.11 GENERATING FUNCTION OF AN ARITHMETIC FUNCTION

To each arithmetic function a we associate the *DIRICHLET SERIES*

$$\sum_{n=1}^{\infty} \frac{a(n)}{n^s} .$$

It is known that there exists $\sigma_c \in \bar{\mathbb{R}}$ such that the series converges if $\mathrm{Re}\, s \geqslant \sigma_c$ and diverges if $\mathrm{Re}\, s < \sigma_c$. If $\sigma_c = +\infty$ the series diverges for all $s \in \mathbb{C}$. But if $\sigma_c = +\infty$ the series defines a holomorphic function of s for $\mathrm{Re}\, s > \sigma_c$. The analytic function thus defined is called the GENERATING FUNCTION *of the arithmetic function a*.

Thus the *RIEMANN ζ-FUNCTION is*, by definition the *generating function of the function z*: for $\mathrm{Re}\, s > 1$ we have:

$$\zeta(s) = \sum_{1}^{\infty} \frac{1}{n^s} .$$

PROPOSITION: *If the arithmetic functions a and b have generating functions, then the function a*b has as generating function the product of the one for a and the one for b.*

In fact, for every value of s for which the Dirichlet series associated with a and b are absolutely convergent, the series associated with $a * b$ is also absolutely convergent and has as sum the product of the sums of the preceding series.

Thus we see, for example, that for $\mathrm{Re}\, s > 1$ we have:

$$\zeta(s) \sum_{n=1}^{\infty} \frac{\mu(n)}{n^s} = 1,$$

which shows that $\zeta(s) \neq 0$ for $\mathrm{Re}\, s > 1$, and that the function μ *has* $1/\zeta$ *as generating function.*

It is also seen that *von Mangoldt's function* Λ *has* $-\zeta'/\zeta$ *as generating function*: for $\mathrm{Re}\, s > 1$ we have:

$$\sum_{n=1}^{\infty} \frac{\Lambda(n)}{n^s} = - \frac{\zeta'(s)}{\zeta(s)}$$

(since $\Lambda * z = \log$). This formula is fundamental in the theory of the distribution of primes.

The results of Section 1.10 give the following:

PROPOSITION: *Let f be a multiplicative arithmetic function. If, for some real* σ,

$$\sum_{p,r} \frac{|f(p^r)|}{p^{r\sigma}} < \infty,$$

where, in the sum, p runs over the set of prime numbers and r runs over $\mathbb{N}^*$, *then the Dirichlet series associated with f,*

$$\sum_{n=1}^{\infty} \frac{f(n)}{n^s},$$

is absolutely convergent for $\mathrm{Re}\, s \geqslant \sigma$, *and for these values of s we have:*

$$\sum_{n=1}^{\infty} \frac{f(n)}{n^s} = \prod_p \left(1 + \sum_{r=1}^{\infty} \frac{f(p^r)}{p^{rs}}\right), \tag{5}$$

and the infinite product, in which p runs over the sequence of prime numbers, is absolutely convergent.

In fact, we have the relation (5) *for every value of s for which the Dirichlet series* $\sum_{n=1} \frac{f(n)}{n^s}$ *is absolutely convergent.*

As an example, by taking $f = z$, *EULER'S FORMULA* is obtained:

$$\text{For } \mathrm{Re}\, s > 1, \quad \zeta(s) = \prod_p \left(1 - \frac{1}{p^s}\right)^{-1}.$$

1.12 IKEHARA'S THEOREM (cf., [E11])

THEOREM: (Ikehara): *Let* $\sum_{n=1}^{+\infty} \frac{a_n}{n^s}$ *be a Dirichlet series for which the coefficients* a_n *are all real* $\geqslant 0$. *Assume that this series is convergent for* $\mathrm{Re}\, s > 1$, *with F(s) as sum, and that there exists* $A > 0$ *such that, for all real t,* $F(s) - \frac{A}{s-1}$ *tends towards a finite limit as* $s \to 1 + it$ *while remaining in the half-plane* $\mathrm{Re}\,s > 1$.

Then when $x \to +\infty$ *we have:*

$$\sum_{n \leqslant x} a_n \sim Ax.$$

It should be noted that the assumptions hold, in particular, when F is holomorphic at every point on the line $\mathrm{Re}\,s = 1$, apart from the point 1, and this point is a simple pole with residue A.

1.13 SELBERG'S SIEVE

Here we shall be content with giving an inequality, due to Selberg, a proof of which the reader will find in Chapter IV of the book by Halberstam and Roth [Ha]; in the Exercises one will find various examples giving applications of this inequality.

Let $\mathrm{A} = \{a_1, \ldots, a_n\}$ be a family of positive integers and P a finite family of prime numbers; (we shall *denote* by $\Pi(P)$ the *product of the elements of* P). For each prime number $p \in P$ one chooses $\omega(p)$ distinct classes $(R_1^{(p)}, \ldots, R_{\omega(p)}^{(p)})$ modulo p. ω is made into a completely multiplicative function by taking:

$$\omega(p) = 0 \text{ if } p \notin P.$$

Then let d be a divisor of $\Pi(P)$; we *denote* by A_d the set of elements of A that are in a class $Q_i^{(p)}$ for at least one p dividing d and one i, $1 \leqslant i \leqslant \omega(p)$. We make the assumption:

$$\left|\mathrm{card}\mathrm{A}_d - \frac{n\omega(d)}{d}\right| \leqslant \omega(d). \tag{H}$$

We *denote* by $S(\mathrm{A}, P)$ the *number of elements of* A *which are not in any of the classes* $R_i^{(p)}$; then for every positive real number z, we have:

$$S(A,P) \leqslant n\left(\sum_{m\leqslant z} \frac{\omega(m)}{m}\right)^{-1} + z^2 \prod_{p\in P}\left(1 - \frac{\omega(p)}{p}\right)^{1-2}$$

In Exercise 1.27 a lower bound will be given for the sum $\sum\limits_{m\leqslant z} \frac{\omega(m)}{m}$ in certain special cases.

BIBLIOGRAPHY

[Har] Chapters I,II,XVI,XVII,XVIII,XIX,XXII; [Bla]; [Ell]; [Hal].

APPENDIX

This Appendix contains a Table of the usual arithmetic functions, giving their definition, their fundamental properties, and the generating function whenever it is simple, and in this case the abscissa of convergence of the associated Dirichlet series.

θ is the upper bound of the real parts of the zeros of the ζ-function. It is known only that $\frac{1}{2} \leqslant \theta \leqslant 1$. The Riemann hypothesis is that $\theta = \frac{1}{2}$.

Definition	Fundamental Properties	Generating Function	Abscissa of Convergence of the Dirichlet Series
$d(n) =$ number of divisors of n	multiplicative $d = z*z$	$\zeta(s)^2$	1
$\varphi(n) =$ sum of divisors of n	multiplicative $\sigma = i*z$	$\zeta(s)\zeta(s-1)$	2
$\varphi(n) =$ number of m's such that $1 \leqslant m \leqslant n$ and $(m,n) = 1$	multiplicative $\varphi = \mu * i$ $\varphi(n) = n \prod_{p/n} \left(1 - \frac{1}{p}\right)$	$\frac{\zeta(s-1)}{\zeta(s)}$	2
$\nu(n) =$ number of *prime* divisors of n	additive		
$\Omega(n) =$ total number of factors in the factorisation of n	completely additive		
$\mu(n) = \begin{cases} (-1)^{\nu(n)} & \text{if } n \text{ has no square factor} \\ 0 & \text{otherwise} \end{cases}$	multiplicative $\mu = z^{-*}$ $\mu * z = e$	$\frac{1}{\zeta(s)}$	θ
$\lambda(n) = (-1)^{\Omega(n)}$	completely multiplicative	$\frac{\zeta(2s)}{\zeta(s)}$	θ
$z(n) = 1$ for all n	completely multiplicative	$\zeta(s)$	1
$i(n) = n$ for all n	completely multiplicative	$\zeta(s-1)$	2
$e(n) = \begin{cases} 1 & \text{for } n = 1 \\ 0 & \text{for } n > 1 \end{cases}$	multiplicative; Unit element of the algebra A	1	$-\infty$
$\Lambda(n) = \begin{cases} \log p & \text{if } n = p \\ 0 & \text{otherwise} \end{cases}$	neither additive nor multiplicative $\Lambda * z = \log$	$-\frac{\zeta'(s)}{\zeta(s)}$	1

PROBLEMS

EXERCISE 1·1: Set $F_n = 2^{2^n} + 1$. Show that F_n divides $F_m - 2$ if n is less than m, and from this deduce that F_n and F_m are relatively prime if $m \neq n$.

From the latter statement deduce a proof of the existence of an infinite number of primes.

EXERCISE 1·2: Let β be a positive real number less than one; show that if the integer N is large enough there exists at least one prime number between βN and N.

EXERCISE 1·3: Show the equivalence of the Prime Number Theorem (Section 1.1) with each of the following assertions:

(i): $\vartheta(x) \sim x$;

(ii): $p_n \sim n\log n$ (p_n being the n-th prime number).

EXERCISE 1·4: Show that the series $\sum_{n=3}^{\infty} p_n^{-1}(\log\log p_n)^{-\alpha}$ converges if and only if the real number α is greater than 1. (p_n is the n-th prime number).

EXERCISE 1·5: SOME SUMS $\sum_{p\leqslant x} f(p)$

I:(1): Given a prime number p and a real number $x \geqslant p$, determine the exponent α_p of p in the decomposition into prime factors of the product $\prod_{n\leqslant x} n$.

Show that:

$$\max\left(\frac{x}{p} - 1, \frac{x}{2p}\right) \leqslant \alpha_p \leqslant \frac{x}{p-1}.$$

I:(2): Show that for all $x > 1$ we have:

$$\sum_{p\leqslant x} \frac{\log p}{p} \leqslant 2\log x,$$

and, for $1 < y \leqslant x$:

$$\sum_{y<p\leqslant x} \frac{1}{p} \leqslant 2\,\frac{\log x}{\log y}\,.$$

(Note that $\sum_{n\leqslant x} \log n \leqslant x\log x$, as well as that for each $p \leqslant x$ $\alpha_p \geqslant \frac{x}{2p}$).

I:(3): By now noticing that:

$$\sum_{n\leqslant x} \log n \geqslant \int_0^{[x]} \log t\,dt \geqslant \int_0^x \log t\,dt - \log x,$$

show that for all $x > 1$:

$$\sum_{p\leqslant x} \frac{\log p}{p-1} \geqslant \log x - 1 - \frac{\log x}{x},$$

and deduce from this the existence of a constant $C \geqslant 0$ such that:

$$\sum_{p\leqslant x} \frac{\log p}{p} \geqslant \log x - C.$$

II:(1): Let $\vartheta(x) = \sum_{p\leqslant x} \log p$. Starting from the fact that $\binom{2m}{m} \leqslant 2^{2m}$, show that for each $m \in \mathbb{N}^*$:

$$\vartheta(2m) - \vartheta(m) < 2m\log 2.$$

From this deduce that for every $x > 1$:

$$\vartheta(x) > 4x\log 2.$$

II:(2): By using one of the inequalities of I(1), show that for every $x > 1$:

$$\sum_{p\leqslant x} \frac{\log p}{p} < \log x + 4\log 2.$$

III: Set $\sum_{p\leqslant x} \frac{\log p}{p} = \Phi(x) = \log x + \eta(x)$, so that by, I(3) and and II(2), $\eta(x)$ is bounded for $x > 1$.

Using the Lemma of Section 1.2 of the Introduction to this Chapter, show that for $x > 2$ we have:

$$\sum_{p\leqslant x} \frac{1}{p} = \frac{\Phi(x)}{\log x} + \int_2^X \frac{\Phi(t)}{t(\log t)^2}\, dt.$$

Deduce from this that there exists a real constant b such that, when $x \to +\infty$:

$$\sum_{p\leqslant x} \frac{1}{p} = \log\log x + b + O\left(\frac{1}{\log x}\right).$$

EXERCISE 1·6: MERTENS' FORMULA

(1): Show that the series:

$$\sum \left(\log \frac{1}{1 - \frac{1}{p^x}} - \frac{1}{p^x}\right)$$

is convergent for all real $x > \frac{1}{2}$, and that if the sum is $g(x)$ then g is continuous on the interval $[\frac{1}{2},+\infty)$.

(2): For $x \geqslant 0$, set:

$$P(x) = \sum_{p \leqslant x} \frac{1}{p} \quad \text{and} \quad N(x) = \sum_{n \leqslant x} \frac{1}{n}.$$

Show that for s real > 0 we have:

$$\log\zeta(1 + s) - g(1 + s) = s\int_0^{+\infty} e^{-st}P(e^t)dt,$$

and

$$\log \frac{1}{1 - e^{-s}} = s\int_0^{+\infty} e^{-st}N(t)dt.$$

(3): By Exercise 1·5 we know that there exists a constant b such that, when $x \to +\infty$:

$$P(x) = \log\log x + b + O\left(\frac{1}{\log x}\right).$$

With γ being Euler's constant, show that when $s \to 0$ through real values greater than zero:

$$s\int_0^{+\infty} e^{-st}(P(e^t) - N(t) + \gamma - b)dt \to 0.$$

From this deduce that $b = \gamma - g(1)$.

(4): From the preceding deduce a proof of Mertens' Formula:

As $x \to +\infty$ we have: $$\prod_{p \leqslant x} \left(1 - \frac{1}{p}\right) \sim \frac{e^{-\gamma}}{\log x}.$$

(Notice that:

$$P(x) + \log\prod_{p \leqslant x} \left(1 - \frac{1}{p}\right) \to -g(1)).$$

EXERCISE 1·7: INTEGERS n SUCH THAT ALL THE PRIME DIVISORS OF n ARE LESS THAN OR EQUAL TO $\sqrt{n}$

Let A be the set of those integers $n > 1$ all of whose prime divis satisfy $p < \sqrt{n}$. For every prime number p set:

$$B_p = \{p, 2p, \ldots, (p-1)p\}.$$

(1): Show that $A = \complement_{\mathbb{N}^*}(\bigsqcup_p B_p)$. Verify that for distinct prime numbers p and q, $B_p \cap B_q = \emptyset$.

(2): Let $Q_A(x) = \mathrm{card}\{n \leqslant x, n \in A\}$. Show that:

$$Q_A(x) = [x] - \sum_{p \leqslant \sqrt{x}} (p-1) - \sum_{x \geqslant p > \sqrt{x}} \left[\frac{x}{p}\right]$$

(3): By using the formula:

$$\sum_{p \leqslant x} \frac{1}{p} = \log\log x + b + O\left(\frac{1}{\log x}\right)$$

find a quantity equivalent to $Q_A(x)$ when $x \to \infty$.

EXERCISE 1·8: A THEOREM OF HARDY AND RAMANUJAN

In this Exercise we will need the results of Exercises 1·5:I(2) and 1·5:III, in particular the formula:

$$\sum_{p \leqslant x} \frac{1}{p} = \log\log x + b + O\left(\frac{1}{\log x}\right),$$

established in Exercise 1 5:III, as well as the fact that $\pi(x) = O\left(\frac{x}{\log x}\right)$

(1): By starting from the property that for each $n \in \mathbb{N}^*$:

$$\nu(n) = \sum_{p/n} 1,$$

show that as $x \to +\infty$ we have:

$$\sum_{n\leqslant x} \nu(n) = x\log\log x + bx + O\left(\frac{x}{\log x}\right) .$$

(2): For $x \geqslant 4$ set $P(x) = \sum_{p\leqslant x} \frac{1}{p}$. Show that for $x \geqslant 4$:

$$\sum_{p_1p_2\leqslant x} \frac{1}{p_1p_2} = 2 \sum_{p\leqslant\sqrt{x}} \frac{1}{p} P\left(\frac{x}{p}\right) - P(\sqrt{x})^2.$$

(3): From the preceding formula deduce that as $x \to +\infty$:

$$\sum_{p_1p_2\leqslant x} \frac{1}{p_1p_2} = (\log\log x)^2 + 2b\log\log x + O(1).$$

(Set $P(x) = \log\log x + b + \eta(x)$ so that there exists $M > 0$ such that $|\eta(x)| \leqslant \frac{M}{\log x}$ for $x \geqslant 2$.

Notice that for $p < x$:

$$\log\log \frac{x}{p} = \log\log x + \log\left(1 - \frac{\log p}{\log x}\right)$$

and that there exists $K > 0$ such that $|\log(1 - u)| \leqslant Ku$ for $0 \leqslant u \leqslant \frac{1}{2}$).

Starting from the fact that for each $n \in \mathbb{N}^*$:

$$\nu(n)^2 = \sum_{\substack{p_1/n \\ p_2/n}} 1,$$

show that when $x \to +\infty$ we have:

$$\sum_{n\leqslant x} \nu(n)^2 = x(\log\log x)^2 + (2b + 1)x\log\log x + O(x).$$

(5): Show that when $x \to +\infty$:

$$\sum_{n \leqslant x} (\nu(n) - \log\log x)^2 = x\log\log x + O(x).$$

(6): Given $\lambda > 0$, we denote by $n_\lambda(x)$, for $x > e$ (so that $\log\log x > 0$), the number of $n \leqslant x$ such that:

$$|\nu(n) - \log\log x| \geqslant \lambda\sqrt{\log\log x}.$$

Show that $\overline{\lim\limits_{x\to+\infty}} \frac{1}{x} n_\lambda(x) \leqslant \frac{1}{\lambda^2}$.

(7): Let g be a positive increasing function on $[a,+\infty)$, where $a > e$, and such that $g(x) \to +\infty$ when $x \to +\infty$.

Denote by $N(x)$, for $x > a$, the number of n's satisfying $a \leqslant n \leqslant x$, such that:

$$|\nu(n) - \log\log n| \geqslant g(n)\sqrt{\log\log n}.$$

(a): Show that if $x > a^{2}$, then for every n satisfying $\sqrt{x} < n < x$ and such that $|\nu(n) - \log\log n| \geqslant g(n)\sqrt{\log\log n}$, we have:

$$|\nu(n) - \log\log x| \geqslant g(\sqrt{x})\sqrt{\log\log x - \log 2} - \log 2.$$

(b): Show that for any $\lambda > 0$, for x large enough:

$$N(x) \leqslant \sqrt{x} + n_\lambda(x).$$

From this deduce that $\frac{N(x)}{x} \to 0$ when $x \to +\infty$.

EXERCISE 1·9: DIFFERENCE BETWEEN TWO CONSECUTIVE PRIME NUMBERS

Set $d_n = p_{n+1} - p_n$ ($n = 1,2,\ldots$), where p_n denotes the n-th prime number.

(a): Show: $$\sum_{2 \leqslant n \leqslant x} \frac{d_n}{\log n} \sim x.$$

(b): From this deduce the inequalities:

$$\liminf_{n\to\infty}\left(\frac{d_n}{\log n}\right) \leqslant 1 \leqslant \limsup_{n\to\infty}\left(\frac{d_n}{\log n}\right) .$$

(In Exercises 1·10 and 1·31 it will be shown that these two equalities are, in fact, strict).

EXERCISE 1·10: Denote by p_n the n-th prime number, and by Π_n the product of the prime numbers less than or equal to p_n.

(a): Let n be an integer strictly greater than 2; show that there exists a strictly positive integer $\ell \leqslant p_n p_{n-1}$ such that all the integers in the interval $[\ell\Pi_{n-2} - p_{n-1} + 1, \ell\Pi_{n-2} + p_{n-1} - 1]$ are divisible by at least one of the prime numbers $p_1,\ldots,p_n$.

(b): From the preceding deduce that:

$$\limsup_{n\to\infty}\left(\frac{d_n}{\log n}\right) \geqslant 2 \quad \text{(with the notations of Exercise 1·9).}$$

EXERCISE 1·11: DIFFERENCE BETWEEN TWO CONSECUTIVE SQUAREFREE INTEGERS

Let Q be the set of integers without square factors. The elements of Q are arranged in an increasing sequence:

$$q_1 = 2,\quad q_2 = 3,\quad q_3 = 5,\quad q_4 = 6,\quad q_5 = 7,\quad q_6 = 10,\ \text{etc..}$$

Let S be the complement of Q in $\mathbb{N}$; the elements of S are arranged into an increasing sequence:

$$s_1 = 0,\quad s_2 = 1,\quad s_3 = 4,\quad s_4 = 8,\ \text{etc..}$$

Recall that (cf., [Har Ch. XVIII) we have:

$$Q(x) = \text{card}\{n \in Q, n \leqslant x\} = \frac{6x}{\pi^2} + O(\sqrt{x}).$$

(1): Show that when $i \to \infty$ we have:

$$q_i = \frac{\pi^2}{6} i + O(\sqrt{i}),$$

and

$$q_{i+1} - q_i = O(\sqrt{i}).$$

(2): Let $n_1, n_2, \ldots, n_k$ be pairwise relatively prime. Show that there exists a number N satisfying:

$$N + j \equiv 0 \pmod{n_j^2} \quad \text{for } 1 \leqslant j \leqslant k.$$

Show that:

$$\overline{\lim}(q_{i+1} - q_i) = +\infty.$$

Find q_i and q_{i+1} such that $q_{i+1} - q_i \geqslant 5$.

(3): Show that:

$$\overline{\lim}\left(\frac{(q_{i+1} - q_i)\log\log q_i}{\log q_i}\right) \geqslant \frac{1}{2}\ .$$

(4): Show that:

$$\underline{\lim}(s_{i+1} - s_i) = 1 \text{ and } \overline{\lim}(s_{i+1} - s_i) = 4.$$

EXERCISE 1·12: Let f be a multiplicative function. Verify that the function g, defined by:

$$g(n) = \begin{cases} 1 & \text{if } f(n) \neq 0, \\ 0 & \text{if } f(n) = 0, \end{cases}$$

is also multiplicative.

EXERCISE 1·13: BELL'S SERIES FOR AN ARITHMETIC FUNCTION

Verify that if $h = f*g$, then for each p the power series $\sum_{r=0}^{+\infty} h(p^r)z^r$ is the formal product of the series:

$$\sum_{r=0}^{+\infty} f(p^r)z^r \quad \text{and} \quad \sum_{r=0}^{+\infty} g(p^r)z^r.$$

If $f(1) \neq 0$, what can be said about the power series $\sum_{r=0}^{+\infty} f^{-*}(p^r)z^r$?

APPLICATION 1: Determine the functions $\lambda * z$ and λ^{-*}.

APPLICATION 2: With f the multiplicative function defined by:

$$f(p^r) = \frac{1.3.5...(2r - 1)}{2.4.6...2r},$$

determine the function $f*f$.

EXERCISE 1·14: With h a completely multiplicative arithmetic function, show that for any arithmetic functions f, g:

$$(fh)*(gh) = (f*g)h,$$

and that for any f such that $f(1) \neq 0$ we have:

$$(fh)^{-*} = f^{-*}h.$$

What does this latter relation yield for $f = z$?

EXERCISE 1·15: Let f be a multiplicative function.

Show that for all square-free n:

$$\sum_{d/n} f(d) = \prod_{p/n} (1 + f(p)),$$

and:

$$\sum_{d/n} \mu(d) f \frac{n}{d} = \prod_{p/n} (f(p) - 1).$$

Show that if $f(p^r) = 0$ when $r > 1$ the first of the above inequalities holds for all $n \in \mathbb{N}^*$.

EXERCISE 1·16: PERIODIC COMPLETELY MULTIPLICATIVE FUNCTIONS

Let f be a periodic completely multiplicative function. Show that f has a period k such that:

$$f(n) = 0 \text{ if } (n,k) > 1 \quad \text{and} \quad f(n) \neq 0 \text{ if } (n,k) = 1$$

(so that f is a Dirichlet character (or a modular character), cf., [Bor]).

(For this, with h a period of f, set:

$$k = \prod_{\substack{p/h \\ f(p)=0}} p^{v_p(h)}, \qquad q = \prod_{\substack{p/h \\ f(p)\neq 0}} p^{v_p(h)},$$

where $v_p(h)$ is the exponent of p in the decomposition of h into prime factors. One would then show that k is still a period of f by considering the product $f(q)f(n + k)$).

EXERCISE 1·17:I: With ℓ a completely additive arithmetic function, show that, for all arithmetic functions f and g:

$$(f*g)\ell = (f\ell)*g + f*(g\ell).$$

II: To each arithmetic function $f \in G$, where G is the group of arithmetic functions invertible with respect to convolution, we associate the arithmetic function:

$$\Lambda_f = (f\log)*f^{-*}.$$

($f\log$ is the product of the functions f and log, that is to say the function $n \mapsto f(n)\log n$).

II:(1): Show that if $h = f*g$ one has $\Lambda_h = \Lambda_f + \Lambda_g$. (Apply Question I above, with $\ell = \log$). This says that the mapping $f \to \Lambda_f$ is a homomorphism of the group G into the additive group of arithmetic functions.

What are the kernel and image of this homomorphism?

What is the function Λ_z?

II:(2): Show that if f is multiplicative then:

$$\Lambda_f(n) = 0 \text{ if } n \text{ is not a power of a prime number.} \qquad (*)$$

(Notice that if $n > 1$ and if n is not a power of a prime number n can be written in the form n_1n_2, with $n_1 > 1$, $n_2 > 1$ and $(n_1,n_2) = 1$, and that all the divisors of n are obtained, each one only once, by forming all the products d_1d_2, where d_1/n and d_2/n).

Show that, if $f(1) = 1$ and if relation (*) holds, then f is multiplicative.

(Notice that if g is the multiplicative function determined by $g(p^r) = f(p^r)$, then $\Lambda_g = \Lambda_f$, and, by using of Question II(1) above, deduce that $g = f$).

EXERCISE 1·18: SQUARE-FREE NUMBERS THAT ARE MULTIPLES OF A GIVEN NUMBER

I: Let E be a finite set of prime numbers that has k elements. Set $P = \prod_{p \in E} p$, and denote by $\mathcal{E}$ the set of integers of the form mq^2, where m,P and $(q,P) = 1$. $\mathcal{E}$ can be empty; in which case $k = 0$ and $P = 1$.

I:(a): Show that for every $x > 0$ the number of elements of

$E < x$ does not exceed $\sqrt{x} \prod_{p \in E} \left(1 + \frac{1}{\sqrt{p}}\right)$ and that:

$$\sum_{\substack{n \in E \\ n > x}} \frac{1}{n} \leq \frac{2^k}{x} + \frac{1}{\sqrt{x}} \prod_{p \in E} \left(1 + \frac{1}{\sqrt{p}}\right) .$$

(Use the fact that, for all $u > 0$,

$$\sum_{n > u} \frac{1}{n^2} \leq \frac{1}{u^2} + \int_u^\infty \frac{dt}{t^2} = \frac{1}{u^2} + \frac{1}{u} ,$$

an inequality that is obtained by majorising the first term of the sum by $1/u^2$ and each of the others by $\int_{n-1}^{n} \frac{dt}{t^2}$).

I:(b): Let χ be the characteristic function of the set of those $n \in \mathbb{N}^*$ which are square-free and prime to P.
to P.

Determine the function $h = \chi * \mu$. (Notice that χ is multiplicative).

Show that the series $\sum_{n=1}^{\infty} \frac{h(n)}{n}$ is absolutely convergent, and determine its sum.

For which values of n is $h(n) \neq 0$? What then is the absolute value of $h(n)$?

I:(c): Deduce from the preceding that for all $x > 0$ we have:

$$\left| \sum_{n \leq x} \chi(n) - \frac{6x}{\pi^2} \prod_{p \in E} \left(1 + \frac{1}{p}\right)^{-1} \right| \leq 2^k + 2\sqrt{x} \prod_{p \in E} \left(1 + \frac{1}{\sqrt{p}}\right)$$

$$\leq 2^k(1 + 2\sqrt{x}).$$

(Use the property $\chi = h * z$).

II: With m an arbitrary integer ≥ 1, let $Q_m(x)$ be the number of square-free $n \in \mathbb{N}^*$ divisible by m and $\leqslant m$.

From result I(c) above, deduce that when $x \to +\infty$:

$$Q_m(x) = \frac{6}{\pi^2}\gamma(m)x + O(\sqrt{x}),$$

where γ is the multipicative function determined by:

$$\gamma(p^r) = \begin{cases} \dfrac{1}{p+1} & \text{for } r = 1, \\ 0 & \text{for } r > 1, \end{cases}$$

and O is uniform with respect to m.

More precisely, there exists a constant $C > 0$ independent of m such that:

$$\left| Q_m(x) - \frac{6}{\pi^2}\gamma(m)x \right| < C\sqrt{x} \quad \text{for all } x > 0.$$

(The case where m does have a square factor is trivial. When m is square-free, the cases where $x \geqslant m$ and where $x < m$ need to be distinguished, the second being treated by a direct majorisation of the first member, without using I(c) above).

EXERCISE 1·19: SQUARE-FREE NUMBERS IN ARITHMETIC PROGRESSIONS

(1): Show that every $n \in \mathbb{N}^*$ can be written uniquely in the form:

$n = m^2 q$ with $m, q \in \mathbb{N}^*$, q square-free.

(2): Verify:

$$\sum_{d^2/n} \mu(d) = \begin{cases} 1 & \text{if } n \text{ is square-free}, \\ 0 & \text{otherwise}. \end{cases}$$

(Note that "d^2/n" is equivalent to "d/m").

(3): Given $k \in \mathbb{N}^*$ and $\ell \in \mathbb{Z}$, let $Q(x;k,\ell)$ be the number of the $n \in \mathbb{N}^*$ which are square-free $\leqslant x$, and congruent to ℓ modulo k.

Assume that (k,ℓ) is square-free. (Otherwise it is clear that $Q(x:k,\ell) = 0$).

Show that when $x \to +\infty$ we have:

$$Q(x:k,\ell) = A_{k,\ell}x + O(\sqrt{x}),$$

where $A_{k,\ell}$ is a constant depending upon k and ℓ and the O is uniform with respect to k and ℓ.

Show that:

$$A_{k,\ell} = \frac{1}{k} \prod_{(p^2,k)|\ell} \left(1 - \frac{(p^2,k)}{p^2}\right)$$

$$= \frac{1}{k}\frac{6}{\pi^2}\left[\prod_{\substack{p|k \\ p\nmid\ell}} \left(1 - \frac{1}{p^2}\right)^{-1}\right]\left[\prod_{\substack{p|(k,\ell) \\ p^2|k}} \left(1 - \frac{1}{p^2}\right)^{-1}\right]$$

$$\times \left[\prod_{\substack{p|(k,\ell) \\ p^2\nmid k}} \left(1 + \frac{1}{p}\right)^{-1}\right]$$

(Write:

$$Q(x:k,\ell) = \sum_{\substack{n\leqslant x \\ n\equiv \ell (\mathrm{mod}\, k)}} \left(\sum_{d^2/n} \mu(d)\right),$$

and reverse the order of summationa.

In order to obtain the definitive expression for $A_{k,\ell}$ it is useful to observe that the characteristic function of the set of the $d \in \mathbb{N}^*$ such that $(d^2,k)/\ell$ is multiplicative).

EXERCISE 1·20: SUMS INVOLVING THE EULER FUNCTION ϕ AND DISTRIBUTION OF IRREDUCIBLE FRACTIONS WITH DENOMINATOR n

(1): Show that $\varphi = \mu * i$ (where $i(n) = n$ for all $n \in \mathbb{N}^*$), and deduce from this that the Euler function φ is multiplicative, and that:

$$\varphi(n) = n \prod_{p|n} \left(1 - \frac{1}{p}\right) .$$

(Regrouping the m's such that $1 \leqslant m \leqslant n$ according to the values of (m,n), we see that $i = z * \varphi$).

(2): By using the property $\varphi = \mu * i$, show that when $x \to \infty$:

$$\sum_{n \leqslant x} \varphi(n) = \frac{3}{\pi^2} x^2 + O(x \log x).$$

(3): Given $m,n \in \mathbb{N}^*$, what is the expression:

$$\sum_{\substack{d|m \\ d|n}} \mu(d)$$

equal to?

(4): Given $x,y > 1$, we denote by $\Phi(x,y)$ the number of pairs $[m,n] \in \mathbb{N}^* \times \mathbb{N}^*$ such that:

$$m \leqslant x,\ n \leqslant y \quad \text{and} \quad (m,n) = 1.$$

Show that:

$$\Phi(x,y) = \sum_{d \leqslant \inf(x,y)} \mu(d) \left[\frac{x}{d}\right] \left[\frac{y}{d}\right],$$

and deduce from this that when $x,y \to \infty$:

$$\frac{1}{xy}\,\Phi(x,y) = \frac{6}{\pi^2} + O\left(\frac{\log \inf(x,y)}{\inf(x,y)}\right).$$

Can the result in (2) above be obtained by starting from this?

(5): Given $n \in \mathbb{N}^*$ and $t > 0$, we denote by $N(n,t)$ the number of irreducible fractions with denominator n belonging to the interval $(0,t]$, in other words the number of the $m \in \mathbb{N}^*$ such that $(m,n) = 1$ and $m \leqslant tn$. Set $B(x) = x - [x] - \frac{1}{2}$.

(a): Show that if $n > 1$, for all $t > 0$:

$$N(n,t) - t\varphi(n) = -\sum_{d/n} \mu(d)B\left(\frac{tn}{d}\right).$$

From this deduce that:

$$|N(n,t) - t\varphi(n)| \leqslant 2^{\nu(n)-1}.$$

(b): From the preceding result deduce that for any $0 < \varepsilon < 1$, there exists $K_\varepsilon > 0$ such that for all $n > 1$ and for all $t > 0$:

$$\left|\frac{1}{\varphi(n)}\,N(n,t) - t\right| \leqslant \frac{K_\varepsilon}{n^{1-\varepsilon}}.$$

EXERCISE 1·21: SMALL VALUES OF $\frac{\varphi(n)}{n}$

We set:

$$\vartheta(x) = \sum_{p\leqslant x} \log p \qquad \text{(CHEBYCHEFF'S FUNCTION)}.$$

We denote by p_k the k-th prime number, and set:

$$N_k = e^{\vartheta(p_k)} = 2.3.5\ldots p_k.$$

(1): With φ the Euler function and $\nu(n)$ the number of prime factors of n, show that:

$$\nu(n) < k \quad\text{and}\quad \frac{\varphi(n)}{n} > \frac{\varphi(N_k)}{N_k} \text{ for } n < N_k:$$

(2): Show that the only numbers N such that:

$$m < N \Rightarrow \frac{\varphi(m)}{m} > \frac{\varphi(N)}{N}$$

are the integers N_k, $k \geqslant 1$.

(3): By Using Mertens' Formula:

$$\prod_{p\leqslant x}\left(1 - \frac{1}{p}\right) \sim \frac{e^{-\gamma}}{\log x},$$

where γ is Euler's constant, show that:

$$\underline{\lim}\left(\frac{\varphi(n)\log\log n}{n}\right) = e^{-\gamma}.$$

(4): Let $\sigma(n) = \sum_{d|n} d$ be the sum of divisors of n.

Show that $\dfrac{\sigma(n)\varphi(n)}{n^2}$ is bounded. For a fixed integer r calculate:

$$\lim_{k\to\infty}\left(\frac{\sigma(N_k^r)}{N_k^r\log\log(N_k^r)}\right),$$

and from this deduce the value of:

$$\overline{\lim}\left(\frac{\sigma(n)}{n\log\log n}\right).$$

EXERCISE 1·22: $\sum_{n\leqslant x} \mu(n) = o(x)$

(1): Let a be an arithmetic function and let $A(x) = \sum_{n\leqslant x} a(n)$.

By using the Lemma in Section 1.2 of the Introduction, show that if the function a is bounded on $\mathbb{N}^*$, then for x tending to infinity:

$$\sum_{n\leqslant x} a(n)\log(n) = A(x)\log(x) + O(x).$$

(2): Show that $\Lambda = \mu * \log$, and, by writing down the value of the function $\mu * \log$, deduce from it that $\mu\log = -\mu * \Lambda$.

(3): Show that for all $x \geqslant 1$:

$$\sum_{n\leqslant x} \mu(n)\log n = -1 - \sum_{n\leqslant x} \mu(n) g\left(\frac{x}{n}\right),$$

where $g(x) = \psi(x) - [x]$.

(4): Set $M(x) = \sum_{n\leqslant x} \mu(n)$.

Show that the Prime Number Theorem implies that we have:

$$M(x) = o(x) \quad (x \to +\infty).$$

(Notice that the Prime Number Theorem is equivalent to the relation $\psi(x) \sim x$).

EXERCISE 1·23: EVALUATION OF THE SUM $\sum_{n\leqslant x} \frac{\mu^2(n)}{\varphi(n)}$

(1): We consider the Dirichlet series:

$$F(s) = \sum_{n=1} \frac{\mu^2(n)}{\varphi(n)n^s},$$

where μ is the Möbius function and φ the Euler function. Determine the abscissa of convergence of this series and put it into the form of an Eulerian product.

(2): Develop as an Euler product the function $G(s) =$

$\dfrac{F(s)}{\zeta(s+1)}$ and show that this product converges uniformly on every compact set K contained in the region Re$s > -\frac{1}{2}$.

Calculate $G(0)$ and show that:

$$G'(0) = \sum_p \frac{\log p}{p(p-1)} \, .$$

(3): Show that $G(s)$ is the sum of a Dirichlet series:

$$G(s) = \sum \frac{g(n)}{n^s} \, ,$$

Calculate the abscissa of absolute convergence as well.

(4): Show that:

$$\sum_{n\leqslant x} \frac{\mu^2(n)}{\varphi(n)} = \sum_{n\leqslant x} g(n) L\left(\frac{x}{n}\right) \, ,$$

with $L(x) = \sum_{n\leqslant x} \dfrac{1}{n}$.

By using $L(x) = \log x + \gamma + \eta(x)$, where γ is Euler's constant and $|\eta(x)| \leqslant \dfrac{1}{x}$ for $x > 1$, show that:

$$\sum_{n\leqslant x} \frac{\mu^2(n)}{\varphi(n)} = \log x + \gamma + \sum_p \frac{\log p}{p(p-1)} + o(1).$$

(The evaluation of this sum is useful in sieve methods).

EXERCISE 1·24: INTEGERS n SUCH THAT $\Omega(n) - \nu(n) = q$ (q GIVEN)

(1): Show that the series $\displaystyle\sum_{n=1}^{+\infty} \frac{z^{\Omega(n)-\nu(n)}}{n^s}$ is absolutely convergent for Re$s > 1$ and $|z| < 2$, its sum being equal to:

$$\frac{\zeta(s)}{\zeta(2s)} \prod \left[\left(1 - \frac{z}{p^s+1}\right)\left(1 - \frac{z}{p^s}\right)^{-1}\right] .$$

(2): Show that the infinite product:

$$\prod \left[\left(1 - \frac{z}{p^s + 1}\right)\left(1 - \frac{z}{p^s}\right)^{-1}\right]$$

is absolutely convergent for $\mathrm{Re}\, s > \frac{1}{2}$ and $|z| < \sqrt{2}$, and defines a function of s and z holomorphic for $\mathrm{Re}\, s > \frac{1}{2}$ and $|z| < \sqrt{2}$.

(3): With $q \geqslant 0$ an integer, we denote by $N_q(x)$ the number of the $n \leqslant x$ such that $\Omega(n) - \nu(n) = q$.

From the preceding deduce that when $x \to +\infty$:

$$N_q(x) = d_q x + o(x),$$

where the numbers d_q are determined by the property, for $|z| < 2$:

$$\sum_{q=0}^{\infty} d_q z^q = \frac{6}{\pi^2} \prod \left(\frac{1 - \frac{z}{p+1}}{1 - \frac{z}{p}}\right).$$

Verify that all the $d_q > 0$ and that $\sum_{q=0}^{\infty} d_q = 1$.

(By a theorem from the theory of functions of several complex variables†, for $|z| < \sqrt{2}$ and $\mathrm{Re}\, s > \frac{1}{2}$ we can write:

$$\prod \left(\frac{1 - \frac{z}{p^s + 1}}{1 - \frac{z}{p^s}}\right) = \sum_{q=0}^{+\infty} A_q(s) z^q,$$

where the A_q are holomorphic for $\mathrm{Re}\, s > \frac{1}{2}$.

---

† If F is a function of two complex variables s and z, holomorphic in s and z for s belonging to a domain D of $\mathbb{C}$ and $|z| < R$, there exist functions A_q holomorphic on D such that, for $s \in D$ and $|z| < R$:

$$F(s,z) = \sum_{q=0}^{\infty} A_q(s) z^q.$$

Notice that for $\text{Re}s > 1$:

$$\sum_{\Omega(n)-\nu(n)=q} \frac{1}{n^s} = \frac{\zeta(s)}{\zeta(2s)} A_q(x),$$

and apply Ikehara's Theorem).

EXERCISE 1·25: THE NUMBER OF n's SUCH THAT $\varphi(n) \leqslant x$

Let φ be the Euler function. For $m \geqslant 1$ set:

$$E_m = \{n \mid \varphi(n) = m\}.$$

(1): Show that for every integer $m \geqslant 1$ the set E_m is finite. Determine E_m when m is odd.

(2): Set $a_m = \text{card}E_m$, and:

$$A(x) = \sum_{m \leqslant x} a_m = \text{card}\{n \mid \varphi(n) \leqslant x\}.$$

Show that there exists a constant C, independent of n, such that:

$$\frac{n}{\varphi(n)} \leqslant C \sum_{d/n} \frac{1}{d} \,.$$

Show that $\frac{n}{\varphi(n)} = O(\log n)$, that $x \leqslant A(x)$, and that $A(x) = O(x\log x)$.

(3): Show that the series:

$$\sum_{n=1}^{\infty} \frac{1}{(\varphi(n))^s} \quad \text{and} \quad \sum_{m=1}^{\infty} \frac{a_m}{m^s}$$

are absolutely convergent for $\text{Re}s > 1$, and that for $\text{Re}s > 1$ the equalities:

$$\sum_{m=1}^{\infty} \frac{a_m}{m^s} = \sum_{n=1} \frac{1}{(\varphi(n))^s} = \zeta(s)f(s)$$

hold, where $\zeta(s)$ denotes the Riemann ζ-function, and:

$$f(s) = \prod_p \left\{1 + \frac{1}{(p-1)^s} - \frac{1}{p^s}\right\} .$$

(4): Show that for every prime p and $\operatorname{Re} s > 0$ we have the inequality:

$$\left|\frac{1}{(p-1)^s} - \frac{1}{p^s}\right| \leqslant \frac{|s|}{(p-1)^{\operatorname{Re} s+1}} .$$

From this deduce that f can be extended to a function $\hat{f}$ holomorphic for $\operatorname{Re} s > 0$. Calculate $\hat{f}(1)$ as a function of $\zeta(2), \zeta(3), \zeta(6)$.

(5): Show that there exists a constant α such that when $x \to \infty$ we have:

$$A(x) \sim \alpha x.$$

EXERCISE 1·26: PARTITIONS

Let $A \subset \mathbb{N}^*$. Denote by $p_A(n)$ the number of partitions of n the summands of which are in A, and to which we associate the power series:

$$F_A(x) = \sum_{n=0}^{\infty} p_A(n)x \ = \prod_{n \in A} (1 - x\ )^{-1},$$

where $p_A(0) = 1$. (Always assuming that $A \subset \mathbb{N}^*$.) By $P(A)$ we denote the property:

PROPERTY: $P(A)$: There exists N such that $n \geqslant N \Rightarrow p_A(N) > 0$.

The object of this Exercise is to characterise the sets A of $\mathbb{N}^*$ with Property $P(A)$.

(1): Show that if $A' \subset A$, $P(A') \Rightarrow P(A)$.

(2): Let $a,b > 0$ be relatively prime integers.

Show that every $n \geqslant ab$ can be written as $n = xa + yb$ with integers $x,y \geqslant 0$. Show that if A contains two relatively prime elements, then Property $P(A)$ holds.

(3): Show that if A is finite and equal to $\{a_1,\dots,a_r\}$ then,

$$p_A(n) = g(n) + O(n^{q-1}),$$

where $g(n)$ is a polynomial of degree $r - 1$, for which the term of highest degree will be given, and q is the largest number of elements which can be taken from A, and whose greatest common divisor (g.c.d.) is greater than one. Use the decomposition into simple elements of $F_A(x)$.

(4): For $A \subset \mathbb{N}^*$ define the properties:

PROPERTY: $P_1(A)$ <=> the g.c.d. of the elements of A is 1;

PROPERTY: $P_2(A)$ <=> For all prime p's there exists $a \in A$ such that $p \nmid a$.

Show that for all $A \subset \mathbb{N}^*$, $P_1(A) \Leftrightarrow P_2(A)$.

Show that for all $A \subset \mathbb{N}^*$, $P(A) \Rightarrow P_1(A)$.

Show that for every finite subset $A \subset \mathbb{N}^*$, $P(A) \Leftrightarrow P_1(A)$.

(5): Let A be an infinite set in $\mathbb{N}^*$ satisfying Property $P_2(A)$.

Show that there exists a finite subset $A' \subset A$ satisfying Property $P_2(A')$.

Show that for every set $A \subset \mathbb{N}^*$ we have $P(A) \Leftrightarrow P_1(A)$.

EXERCISE 1·27: LOWER BOUND FOR A SUM APPEARING IN THE SELBERG SIEVE

(2): Let $N \geqslant 2$ be an integer, Q a set of prime numbers less than or equal to N, and $A(N,Q)$ the set of those integers less than or equal to N that contain only prime factors belonging to Q.

Show that there exists a positive constant not depending on N such that:

$$\sum_{m \in A(N,Q)} \frac{1}{m} \quad B \prod_{p \in Q} \left(1 - \frac{1}{p}\right)^{-1}.$$

(One can induct on the cardinality of Q, after doing the case where Q is formed by all the prime numbers less than or equal to N).

(b): Let $z \geqslant 4$ be a real number, P be the set of prime numbers less than or equal to z, P_1 a subset of P, and P_2 the complement of P_1 in P; P_i $(i = 1,2)$ will denote the product of the elements of P_i, and ω will denote the completely multiplicative function determined by:

$$\omega(p) = \begin{cases} 1 \text{ if } p \text{ divides } P_1, \\ 2 \text{ if } p \text{ divides } P_2, \\ 0 \text{ if } p > z. \end{cases}$$

Show that there exists an *absolute* constant C such that:

$$\sum_{m \leqslant z} \frac{\omega(m)}{m} \geqslant \sum_{k \leqslant \sqrt{z}} \frac{1}{k} \sum_{\substack{\ell \leqslant \sqrt{z} \\ (\ell, P_1)=1}} \frac{1}{\ell} \geqslant C(\log z)^2 \prod_{p/P_1} \left(1 - \frac{1}{p}\right).$$

EXERCISE 1·28: Let x be a positive real number > 8, and m a non-zero even integer with absolute value less than x. By using Selberg's inequality prove that there exist two absolute constants D,E such that we have:

$$\text{card}\{p \mid p \leqslant x,\ p \text{ and } |p + m| \text{ prime numbers}\} \leqslant$$

$$\leqslant D \frac{x}{\log^2 x} \prod_{p \nmid m} \left(1 - \frac{1}{p}\right)^{-1}$$

$$\leqslant E \frac{x}{\log^2 x} \prod_{p/m} \left(1 + \frac{1}{p}\right).$$

EXERCISE 1·29: Deduce from Exercise 1·28 that the series of the reciprocals of the prime numbers p such that $p + 2$ is a prime number converges.

EXERCISE 1·30: GOLDBACH'S PROBLEM

Let $r(N)$ be the number of representations of an integer $N > 1$ as a sum of two prime numbers.

(a): Using Exercise 1·28, prove that there exists F not depending on n, such that:

$$r(N) \leqslant FN(\log N)^{-2} \sum_{d/N} \frac{1}{d}.$$

(b): From this deduce that there exists a constant F' such that:

$$\sum_{N \leqslant x} r^2(N) \leqslant F' x^3 (\log x)^{-4} \quad \text{for all } x > 4.$$

(c): Deduce from the Prime Number Theorem that there exists a positive constant F'' such that:

$$\sum_{N \leqslant x} r(N) \geqslant F'' x^2 (\log x)^{-2} \quad \text{for all } x > 4.$$

(d): Let $M(x)$ be the number of integers less than or equal to x which are the sum of two primes; prove that there exists a positive constant C such that:

$$M(x) \geqslant Cx \quad \text{for all } x \geqslant 4.$$

(e): Deduce from this that the sequence formed by 0,1, and the prime numbers, is a basis of the integers.

EXERCISE 1·31: SMALL DIFFERENCES BETWEEN CONSECUTIVE PRIME NUMBERS

We return to the notations of Exercise 1·9. By a and A we denote two positive real numbers.

(a): Prove the inequality:

$$\limsup_{x\to\infty}\left(\frac{\log x}{x}\right).\mathrm{card}\{i \mid p_{i+1} \leqslant x, d_i \geqslant A\log x\} \leqslant A^{-1}.$$

(b): $M(x)$ denotes the number of integers i such that p_i lies between $\frac{1}{2}x$ and x, and d_i between $a\log x$ and $A\log x$; by using Exercise 1·28 prove that there exists a constant F such that for all $x \geqslant 2$ we have:

$$M(x) \leqslant F\,\frac{x}{\log^2 x}\sum_{a\log x<m\leqslant A\log x}\ \prod_{p/m}\left(1+\frac{1}{p}\right).$$

(c): Deduce from this the existence of a constant F' such that:

$$\limsup_{x\to\infty}\left(\frac{M(x)\log x}{x}\right) \leqslant F'(A-a).$$

(d): Write:

$$S(x) = \sum_{\frac{1}{2}x<p_i\leqslant x} d_i$$

$$= \sum_{d_i\leqslant a\log x} d_i + \sum_{a\log x<d_i\leqslant A\log x} d_i + \sum_{d_i>A\log x} d_i$$

Show that the hypothesis: "The first sum of the right-hand side is empty for x large enough" implies:

$$\liminf_{x\to\infty} \frac{S(x)}{x} \geq \tfrac{1}{2}A - F'(A - a)^2.$$

(e): By choosing $a < 1$ and A so that

$$\tfrac{1}{2}A - F'(A - a)^2 > \tfrac{1}{2}$$

holds, prove that:

$$\liminf_{i\to\infty} \left(\frac{d_i}{\log i}\right) \leq a < 1.$$

SOLUTIONS

SOLUTION 1·1: Let us write $m = n + k$, where $k \in \mathbb{N}^*$, and let us set $u = 2^{2^n}$. We have:

$$\frac{F_{n+k} - 2}{F_n} = \frac{2^{2^{n+k}} - 1}{2^{2^n} + 1} = \frac{u^{2^k} - 1}{u + 1} = u^{2^k - 1} - u^{2^k - 2} + \cdots - 1,$$

hence F_n divides $F_m - 2$. Let d be the g.c.d. of F_n and F_m; $d|F_n$, hence $d|F_m - 2$ and $d|F_m$, so $d|2$. Since F_n and F_m are odd, $d = 1$, and therefore F_n and F_m are relatively prime.

The mapping of $\mathbb{N}^*$ into the set of prime numbers which assigns to each integer n the smallest prime factor of F_n is therefore injective, and so there are infinitely many prime numbers.

SOLUTION 1·2: Let us fix real $\beta < 1$; by the Prime Number Theorem:

$$\pi(N) \sim \frac{N}{\log N} \quad \text{and} \quad \pi(\beta N) \sim \frac{\beta N}{\log N + \log\beta} \sim \frac{\beta N}{\log N}.$$

If N is large enough we have $\pi(N) > \pi(\beta N)$, and hence there exists a prime number between βN and N.

SOLUTION 1·3: (i): We have the following two relations:

$$\vartheta(x) = \sum_{p \leqslant x} \log p \leqslant \log x \,.\, \sum_{p \leqslant x} 1 = \pi(x)\log x; \tag{1}$$

$$\text{For all } \delta \in (0,1)\text{:}\quad \vartheta(x) \geqslant \sum_{x^{\delta} < p \leqslant x} \log p$$

$$\geqslant \delta\log x(\pi(x) - \pi(x^{\delta}))$$

$$\geqslant \delta\pi(x)\log x - x^{\delta}\log x. \tag{2}$$

Assuming the Prime Number Theorem we deduce from these relations that:

$$\limsup \frac{\vartheta(x)}{x} \leqslant 1 \quad \text{and} \quad \liminf \frac{\vartheta(x)}{x} \geqslant \delta$$

(for all $\delta \in (0,1)$).

Hence we certainly have:

$$\liminf \frac{\vartheta(x)}{x} \geqslant 1,$$

and therefore $\vartheta(x) \sim x$.

Conversely, if we start from the relation $\vartheta(x) \sim x$, using (1) we find:

$$\liminf \frac{\pi(x)\log x}{x} \geqslant 1,$$

whence $x^{\delta} = o(\pi(x))$, and from (2):

$$\limsup \frac{\pi(x)\log x}{x} \leqslant \frac{1}{\delta},$$

and thus the Prime Number Theorem.

(ii): For each $n \geqslant 1$ we have $\pi(p_n) = n$.

If the Prime Number Theorem is assumed, we deduce that when $n \to \infty$:

$$n \sim \frac{p_n}{\log p_n}.$$

This implies:

$$\log n \sim \log p_n \quad \text{and} \quad p_n \sim n\log p_n \sim n\log n.$$

Let us now observe that for all $x \geq 2$:

$$p_{\pi(x)} \leq x \leq p_{\pi(x)+1}$$

If, for "infinite" n, we assume $p_n \sim n\log n$, we deduce from this, for "infinite" x, that the extreme terms are equivalent to $\pi(x)\log\pi(x)$, and consequently:

$$x \sim \pi(x)\log\pi(x).$$

This implies:

$$\log x \sim \log\pi(x) \quad \text{and} \quad \pi(x) \sim \frac{x}{\log\pi(x)} \sim \frac{x}{\log x}.$$

SOLUTION 1·4: By the Prime Number Theorem and by Exercise 1·3, we have $p_n \sim n\log n$. As the given series has positive terms, the general term can be replaced by an equivalent one, which comes down to studying the series with general term $u_n = (n\log n(\log\log n)^\alpha)^{-1}$; by comparing with the integral:

$$\int_3^\infty \frac{dt}{t\log t(\log\log t)^\alpha},$$

we see that the given series converges if and only if $\alpha > 1$.

SOLUTION 1·5: I:(1): The exponent of p in the decomposition of an integer n into prime factors is clearly equal to:

$$\sum_{\substack{r\geqslant 1\\ p^r/n}} 1.$$

(The sum being zero if it is empty, i.e., if $p\nmid n$).

The exponent of p in the decomposition into prime factors of the product $\prod_{n\leqslant x} n$ is therefore

$$\alpha_p = \sum_{n\leqslant x}\left(\sum_{\substack{r\geqslant 1\\ p^r/n}} 1\right) = \sum_{\substack{r\geqslant 1\\ p^r\leqslant x}}\left(\sum_{\substack{n\leqslant x\\ p^r/n}} 1\right) = \sum_{\substack{r\geqslant 1\\ p^r\leqslant x}}\left[\frac{x}{p^r}\right].$$

This is equal at most to $\sum_{r=1}^{+\infty} \frac{x}{p^r} = \frac{x}{p-1}$ and at least to $\left[\frac{x}{p}\right]$.

We have:

$$\left[\frac{x}{p}\right] \geqslant \frac{x}{p} - 1, \quad \text{also} \quad \left[\frac{x}{p}\right] \geqslant \frac{x}{2p},$$

for, if $u \geqslant 1$, we have $u \leqslant [u] + 1 \leqslant 2[u]$, and consequently $[u] \geqslant \frac{1}{2}u$.

I:(2): It is clear that we have $\sum_{n\leqslant x} \log n \leqslant x\log x$, since, for each $n \leqslant x$, $\log n \leqslant \log x$. However, by what has already been shown, we have:

$$\sum_{n\leqslant x} \log n = \log \prod_{n\leqslant x} n = \sum_{p\leqslant x} \alpha_p \log p$$

$$\geqslant \sum_{p\leqslant x} \frac{x}{2p} \log p = \tfrac{1}{2}x \sum_{p\leqslant x} \frac{\log p}{p}.$$

Hence we have:

$$\tfrac{1}{2}x \sum_{p\leqslant x} \frac{\log p}{p} \leqslant x\log x,$$

whence:

$$\sum_{p\leqslant x} \frac{\log p}{p} \leqslant 2\log x.$$

If $1 \leqslant y \leqslant x$ one can also write:

$$\sum_{n\leqslant x} \log n \geqslant \sum_{y<p\leqslant x} \frac{x}{2p} \log p \geqslant \tfrac{1}{2}x\log y \sum_{y<p\leqslant x} \frac{1}{p},$$

and from this it follows that:

$$\tfrac{1}{2}x\log y \sum_{y<p\leqslant x} \frac{1}{p} \leqslant x\log x,$$

whence:

$$\sum_{y<p\leqslant x} \frac{1}{p} < 2\,\frac{\log x}{\log y}.$$

I:(3): For each $n \geqslant 1$ we have:

$$\int_{n-1}^{n} \log t dt \leqslant \log n.$$

Consequently:

$$\sum_{n\leqslant x} \log n \geqslant \int_0^{[x]} \log t dt = \int_0^{x} \log t dt - \int_{[x]}^{x} \log t dt$$

$$\geqslant \int_0^{x} \log t dt - \log x = (x - 1)\log x - x.$$

For each $p < x$, $\alpha_p < \frac{x}{p - 1}$, we have:

$$\sum_{n\leqslant x} \log n = \sum_{p\leqslant x} \alpha_p \log p \leqslant x \sum_{p\leqslant x} \frac{\log p}{p - 1}.$$

Hence:

$$(x - 1)\log x - x \leqslant x \sum_{p\leqslant x} \frac{\log p}{p - 1},$$

whence:

$$\sum_{p\leqslant x} \frac{\log p}{p - 1} \geqslant \log x - 1 - \frac{\log x}{x}.$$

Since $\frac{1}{p} = \frac{1}{p - 1} \quad \frac{1}{p(p - 1)}$,

$$\sum_{p\leqslant x} \frac{\log p}{p} = \sum_{p\leqslant x} \frac{\log p}{p - 1} - \sum_{p\leqslant x} \frac{\log p}{p(p - 1)}$$

$$\geqslant \log x - 1 - \frac{\log x}{x} - \sum \frac{\log p}{p(p - 1)}$$

$$\geqslant \log x - C,$$

where:

$$C = 1 + \frac{1}{e} + \sum \frac{\log p}{p(p - 1)}$$

(as $\frac{\log x}{x}$ has a maximum when $x = e$).

II:(1): Every prime number p satisfying $m \leqslant p \leqslant 2m$ divides $(2m)!$ but not $m!$, and so divides $\binom{2m}{m} = \frac{(2m)!}{(m!)^2}$. Consequently:

$$\log\binom{2m}{m} \geqslant \sum_{m<p\leqslant 2m} \log p = \vartheta(2m) - \vartheta(m).$$

But:

$$2^{2m} = \sum_{k=0}^{2m} \binom{2m}{k} > \binom{2m}{m},$$

and so:

$$\log\binom{2m}{m} < 2m\log 2.$$

Hence we have:

$$\vartheta(2m) - \vartheta(m) < 2m\log 2.$$

From this, for all $k \geqslant 1$, it follows that:

$$\theta(2^k) = \theta(2^k) - \theta(1) = \sum_{j=1}^{k} (\theta(2^j) - \theta(2^{j-1}))$$

$$< \sum_{j=1}^{k} 2^j \log 2 < 2^{k+1}\log 2.$$

Given $x > 1$, there exists $h \geqslant 0$ such that $2^h \leqslant x \leqslant 2^{h+1}$, and we have

$$\theta(x) \leqslant \theta(2^{h+1}) \leqslant 2^{h+2}\log 2 < 4x \log 2 .$$

II: (2): Given $x > 1$, since for each $p \leqslant x$ we have: $\alpha_p \geqslant \frac{x}{p} - 1$,

we obtain:

$$\sum_{n \leqslant x} \log n = \sum_{p \leqslant x} \alpha_p \log p \geqslant \sum_{p \leqslant x} \left(\frac{x}{p} - 1\right) \log p = x \sum_{p \leqslant x} \frac{\log p}{p} - \theta(x).$$

Therefore:

$$x \sum_{p \leqslant x} \frac{\log p}{p} \leqslant \sum_{n \leqslant x} \log n + \theta(x) \leqslant x\log x + 4x\log 2,$$

whence:

$$\sum_{p \leqslant x} \frac{\log p}{p} \leqslant \log x + 4\log 2.$$

III: Using the Lemma of Section 1.2 of the Introduction, for $x > 2$ we see that:

$$\sum_{p \leqslant x} \frac{1}{p} = \frac{1}{2} + \sum_{2 < p \leqslant x} \frac{\log p}{p} \frac{1}{\log p} = \frac{1}{2} + \frac{\Phi(x)}{\log x} - \frac{\Phi(2)}{\log 2} - \int_2^x \Phi(t) \frac{d}{dt}\left(\frac{1}{\log t}\right) dt$$

$$= \frac{\Phi(x)}{\log x} + \int_2^x \frac{\Phi(t)}{t(\log t)^2}\, dt$$

$$= 1 + \frac{\eta(x)}{\log x} + \int_2^x \frac{dt}{t\log t} + \int_2^x \frac{\eta(t)}{t(\log t)^2}\, dt =$$

(Contd) $$= 1 + \frac{\eta(x)}{\log x} + \log\log x - \log\log 2 + \int_2^x \frac{\eta(t)}{t(\log t)^2}\,dt.$$

Since, by I(3) and II(2) above, $\eta(x)$ is bounded for $x > 1$, the integral $\int_2^{+\infty} \frac{\eta(t)}{t(\log t)^2}\,dt$ is convergent, and we can write:

$$\sum_{p \leqslant x} \frac{1}{p} = \log\log x + 1 - \log\log 2 + \int_2^{+\infty} \frac{\eta(t)}{t(\log t)^2}\,dt + \frac{\eta(x)}{\log x}$$

$$- \int_2^{+\infty} \frac{\eta(t)}{t(\log t)^2}\,dt.$$

If $|\eta(x)| \leqslant M$ for all $x > 1$, we see that:

$$\left|\frac{\eta(x)}{\log x}\right| \leqslant \frac{M}{\log x} \quad \text{and} \quad \left|\int_x^{+\infty} \frac{\eta(t)dt}{t(\log t)^2}\right| \leqslant M\int_x^{+\infty} \frac{dt}{t(\log t)^2} = \frac{M}{\log x}\,.$$

One obviously obtains the formula indicated, with

$$b = 1 - \log\log 2 + \int_2^{+\infty} \frac{\eta(t)}{t(\log t)^2}\,dt.$$

EXERCISE 1·6: (1): The series:

$$\sum \left(\log \frac{1}{1 - \frac{1}{p^x}} - \frac{1}{p^x}\right)$$

is convergent for real $x > \frac{1}{2}$, because when $p \to +\infty$:

$$\log \frac{1}{1 - \frac{1}{p^x}} - \frac{1}{p^x} \sim \frac{1}{2p^{2x}}\,.$$

For all $x_0 > \frac{1}{2}$, there is normal, hence uniform, convergence for $x \geqslant x_0$. In fact, $\log \frac{1}{1 - u} - u$ is a positive, increasing function of u for $0 \leqslant u < 1$. Consequently, for any p, we have for $x \geqslant x_0$:

$$0 \leqslant \log \frac{1}{1 - \frac{1}{p^x}} - \frac{1}{p^x} \leqslant \log \frac{1}{1 - \frac{1}{p^{x_0}}} - \frac{1}{p^{x_0}} .$$

From this it follows that if $g(x)$ is the sum of the series, the function g is continuous on the interval $]\frac{1}{2},+\infty[$.

(2): For real $s > 0$ we have:

$$\zeta(1 + s) = \prod \frac{1}{1 - \frac{1}{p^{1+s}}} ,$$

and consequently:

$$\log\zeta(1 + s) = \sum \log \frac{1}{1 - \frac{1}{p^{1+s}}} .$$

From this it follows that:

$$\log\zeta(1 + s) - g(1 + s) = \sum \frac{1}{p^{1+s}} .$$

By applying the Lemma of Section 1.2 of the Introduction, we see that, for all $T > 0$:

$$\sum_{p \leqslant e^T} \frac{1}{p^{1+s}} = \sum_{0<\log p\leqslant T} \frac{1}{p} e^{-s\log p} = P(e^T)e^{-sT} + \int_0^T se^{-st}P(e^t)dt.$$

When $T \to +\infty$ the first sum tends to $\sum \frac{1}{p^{1+s}}$, and, since $P(e^T) \sim \log T$, $P(e^T)e^{-sT} \to 0$. Consequently, the integral $\int_0^{+\infty} se^{-st}P(e^t)dt$ is convergent and equal to $\sum \frac{1}{p^{1+s}}$, that is to say, to $\log\zeta(1 + s) - g(1 + s)$.

Furthermore, for real $s > 0$ we have:

$$\log \frac{1}{1 - e^{-s}} = \sum_{n=1}^{+\infty} \frac{1}{n} e^{-ns}.$$

Moreover as before, we have for $T > 0$:

$$\sum_{n \leqslant T} \frac{1}{n} e^{-ns} = \sum_{0<n\leqslant T} \frac{1}{n} e^{-ns} = N(T)e^{-sT} + \int_0^T se^{-st} N(t)dt,$$

so, by letting $T \to +\infty$, we obtain in the limit:

$$\log \frac{1}{1 - e^{-s}} = \int_0^{+\infty} se^{-st} N(t)dt.$$

(3): When $t \to +\infty$ we have:

$$P(e^t) = \log t + b + O\left(\frac{1}{t}\right) \quad \text{and} \quad N(t) = \log t + \gamma + o(1).$$

Consequently:

$$P(e^t) - N(t) + \gamma - b \to 0.$$

Given $\varepsilon > 0$, there exists $t_0 > 0$ such that:

$$|P(e^t) - N(t) + \gamma - b| \leqslant \varepsilon \quad \text{for} \quad t \geqslant t_0.$$

Moreover, for $0 \leqslant t \leqslant t_0$, we clearly have:

$$0 \leqslant P(e^t) \leqslant P(e^{t_0}) \quad \text{and} \quad 0 \leqslant N(t) \leqslant N(t_0),$$

and consequently:

$$|P(e^t) - N(t) + \gamma - b| \leqslant P(e^{t_0}) + N(t_0) + |\gamma - b| = M.$$

Then for all real $s > 0$:

$$\left| s\int_0^{+\infty} e^{-st}(P(e^t) - N(t) + \gamma - b)dt \right| \leqslant \qquad \text{(Contd)}$$

(Contd) $$\leqslant Ms\int_0^{t_0} e^{-st}dt + \varepsilon s\int_{t_0}^{+\infty} e^{-st}dt$$

$$\leqslant M(1 - e^{-st_0}) + \varepsilon e^{-st_0},$$

and from this we have:

$$\overline{\lim_{\substack{s\to 0\\ s>0}}}\left| s\int_0^{+\infty} e^{-st}(P(e^t) - N(t) + \gamma - b)dt\right| \leqslant \varepsilon.$$

As this is true for any $\varepsilon > 0$, we see that:

$$s\int_0^{+} e^{-st}(P(e^t) - N(t) + \gamma - b)dt \to 0$$

when $s \to 0$ through real positive values.

The formulas established in the preceding paragraph show that the expression under consideration is equal to:

$$\log\zeta(1 + s) - g(1 + s) - \log\frac{1}{1 - e^{-s}} + \gamma - b,$$

that is to say:

$$\log((1 - e^{-s})\zeta(1 + s)) - g(1 + s) + \gamma - b,$$

which tends to $\gamma - b - g(1)$ when $s \to 1$, because $(1 - e^{-s}) \sim s$ and g is continuous at the point 1. Hence we have $\gamma - b - g(1) = 0$, whence $b = \gamma - g(1)$.

(4): Since, for $x > 0$, we have:

$$P(x) + \log\prod_{p\leqslant x}\left(1 - \frac{1}{p}\right) = -\sum_{p\leqslant x}\left(\log\frac{1}{1 - \frac{1}{p}} - \frac{1}{p}\right),$$

when $x \to +\infty$ we see that:

$$P(x) + \log \prod_{p \leq x} \left(1 - \frac{1}{p}\right) \to - g(1),$$

which can be written:

$$\log \prod_{p \leq x} \left(1 - \frac{1}{p}\right) = - P(x) - g(1) + o(1)$$

$$= - \log\log x - b - g(1) + o(1)$$

$$= - \log\log x - \gamma + o(1).$$

Thus one obtains the well known Mertens' Formula:

$$\prod_{p \leq x} \left(1 - \frac{1}{p}\right) \sim \frac{e^{-\gamma}}{\log x} .$$

SOLUTION 1·7: (1): $n \notin A$ if and only if there exists a prime number p, $p > \sqrt{n}$, which divides n. Then we have:

$$1 \leq \frac{n}{p} < p, \text{ that is to say, } n = ap \text{ with } 1 \leq a \leq p - 1,$$

hence $n \in B_p$.

Let us assume $p < q$, and that $x \in B_p \cap B_q$. We must have:

$$x = ap = bq \quad \text{with } 1 \leq a \leq p - 1 \text{ and } 1 \leq b \leq q - 1;$$

but q divides ap, hence it divides a, therefore $q \leq a < p$, which contradicts $p < q$.

(2): Let us define $Q_{B_p}(x) = \operatorname{card}\{n \leq x, n \in B_p\}$. We have:

$$Q_{B_p}(x) = \min\left(\left[\frac{x}{p}\right], p - 1\right) = \begin{cases} \left[\frac{x}{p}\right] & \text{if } x < p^2, \\ p - 1 & \text{if } x \geq p^2. \end{cases}$$

By part (1) above,

$$Q_A[x] = [x] - \sum_{p\,\text{prime}} Q_{B_p}(x),$$

$$Q_A[x] = [x] - \sum_{p \leq \sqrt{x}} (p - 1) - \sum_{\sqrt{x} < p \leq x} \left[\frac{x}{p}\right].$$

(3): Let $\pi(x)$ be the number of prime numbers less than or equal to x. We know that $\pi(x) \sim \dfrac{x}{\log x}$. We have then that:

$$\sum_{p \leq \sqrt{x}} (p - 1) \leqslant \sqrt{x}\,\pi(\sqrt{x}) = O\left(\frac{x}{\log x}\right).$$

On the other hand:

$$\sum_{\sqrt{x} < p \leq x} \left[\frac{x}{p}\right] = \sum_{\sqrt{x} < p \leq x} \frac{x}{p} + O(\pi(x))$$

$$= x\left[\log\log x + b - \log\log\sqrt{x} - b + O\left(\frac{1}{\log x}\right)\right] + O\left(\frac{x}{\log x}\right)$$

$$= x\log 2 + O\left(\frac{x}{\log x}\right).$$

Finally, we have:

$$Q_A(x) = x + O(1) + O\left(\frac{x}{\log x}\right) - x\log 2 + O\left(\frac{x}{\log x}\right),$$

and:

$$Q_A(x) \sim x(1 - \log 2).$$

SOLUTION 1·8: (1): By the definition of the function ν, for all $n \in \mathbb{N}^*$:

$$\nu(n) = \sum_{p/n} 1. \tag{1}$$

Consequently, for $x \geqslant 1$:

$$\sum_{n\leqslant x} \nu(n) = \sum_{n\leqslant x} \left(\sum_{p/n} 1\right) = \sum_{p\leqslant x} \left(\sum_{\substack{n\leqslant x\\ p/n}} 1\right)$$

$$= \sum_{p\leqslant x} \left[\frac{x}{p}\right] = x \sum_{p\leqslant x} \frac{1}{p} + O(\pi(x)),$$

since, for each $p \leqslant x$, $\left[\frac{x}{p}\right]$ differs from $\frac{x}{p}$ by at most one.

Assuming the formula cited in the Problem, and that $\pi(x) = O\left(\frac{x}{\log x}\right)$, we obtain:

$$\sum_{n\leqslant x} \nu(n) = x\log\log x + bx + O\left(\frac{x}{\log x}\right). \tag{2}$$

(2): The condition $p_1p_2 \leqslant x$ implies:

$$p_1 \leqslant \sqrt{x} \quad or \quad p_2 \leqslant \sqrt{x}.$$

If one adds up the sum of the terms for which $p_1 \leqslant \sqrt{x}$ and the sum of those for which $p_2 \leqslant \sqrt{x}$, then the terms for which:

$$p_1 \leqslant \sqrt{x} \quad and \quad p_2 \leqslant \sqrt{x}$$

have been counted twice.

We have:

$$\sum_{\substack{p_1p_2\leqslant x\\ p_1\leqslant\sqrt{x}}} \frac{1}{p_1p_2} = \sum_{p_1\leqslant\sqrt{x}} \frac{1}{p_1} \sum_{p_2\leqslant\frac{x}{p_1}} \frac{1}{p_2} = \sum_{p\leqslant\sqrt{x}} \frac{1}{p} P\left(\frac{x}{p}\right),$$

and the same relation with p_1 and p_2 interchanged. On the other hand:

$$\sum_{\substack{p_1\leqslant\sqrt{x}\\ p_2\leqslant\sqrt{x}}} \frac{1}{p_1p_2} = \left(\sum_{p\leqslant\sqrt{x}} \frac{1}{p_1}\right)\left(\sum_{p\leqslant\sqrt{x}} \frac{1}{p_2}\right) = P(\sqrt{x})^2.$$

Thus one obtains:

$$\sum_{p_1p_2\leqslant x} \frac{1}{p_1p_2} = 2 \sum_{p\leqslant\sqrt{x}} \frac{1}{p} P\left(\frac{x}{p}\right) - P(\sqrt{x})^2. \tag{3}$$

(3): By expressing the function P as is done in the statement of the problem we have:

$$\sum_{p\leqslant\sqrt{x}} \frac{1}{p} P\left(\frac{x}{p}\right) = \sum_{p\leqslant\sqrt{x}} \frac{1}{p}\left(\log\log\left(\frac{x}{p}\right) + b + \eta\left(\frac{x}{p}\right)\right)$$

$$= \sum_{p\leqslant\sqrt{x}} \frac{1}{p}\left(\log\log x + \log\left(1 - \frac{\log p}{\log x}\right) + b + \eta\left(\frac{x}{p}\right)\right),$$

because, for $p < x$:

$$\log\log\left(\frac{x}{p}\right) = \log(\log x - \log p) = \log\left(\log x\left(1 - \frac{\log p}{\log x}\right)\right).$$

This gives:

$$\sum_{p\leqslant\sqrt{x}} \frac{1}{p} P\left(\frac{x}{p}\right) = (\log\log x + b) \sum_{p\leqslant\sqrt{x}} \frac{1}{p} + \sum_{p\leqslant\sqrt{x}} \frac{1}{p}\left(\log 1 - \frac{\log p}{\log x}\right)$$

$$+ \sum_{p\leqslant\sqrt{x}} \frac{1}{p} \eta\left(\frac{x}{p}\right)$$

$$= (\log\log x + b)P(\sqrt{x}) + \sum_{p\leqslant\sqrt{x}} \frac{1}{p} \log\left(1 - \frac{\log p}{\log x}\right)$$

$$+ \sum_{p\leqslant\sqrt{x}} \frac{1}{p} \eta\left(\frac{x}{p}\right).$$

As we have:

$$P(\sqrt{x}) = \log\log(\sqrt{x}) + b + O\left(\frac{1}{\log x}\right)$$

$$= \log\log x - \log 2 + b + O\left(\frac{1}{\log x}\right),$$

we find that:

$$(\log\log x + b)P(\sqrt{x}) = (\log\log x)^2 + (2b - \log 2)\log\log x + O(1).$$

On the other hand, for $|u| < 1$ it is known that:

$$\log(1 - u) = -\sum_{k=1}^{+\infty} \frac{u^k}{k} = -u \sum_{k=1}^{+\infty} \frac{u^{k-1}}{k}.$$

Consequently, for $0 \leqslant u \leqslant \frac{1}{2}$:

$$|\log(1 - u)| \leqslant u \sum_{k=1}^{+\infty} \frac{1}{k2^{k-1}} = Ku.$$

Since, for $p \leqslant \sqrt{x}$, $0 \leqslant \dfrac{\log p}{\log x} \leqslant \frac{1}{2}$, we see that:

$$\left|\sum_{p\leqslant\sqrt{x}} \frac{1}{p} \log\left(1 - \frac{\log p}{\log x}\right)\right| \leqslant K \sum_{p\leqslant\sqrt{x}} \frac{\log p}{p\log x}$$

$$\leqslant \frac{K}{\log x} \sum_{p\leqslant\sqrt{x}} \frac{\log p}{p} \leqslant K,$$

since, by Exercise 1•5:I(2):

$$\sum_{p\leqslant\sqrt{x}} \frac{\log p}{p} \leqslant \log x.$$

For $p \leqslant \sqrt{x}$ we also have $\dfrac{x}{p} \geqslant \sqrt{x}$, and consequently:

$$\left|\eta\left(\frac{x}{p}\right)\right| \leqslant \frac{M}{\log \dfrac{x}{p}} \leqslant \frac{2M}{\log x}.$$

Therefore:

$$\left|\sum_{p\leqslant\sqrt{x}} \frac{1}{p}\, \eta\left(\frac{x}{p}\right)\right| \leqslant \frac{2M}{\log x} P(\sqrt{x}) = O\left(\frac{\log\log x}{x}\right).$$

Finally, we have:

$$\sum_{p\leqslant\sqrt{x}} \frac{1}{p} P\left(\frac{x}{p}\right) = (\log\log x)^2 + (2b - \log 2)\log\log x + O(1). \qquad (4)$$

Moreover, we have:

$$P(\sqrt{x})^2 = \left\{\log\log x + b - \log 2 + O\left(\frac{1}{\log x}\right)\right\}^2$$

$$= (\log\log x)^2 + 2(b - \log 2)\log\log x + O(1). \qquad (5)$$

Equations (3),(4),(5) give:

$$\sum_{p_1p_2 \leqslant x} \frac{1}{p_1p_2} = (\log\log x)^2 + 2b\log\log x + O(1). \qquad (6)$$

(4): From Equation (1) we deduce that for each $n \in \mathbb{N}^*$ we have:

$$\nu(n)^2 = \left(\sum_{p_1/n} 1\right)\left(\sum_{p_2/n} 1\right) = \sum_{\substack{p_1/n \\ p_2/n}} 1$$

Consequently, for $x \geqslant 1$:

$$\sum_{n \leqslant x} \nu(n)^2 = \sum_{n \leqslant x} \left(\sum_{\substack{p_1/n \\ p_2/n}} 1\right) = \sum_{\substack{p_1 \leqslant x \\ p_2 \leqslant x}} \left(\sum_{\substack{n \leqslant x \\ p_1/n \\ p_2/n}} 1\right).$$

If $p_1 \neq p_2$, "p_1/n and p_2/n" is equivalent to "p_1p_1/n", and we have:

$$\sum_{\substack{n \leqslant x \\ p_1/n \\ p_2/n}} = \begin{cases} 0 & \text{if } p_1p_2 > x, \\ \left[\dfrac{x}{p_1p_2}\right] & \text{if } p_1p_2 < x. \end{cases}$$

Thus:

$$\sum_{n \leqslant x} \nu(n)^2 = \sum_{\substack{p_1p_2 \leqslant x \\ p_1 \neq p_2}} \left[\frac{x}{p_1p_2}\right] + \sum_{p \leqslant x} \left[\frac{x}{p}\right].$$

We have already seen that:

$$\sum_{p \leqslant x} \left[\frac{x}{p}\right] = x \sum_{p \leqslant x} \frac{1}{p} + O(\pi(x)) = x\log\log x + bx + O\left(\frac{x}{\log x}\right).$$

It is sufficient to write here that:

$$\sum_{p\leqslant x} \left[\frac{x}{p}\right] = x\log\log x + O(x).$$

On the other hand, the number of pairs $[p_1,p_2]$ such that $p_1p_2 \leqslant x$ and $p_1 \neq p_2$ is less than $2x$, for each integer of the form p_1p_2 with $p_1 \neq p_2$ is written in this way in precisely two ways. Since, for each pair, the difference between $\left[\frac{x}{p_1p_2}\right]$ and $\frac{x}{p_1p_2}$ is at most equal to one, we have:

$$\sum_{\substack{p_1p_2\leqslant x\\ p_1\neq p_2}} \left[\frac{x}{p_1p_2}\right] = x \sum_{\substack{p_1p_2\leqslant x\\ p_1\neq p_2}} \frac{1}{p_1p_2} + O(x)$$

$$= x \sum_{p_1p_2\leqslant x} \frac{1}{p_1p_2} - x \sum_{p^2\leqslant x} \frac{1}{p^2} + O(x)$$

$$= x \sum_{p_1p_2\leqslant x} \frac{1}{p_1p_2} + O(x),$$

whence, on taking account of (6):

$$\sum_{\substack{p_1p_2\leqslant x\\ p_1\neq p_2}} \left[\frac{x}{p_1p_2}\right] = x(\log\log x)^2 + 2bx\log\log x + O(x).$$

Thus we arrive at:

$$\sum_{n\leqslant x} \nu(n)^2 = x(\log\log x)^2 + (2b + 1)x\log\log x + O(x). \tag{7}$$

(5): For each $n \leqslant x$ (where $x > 1$) we have:

$$(\nu(n) - \log\log x)^2 = \nu(n)^2 - 2\nu(n)\log\log x + (\log\log x)^2.$$

Consequently:

$$\sum_{n \leqslant x} (\nu(n) - \log\log x)^2$$

$$= \sum_{n \leqslant x} \nu(n)^2 - 2\left(\sum_{n \leqslant x} \nu(n)\right)\log\log x + [x](\log\log x)^2.$$

By using Equations (7) and (2), we deduce that when $x \to +\infty$

$$\sum_{n \leqslant x} (\nu(n) - \log\log x)^2 = x\log\log x + O(x). \tag{8}$$

(6): It is clear, for $x \geqslant e$, that:

$$\sum_{n \leqslant x} (\nu(n) - \log\log x)^2 \geqslant \lambda^2 n_\lambda(x)\log\log x,$$

whence:

$$\frac{1}{x} n_\lambda(x) \leqslant \frac{1}{\lambda^2 x\log\log x} \sum_{n \leqslant x} (\nu(n) - \log\log x)^2.$$

Using Equation (8) we get:

$$\overline{\lim_{x \to +\infty}} \frac{1}{x} n_\lambda(x) \leqslant \frac{1}{\lambda^2} .$$

(7):(a): For all n satisfying $\sqrt{x} < n \leqslant x$, we have:

$$\log\log x - \log 2 < \log\log n \leqslant \log\log x.$$

From this it follows that:

$$|\nu(n) - \log\log x| \geqslant |\nu(n) - \log\log n| - |\log\log n - \log\log x|$$

$$\geqslant |\nu(n) - \log\log n| - \log 2$$

and:

$$g(n)\sqrt{\log\log n} \geqslant g(\sqrt{x})\sqrt{\log\log x - \log 2}$$

(since $x > a^2 > e^2$, $\log\log x > \log 2$). This certainly gives the result in the Problem.

(7):(b): We have:

$$g(\sqrt{x})\sqrt{\log\log x - \log 2} - \log 2$$
$$= \sqrt{\log\log x}\left(g(\sqrt{x})\sqrt{1 - \frac{\log 2}{\log\log x}} - \frac{\log 2}{\sqrt{\log\log x}}\right)$$

For any $\lambda > 0$ the expression between parentheses is $\geqslant$ for x large enough. Then for n such that $\sqrt{x} < n \leqslant x$, and for which

$$|\nu(n) - \log\log n| \geqslant g(n)\sqrt{\log\log n},$$

we have:

$$|\nu(n) - \log\log x| \geqslant \lambda\sqrt{\log\log x}.$$

The number of such n's is therefore at most equal to $n_\lambda(x)$, and consequently:

$$N(x) \leqslant \sqrt{x} + n_\lambda(x),$$

whence:

$$\frac{N(x)}{x} \leqslant \sqrt{x} + \frac{1}{x} n_\lambda(x).$$

From this it follows, by what was shown in question (6) above, that:

$$\overline{\lim_{x \to +\infty}} \frac{N(x)}{x} \leqslant \frac{1}{\lambda^2}.$$

As this is true for any $\lambda > 0$, we see that $\frac{N(x)}{x} \to 0$ when $x \to +\infty$.

SOLUTION 1·9: (a): We have:

$$\sum_{2\leqslant n\leqslant x} \frac{d_n}{\log n} \geqslant \sum_{2\leqslant n\leqslant x} \frac{d_n}{\log x} = \frac{1}{\log x} \sum_{n=2}^{[x]} (p_{n+1} - p_n)$$

$$= \frac{p_{[x]+1} - 3}{\log x}.$$

By Exercise 1·3(ii) this expression is equivalent to:

$$\frac{[x + 1]\log[x + 1]}{\log x},$$

that is to say to x, when $x \to \infty$; hence:

$$\liminf_{x\to\infty} \frac{1}{x} \sum_{2\leqslant n\leqslant x} \frac{d_n}{\log n} \geqslant 1.$$

For every positive number $\alpha < 1$ we have:

$$\sum_{2\leqslant n\leqslant x} \frac{d_n}{\log n} = \sum_{2\leqslant n\leqslant x^\alpha} \frac{d_n}{\log n} + \sum_{x^\alpha\leqslant n\leqslant x} \frac{d_n}{\log n}$$

$$\leqslant \frac{1}{\log 2} \sum_{n\leqslant x^\alpha} d_n + \sum_{x^\alpha<n\leqslant x} \frac{d_n}{\alpha\log x}$$

$$\leqslant \frac{1}{\log 2} (p_{[x^\alpha]+1} - 3) + \frac{1}{\alpha\log x} (p_{[x+1]} - p_{[x^\alpha+1]}).$$

From this one deduces (as above) the relation:

$$\limsup_{x\to\infty} \frac{1}{x} \sum_{2\leqslant n\leqslant x} \frac{d_n}{\log n} \leqslant \frac{1}{\alpha}.$$

Since this relation is true for all $\alpha \in (0,1)$, we have:

$$\sum_{2\leqslant n\leqslant x} \frac{d_n}{\log x} \sim x.$$

(b): Let us assume that we have the relation:

$$\liminf_{x\to\infty} \frac{d_n}{\log n} = 1 + a > 1.$$

We would be able to find an integer n_0 such that for every $n > n_0$, $\frac{d_n}{\log n} \geqslant 1 + \frac{1}{2}a$. Then we would have:

$$\sum_{2\leqslant n\leqslant x} \frac{d_n}{\log n} = \sum_{2\leqslant n\leqslant n_0} \frac{d_n}{\log n} + \sum_{n_0<n\leqslant x} \frac{d_n}{\log n}$$

$$\geqslant C_0 + (1 + \tfrac{1}{2}a)[x - n_0].$$

From this one would deduce:

$$\liminf_{x\to\infty} \frac{1}{x} \sum_{2\leqslant n\leqslant x} \frac{d_n}{\log n} \geqslant 1 + \frac{a}{2},$$

which contradicts (a) preceding.

In a similar way we can prove the relation:

$$\limsup_{x\to\infty} \frac{d_n}{\log n} \geqslant 1.$$

SOLUTION 1·10: (a): For every integer ℓ, each of the numbers:

$$\ell\Pi_{n-2} - p_{n-1} + 1,\ \ell\Pi_{n-2} - p_{n-1} + 2,\ \ldots,$$

$$\ell\Pi_{n-2} - 2,\ \ell\Pi_{n-2},\ \ell\Pi_{n-2} + 2,\ \ldots,\ \ell\Pi_{n-2} + p_{n-1} - 1,$$

is divisible by at least one of the numbers $p_1,\ldots,p_{n-2}$. One then chooses an integer ℓ such that:

$$\ell\Pi_{n-2} \equiv 1 \pmod{p_{n-1}} \quad\text{and}\quad \ell\Pi_{n-2} \equiv -1 \pmod{p_n}.$$

By the Chinese Remainder Theorem (cf., [Har] Theorem 121) it is known that there exists such a positive integer $\ell \leqslant p_n p_{n-1}$ (it is then unique). Then the number $\ell\Pi_{n-2} - 1$ is divisible by p_{n-1} and $\ell\Pi_{n-2} + 1$ is divisible by p_n. (When $n = 5$, one has $\ell = 4$).

(b): We have $\log\Pi_n = \vartheta(p_n)$, where ϑ is the Chebycheff function. We know that $\vartheta(x) \sim x$, so it follows, using Exercise 1·3, that for n large enough $\Pi_{n-2} > 2p_n$, whence $\ell\Pi_{n-2} - p_{n-} > p_n$, and the integers in the interval $[\ell\Pi_{n-2} - p_{n-1} + 1, \ell\Pi_{n-2} + p_{n-1} - 1]$ are all composite. There exists $\lambda = \lambda(n)$ such that:

$$p_\lambda \leqslant \ell\Pi_{n-2} - p_{n-1} < \ell\Pi_{n-2} + p_{n-1} \leqslant p_{\lambda+1}.$$

We have $\lambda \geqslant n$, $d_\lambda \geqslant 2p_{n-1}$, and $p_\lambda \leqslant \Pi_n$; hence:

$$\log p_\lambda \leqslant \log\Pi_n = \vartheta(p_n).$$

Using Exercise 1·3, for any $\varepsilon > 0$ and n large enough, we have:

$$\frac{d_{\lambda(n)}}{\log\lambda(n)} \geqslant (1 - \varepsilon)\frac{d_{\lambda(n)}}{\log p_{\lambda(n)}} \geqslant (1 - \varepsilon)\frac{d_{\lambda(n)}}{\vartheta(p_n)}$$

$$\geqslant (1 - 2\varepsilon)\frac{2p_{n-1}}{p_n} \geqslant 2(1 - 3\varepsilon)$$

Hence:

$$\limsup_{n\to\infty}\left(\frac{d_{\lambda(n)}}{\log\lambda(n)}\right) \geqslant 2,$$

and *a fortiori*:

$$\limsup_{N\to\infty}\left(\frac{d_N}{\log N}\right) \geqslant 2 > 1.$$

SOLUTION 1·11: (1): The formula given in the Problem yields:

$$i = Q(q_i) = \frac{6q_i}{\pi^2} + O(\sqrt{q_i}).$$

From this we deduce that $i \sim \frac{6q_i}{\pi^2}$, whence $O(\sqrt{q_i}) = O(\sqrt{i})$, which gives:

$$q_i = \frac{\pi^2}{6} i + O(\sqrt{i}).$$

Next, we have:

$$q_{i+1} - q_i = \frac{\pi^2}{6}(i + 1 - i) + O(\sqrt{i + 1}) + O(\sqrt{i})$$

$$= O(\sqrt{i}) = O(\sqrt{q_i}),$$

since $i \sim \frac{6}{\pi^2} q_i$.

(2): If the numbers $n_1, n_2, \ldots, n_k$ are pairwise relatively prime, the same holds for the number $n_1^2, n_2^2, \ldots, n_k^2$, and by the "Chinese Remainder Theorem" the congruences:

$$N \equiv a_j \bmod n_j^2, \quad j = 1,2,\ldots,k,$$

have an unique solution N satisfying $0 < N \leqslant \prod_{j=1}^{k} n_j^2$, for any a_j. That is true, in particular, if $a_j = -j$.

Let us assume N to be chosen so that $N + j$ will be divisible by n_j^2 for $1 \leqslant j \leqslant k$. Let q_i be the largest element of Q such that $q_i \leqslant N$. Then $q_{i+1} > N + k$, and $q_{i+1} - q_i > k + 1$. As k can be chosen as large as one likes, from the latter statement it follows that:

$$\overline{\lim}(q_{i+1} - q_i) = +\infty.$$

In order to find q_i such that $q_{i+1} - q_i \geqslant 5$ we solve the congruences:

$$N \equiv 0 \bmod 4, \quad N \equiv -1 \bmod 9, \quad N \equiv -2 \bmod 25, \quad N \equiv -3 \bmod 49.$$

$N \equiv 0 \pmod 4$ and $N \equiv -2 \pmod{25}$ are equivalent to $N = 48$ modulo 100, or $N = 48 + 100k$. The equation:

$$48 + 100k \equiv -3 \bmod 49, \quad \text{or} \quad 2k \equiv -2 \bmod 49,$$

has $k = 48$ as solution. It remains to solve:

$$4848 + 4900k' \equiv -1 \bmod 9, \quad \text{or} \quad 4k' \equiv 2 \bmod 9, \quad \text{or} \quad k' = 5.$$

The smallest solution of the congruences required is therefore $N = 29{,}348$. The numbers $N, N+1, N+2, N+3, N+4$ have square factors. Hence we have:

$q_i = 29{,}347$, which is a prime number, and

and

$$q_{i+1} = 29{,}353 = 149 \times 197.$$

REMARK: The smallest solution of $q_{i+1} - q_i \geqslant 5$ is $q_i = 241$ and $q_{i+1} = 246$.

(3): The congruence:

$$N \equiv -j \bmod p_j^2 \quad \text{for } j = 1, 2, \ldots, k,$$

where p_j is the j-th prime number, has a solution $N \leqslant \prod_{i=1}^{k} p_i^2$. Let q_i be the largest element of Q such that $q_i \leqslant N$; then $q_{i+1} - q_i \geqslant k + 1$. We now have:

$$\log q_i \leqslant \log N \leqslant 2 \sum_{i=1}^{k} \log p_i = 2\vartheta(p_k),$$

where ϑ is the Chebycheff function. But we have:

$$\vartheta(p_k) \sim p_k \sim k \log k,$$

$$\log \vartheta(p_k) \sim \log k + \log\log k \sim \log k,$$

and:

$$k \sim \frac{\vartheta(p_k)}{\log k} \sim \frac{\vartheta(p_k)}{\log\vartheta(p_k)} \geqslant \frac{\frac{1}{2}\log q_i}{\log(\frac{1}{2}\log q_i)} \sim \frac{\frac{1}{2}\log q_i}{\log\log q_i},$$

because the function $\frac{x}{\log x}$ is increasing. With the formula:

$$q_{i+1} - q_i \geqslant k + 1,$$

this proves:

$$\overline{\lim}\left(\frac{(q_{i+1} - q_i)\log\log q_i}{\log q_i}\right) \geqslant \frac{1}{2}.$$

(4): As $\overline{\lim}(q_{i+1} - q_i) = +\infty$, this implies:

$$\underline{\lim}(s_{i+1} - s_i) = 1.$$

Let us now show that $\overline{\lim}(s_{i+1} - s_i) = 4$. As the multiples of 4 are in S, we always have $s_{i+1} - s_i \leqslant 4$. Let us now assume that for $i \geqslant i_0$ one has $s_{i+1} - s_i \leqslant 3$. When s is a multiple of 4 let $s_i = 4n$; this would imply $s_{i+1} \leqslant 4n + 3$, and amongst the four numbers $4n, 4n + 1, 4n + 2, 4n + 3$, at least two would be in S.

The density of the numbers in S would then be greater or equal to one half. Now this density, by the Formula of Question (1) above, is $1 - \frac{6}{\pi^2} = 0.39,\ldots$.

SOLUTION 1·12: (1): We have $g(1) = 1$, since $f(1) = 1 \neq 0$.

(2): Let m and n be such that $(m,n) = 1$. We have $f(mn) = f(m)f(n)$. If $f(m) \neq 0$ and $f(n) \neq 0$, then $f(mn) \neq 0$. We have then that $g(m) = 1$, $g(n) = 1$, and $g(mn) = 1$. If $f(m) = 0$ or $f(n) = 0$, then $f(mn) = 0$. Then $g(m) = 0$ or $g(n) = 0$, and $g(mn) = 0$. In all cases, $g(mn) = g(m)g(n)$.

SOLUTION 1·13: We know that:

$$h(p^r) = \sum_{d/p^r} f(d)g\left(\frac{p^r}{d}\right).$$

As the divisors of p^r are the numbers p^j, where $0 \leqslant j \leqslant r$, we have:

$$h(p^r) = \sum_{j=0}^{r} f(p^r)g(p^{r-j}).$$

This shows that the power series $\sum_{r=0}^{\infty} h(p^r)z^r$ is the product of the power series:

$$\sum_{r=0}^{+\infty} f(p^r)z^r \quad \text{and} \quad \sum_{r=0}^{+\infty} g(p^r)z^r.$$

If $f(1) \neq 0$ it is seen that the formal product of the power series:

$$\sum_{r=0}^{+\infty} f(p^r)z^r \quad \text{and} \quad \sum_{r=0}^{+\infty} f^{-*}(p^r)z^r$$

is the series $\sum_{r=0}^{+\infty} e(p^r)z^r$, which reduces to its constant term, i.e. one. Hence, the series $\sum_{r=0}^{+\infty} f^{-*}(p^r)z^r$ is the formal quotient of one by the series $\sum_{r=0}^{+\infty} f(p^r)z^r$.

APPLICATION 1: As the functions λ and z are multiplicative, the function $\lambda * z$ is also multiplicative. In order to determine it it is sufficient to determine its values for the numbers p^r.

Since $\lambda(p^r) = (-1)^r$ and $z(p^r) = 1$, the series $\sum_{r=0}^{+\infty} \lambda(p^r)z^r$ and $\sum_{r=0}^{+\infty} z(p^r)z^r$ are absolutely convergent for $|z| < 1$, with $\frac{1}{1+z}$ and

and $\frac{1}{1 - z}$ as sums. The product series is therefore the expansion of $\frac{1}{1 - z^2}$, that is to say, $1 + z^2 + z^4 + \cdots + z^{2k} + \cdots$. Hence we have:

$$(\lambda * z)(p^r) = \begin{cases} 1 & \text{if } r \text{ is even,} \\ 0 & \text{if } r \text{ is odd.} \end{cases}$$

It follows from this that $\lambda * z$ is the function equal to one for the squares and zero for the non-squares.

The function λ^{-*} is multiplicative, and $\sum_{r=0}^{+\infty} \lambda^{-*}(p^r)z^r$ is the expansion, in a neighbourhood of the origin, of $1/\left(\frac{1}{1 + z}\right) = 1 + z$. Hence we have:

$$\lambda^{-*}(p^r) = \begin{cases} 1 & \text{for } r = 1, \\ 0 & \text{for } r \quad 1. \end{cases}$$

Therefore:

$$\lambda^{-*}(n) = \begin{cases} 1 & \text{if } n \text{ is squarefree} \\ 0 & \text{otherwise,} \end{cases}$$

(i.e., $\lambda^{-*} = |\mu|$ or μ^2).

APPLICATION 2: For the function under consideration,

$$\sum_{r=0}^{+\infty} f(p^r)z^r = 1 + \sum_{r=1}^{\infty} \frac{1.3.5...(2r - 1)}{2.4.6...2r} z^r$$

is the expansion in a neighbourhood of the origin of $\frac{1}{\sqrt{1 - z}}$ (the branch equal to one for $z = 0$).

If $g = f*f$ the series $\sum_{r=0}^{+\infty} g(p^r)z^r$ is the expansion of $\left(\frac{1}{\sqrt{1 - z}}\right)^2 = \frac{1}{1 - z}$, that is to say, $\sum_{r=0}^{+\infty} z^r$. Therefore we have $g(p^r) = 1$ for all r. As g is multiplicative, $g(n) = 1$ for all n, i.e., $g = z$. Thus $f*f = z$.

SOLUTION 1·14: Let $\Phi = f*g$, $\Psi = (fh)*(gh)$. By the definition of convolution, we have:

$$\Psi(n) = \sum_{d/n} f(d)h(d)g\left(\frac{n}{d}\right)h\left(\frac{n}{d}\right) .$$

Since, for each d, $h(d)h\left(\frac{n}{d}\right) = h(n)$, this gives:

$$\Psi(n) = h(n) \sum_{d/n} f(d)g\left(\frac{n}{d}\right) = h(n)\Phi(n).$$

If $f(1) \neq 0$, by taking $g = f^{-*}$ one obtains:

$$(fh)*(f^{-*}h) = (f*f^{-*})h = eh = e,$$

(since $h(1) = 1$ and $e(n) = 0$ for $n > 1$). This shows that $f^{-*}h = (fh)^{-*}$. If $f = z$, one has $f^{-*} = \mu$. Thus we see that $h^{-*} = \mu h$ (as $zh = h$).

SOLUTION 1·15: (1): We have $\sum_{d/n} f(d) = (z*f)(n)$. However, z and f are multiplicative, so $z*f$ is multiplicative, and for n square-free we have:

$$(z*f)(n) = \prod_{p/n} (z*f)(p).$$

With the divisors of p being one and p, we have:

$$(z*f)(p) = z(1)f(p) + z(p)f(1) = f(p) + 1,$$

whence the result indicated.

For arbitrary n:

$$(z*f)(n) = \prod_{p^r//n} (z*f)(p).$$

As the divisors of p^r are the numbers p^j, where $0 \leqslant j \leqslant r$, we have:

$$(z*f)(p^r) = (f*z)(p^r) = \sum_{j=0}^{r} f(p^j)z(p^{r-j}) = \sum_{j=0}^{r} f(p^j).$$

If $f(p^r) = 0$, when $r > 1$ this reduces to $1 + f(p)$. We have therefore:

$$(z*f)(n) = \prod_{p/n} (1 + f(p)) \quad \text{for all } n \in \mathbb{N}^*.$$

(2): One has:

$$\sum_{d/n} \mu(d)f\left(\frac{n}{d}\right) = (\mu*f)(n).$$

f is multiplicative, and consequently, for n square-free

$$(\mu*f)(n) = \prod_{p/n} (\mu*f)(p).$$

But:

$$(\mu*f)(p) = \mu(1)f(p) + \mu(p)f(1) = f(p) - 1.$$

SOLUTION 1·16: We have:

$$kq = \prod_{p/n} p^{v_p(h)} = h.$$

For any $n \in \mathbb{N}^*$ we have:

$$f(q)f(n + k) = f(qn + qk) = f(qn + h) = f(qn) = f(q)f(n).$$

But $f(q) \neq 0$ since for every p dividing q, $f(p) \neq 0$. Therefore this implies $f(n + k) = f(n)$.

If $(n,k) > 1$ there exists a p which divides n and k. Since p divides k, $f(p) = 0$. Consequently $f(n) = 0$.

If $(n,k) = 1$ there exists $m \in \mathbb{N}^*$ such that $mn \equiv 1 \pmod{k}$.

We have:

$$f(m)f(n) = f(mn) = f(1).$$

Therefore one cannot have $f(n) = 0$.

SOLUTION 1·17: I: We have:

$$(f*g)(n)\ell(n) = \left(\sum_{d_1d_2=n} f(d_1)g(d_2)\right)\ell(n)$$

$$= \sum_{d_1d_2=n} f(d_1)g(d_2)\ell(n).$$

On replacing $\ell(n)$ by $\ell(d_1) + \ell(d_2)$ we obtain:

$$(f*g)(n)\ell(n) = \sum_{d_1d_2=n} f(d_1)\ell(d_1)g(d_2)$$

$$+ \sum_{d_1d_2=n} f(d_1)g(d_2)\ell(d_2)$$

$$= ((f\ell)*g)(n) + (f*(g\ell))(n).$$

II: One sees immediately that the definition given for Λ_f is equivalent to defining it by the condition $\Lambda_f * f = f\log$.

II:(1): By taking $\ell = \log$ in the result of Question I above, we obtain:

$$h\log = (f\log)*g + f*(g\log) = (\Lambda_f*f)*g + f*(\Lambda_g*g)$$

$$= \Lambda_f*(f*g) + \Lambda_g*(f*g) \quad \text{the convolution being commutative and associative}$$

$$= \Lambda_f*h + \Lambda_g*h = (\Lambda_f + \Lambda_g)*h$$

By the above remark, this shows that $\Lambda_h = \Lambda_f + \Lambda_g$.

It is clear that we have $\Lambda_f = 0$ if and only if $f\log = 0$, in other words $f(n)\log n = 0$ for all $n \in \mathbb{N}^*$. Since $\log 1 = 0$ and

$\log n \neq 0$ for $n > 1$, this is equivalent to:

$$f(n) = 0 \quad \text{for all } n > 1,$$

which comes down to saying that f is of the form λe, where λ is a non-zero complex constant (by hypothesis $f(1) \neq 0$). The kernel of the homomorphism $f \to \Lambda_f$ is therefore the set of functions λe, where $\lambda \in \mathbb{C}^*$. One sees that the image is the set of arithmetic functions which are zero for $n = 1$.

Indeed, on the one hand we always have:

$$\Lambda_f(1) = (f(1)\log 1)f^{-*}(1) = 0 \ .$$

On the other hand, if g is an arbitrary arithmetic function satisfying $g(1) = 0$ one can determine a function f by arbitrarily choosing a non-zero value for $f(1)$ and determining $f(n)$ for $n > 1$ by the recurrence relation:

$$f(n) = \frac{1}{\log n} \sum_{\substack{d/n \\ d>1}} g(d) f\left(\frac{n}{d}\right) .$$

Thus:

$$f(n)\log n = \sum_{d/n} g(d) f\left(\frac{n}{d}\right) \quad \text{for every } n \in \mathbb{N}^*,$$

in other words $g * f = f\log$, which shows that $\Lambda_f = g$.

It is known that von Mangoldt's function Λ satisfies $\Lambda * z = \log$. This can be written $\Lambda * z = z\log$, and from this it follows that $\Lambda_z = \Lambda$.

II:(2): Let us assume f to be multiplicative. We already know that $\Lambda_f(1) = 0$. If, now, $n > 1$ and is not a power of a prime, there exist $n_1, n_2 > 1$ such that $(n_1, n_2) = 1$ and $n_1 n_2 = n$. By the definition of Λ_f we have:

$$\Lambda_f(n) = \sum_{d/n} f(d)\log d . f^{-*}\left(\frac{n}{d}\right) =$$

(Contd)

(Contd)
$$= \sum_{\substack{d_1/n_1 \\ d_2/n_2}} f(d_1d_2)\log(d_1d_2)f^{-*}\left(\frac{n}{d_1d_2}\right)$$

$$= \sum_{\substack{d_1/n_2 \\ d_2/n_2}} f(d_1)f(d_2)(\log d_1 + \log d_2)f^{-*}\left(\frac{n_1}{d_2}\right)f^{-*}\left(\frac{n_1}{d_2}\right)$$

$$= \sum_{\substack{d_1/n_1 \\ d_2/n_2}} f(d_1)\log d_1 . f^{-*}\left(\frac{n_1}{d_1}\right)f(d_2)f^{-*}\left(\frac{n_2}{d_2}\right)$$

$$+ \sum_{\substack{d_1/n_2 \\ d_2/n_2}} f(d_1)f^{-*}\left(\frac{n_1}{d_1}\right)f(d_2)\log d_2 . f^{-*}\left(\frac{n_2}{d_2}\right)$$

$$= \left(\sum_{d_1/n_1} f(d_1)\log d_1 . f^{-*}\left(\frac{n_1}{d_1}\right)\right)\left(\sum_{d_2/n_2} f(d_2)f^{-*}\left(\frac{n_2}{d_2}\right)\right)$$

$$+ \left(\sum_{d_1/n_1} f(d_1)f^{-*}\left(\frac{n_1}{d_1}\right)\right)\left(\sum_{d_2/n_2} f(d_2)\log d_2 . f^{-*}\left(\frac{n_2}{d_2}\right)\right)$$

$$= \Lambda_f(n_1)e(n_2) + e(n_1)\Lambda_f(n_2)$$

$$= 0,$$

since $e(n_1) = e(n_2) = 0$.

Let us now assume that $f(1) = 1$ and that $\Lambda_f(n) = 0$ whenever n is not a power of a prime number. Let g be the multiplicative function determined by:

$$g(p^r) = f(p^r) \text{ for all prime numbers } p \text{ and all } r \in \mathbb{N}^*.$$

We see that $\Lambda_g = \Lambda_f$.

In fact, first of all, if n is not a power of a prime number, $\Lambda_f(n) = 0$ and $\Lambda_g(n) = 0$.

Next, given a prime number p it follows from the relations $\Lambda_f * f = f\log$ and $\Lambda_g * g = g\log$ that for all $r \geq 1$:

$$\sum_{j=1}^{r} \Lambda_f(p^j)f(p^{r-j}) = rf(p^r)\log p \quad (\Lambda_f(p^0) = \Lambda_f(1) = 0),$$

and similarly:

$$\sum_{j=1}^{r} \Lambda_g(p^j)g(p^{r-j}) = rg(p^r)\log p,$$

that is to say:

$$\sum_{j=1}^{r} \Lambda_g(p^j)f(p^{r-j}) = rf(p^r)\log p.$$

By induction on r, this shows that:

$$\Lambda_g(p^r) = \Lambda_f(p^r) \quad \text{for all } r \geq 1.$$

As $\Lambda_f = \Lambda_g$, the function $f * g^{-*}$ belongs to the kernel of the homomorphism considered in Question II(1). Hence, $f * g^{-*} = \lambda e$, with $\lambda \in \mathbb{C}^*$. Since $(f * g^{-*})(1) = f(1)g^{-*}(1) = 1$, we have $\lambda = 1$. Hence $f * g^{-*} = e$, which shows that $f = g$. Since g is multiplicative, f is multiplicative.

SOLUTION 1·18: I: Let us first notice that each $n \in E$ can be written *uniquely* in the form:

$$n = mq^2 \quad \text{with } m/P \text{ and } (q,P) = 1,$$

for this equality implies $(n,P) = m$.

I:(a): The number of elements of E, $\leq x$, corresponding to a given m is at most equal to the number of $q \in \mathbb{N}$ such that $q^2 \leq \frac{x}{m}$, hence at most to $\sqrt{\frac{x}{m}}$. Consequently, the number of elements of $E \leq x$ is at most $\sqrt{x} \sum_{m/P} \frac{1}{\sqrt{m}}$.

The sum $\sum_{m/P} \frac{1}{\sqrt{m}}$ is the value at P of the convolution of the function z and the function $n \to \frac{1}{\sqrt{n}}$. This convolution is a multiplicative function whose value at each prime number p is $1 + \frac{1}{\sqrt{p}}$. As P is squarefree, we have:

$$\sum_{m/P} \frac{1}{\sqrt{m}} = \prod_{p/P} \left(1 + \frac{1}{\sqrt{p}}\right) = \prod_{p \in E} \left(1 + \frac{1}{\sqrt{p}}\right) .$$

The number of elements of $E \leqslant x$ therefore is at most:

$$\sqrt{x} \prod_{p \in E} \left(1 + \frac{1}{\sqrt{p}}\right) .$$

The $n \in E$ such that $n > x$ are of the form mq^2, with m/P and $q > \sqrt{\frac{x}{m}}$, and we have:

$$\sum_{\substack{n \in E \\ n > x}} \frac{1}{n} \leqslant \sum_{m/P} \frac{1}{m} \left(\sum_{q > \sqrt{\frac{x}{m}}} \frac{1}{q^2} \right) .$$

As:

$$\sum_{q > \sqrt{\frac{x}{m}}} \frac{1}{q^2} \leqslant \frac{m}{x} + \sqrt{\frac{m}{x}} ,$$

this gives:

$$\sum_{\substack{n \in E \\ n > x}} \frac{1}{n} \leqslant \frac{1}{x} \sum_{m/P} 1 + \frac{1}{\sqrt{x}} \sum_{m/P} \frac{1}{\sqrt{m}} .$$

The last sum has already been calculated. On the other hand, the number of divisors of P is the number of subsets of E, i.e. 2^k. Thus we obtain:

$$\sum_{\substack{n \in E \\ n > x}} \frac{1}{n} \leqslant \frac{2^k}{x} + \frac{1}{\sqrt{x}} \prod_{p \in E} \left(1 + \frac{1}{\sqrt{p}}\right) .$$

I:(b): The function χ is multiplicative, for if $(n_1, n_2) = 1$,

n_1n_2 is square-free and prime to P if and only if this is true for n_1 and n_2.

The function h is multiplicative, being the convolution of two multiplicative functions. In order to determine it, it suffices to determine $h(p^r)$. We have:

$$h(p^r) = \sum_{j=0}^{r} \mu(p^j)\chi(p^{r-j}) = \chi(p^r) - \chi(p^{r-1}),$$

which gives:

$$h(p^r) = 0 \quad \text{for } r > 2,$$

$$h(p) = \begin{cases} -1 & \text{if } p \in E, \\ 0 & \text{if } p \notin E, \end{cases} \qquad \text{and} \qquad h(p^2) = \begin{cases} 0 & \text{if } p \in E, \\ -1 & \text{if } p \notin E. \end{cases}$$

We see that:

$$\sum_{r=1}^{+\infty} \frac{|h(p^r)|}{p^r} = \begin{cases} \dfrac{1}{p} & \text{if } p \in E, \\ \dfrac{1}{p^2} & \text{if } p \notin E, \end{cases}$$

so that:

$$\sum_{\substack{p,r \\ r \geq 1}} \frac{|h(p^r)|}{p} < +\infty.$$

Consequently the series $\sum_{n=1}^{+\infty} \frac{h(n)}{n}$ is absolutely convergent, and we have:

$$\sum_{n=1}^{+\infty} \frac{h(n)}{n} = \prod_p \left(1 + \sum_{r=1}^{+\infty} \frac{h(p^r)}{r}\right) = \left(\prod_{p\in E}\left(1 - \frac{1}{p}\right)\right)\left(\prod_{p\notin E}\left(1 - \frac{1}{p^2}\right)\right)$$

$$= \left(\prod_p \left(1 - \frac{1}{p^2}\right)\right)\left(\prod_{p\in E}\left(1 + \frac{1}{p}\right)^{-1}\right) = \frac{6}{\pi^2} \prod_{p\in E}\left(1 + \frac{1}{p}\right)^{-1}$$

On the other hand, $h(n)$ is non-zero if and only if the decomposition of n into prime factors does not contain numbers of E with exponents greater than one, nor numbers not belonging to E with exponents $\neq 2$, in other words if and only if it is of the form mq^2, with m/P and q square-free and prime to P. Furthermore, we see that when $h(n) \neq 0$, $|h(n)| = 1$.

Notice that the set of n's such that $h(n) \neq 0$ is a subset of E.

I:(c): As $h = \chi * \mu$, we have $\chi = h * z$, and consequently, for all $x > 0$:

$$\sum_{n \leqslant x} \chi(n) = \sum_{n \leqslant x} h(n)\left[\frac{x}{n}\right] .$$

Replacing $h(n)\left[\frac{x}{n}\right]$ by $h(n)\,\frac{x}{n}$ the error has absolute value equal at most to $|h(n)|$. Hence, replacing $\sum_{n \leqslant x} h(n)\left[\frac{x}{n}\right]$ by $x \sum_{n \leqslant x} \frac{h(n)}{n}$, one is off in absolute value by at most $\sum_{n \leqslant x} |h(n)|$, hence at most by the number of elements of $E \leqslant x$, and consequently by $\sqrt{x} \prod_{p \in E} \left(1 + \frac{1}{\sqrt{p}}\right)$.

Next, replacing $x \sum_{n \leqslant x} \frac{h(n)}{n}$ by $\sum_{n=1}^{+\infty} \frac{h(n)}{n}$, we introduce a new error with absolute value at most equal to $x\left|\sum_{n>x} \frac{h(n)}{n}\right|$, hence to $x \sum_{n>x} \frac{|h(n)|}{n}$, and consequently at most equal to $\sum_{\substack{n \in E \\ n>x}} \frac{1}{n}$, hence to $2^k + \sqrt{x} \prod_{p \in E} \left(1 + \frac{1}{\sqrt{p}}\right)$.

Finally, observe that:

$$\left|\sum_{n \leqslant x} \chi(n) - x \sum_{n=1}^{+\infty} \frac{h(n)}{n}\right| \leqslant 2^k + 2\sqrt{x} \prod_{p \in E} \left(1 + \frac{1}{\sqrt{p}}\right) .$$

Using the value of $\sum_{n=1} \frac{h(n)}{n}$, this yields:

$$\left|\sum_{n\leqslant x} \chi(n) - \frac{6}{\pi^2} x \prod_{p\in E} \left(1 + \frac{1}{p}\right)^{-1}\right|$$

$$\leqslant 2^k + 2\sqrt{x} \prod_{p\in E} \left(1 + \frac{1}{\sqrt{p}}\right) \leqslant 2^k(1 + 2\sqrt{x})$$

II: If m is square-free, $Q_m(x)$ is clearly 0. As $\gamma(m) = 0$, we have:

$$Q_m(x) - \frac{6}{\pi^2} \gamma(m)x = 0.$$

Let us assume, therefore, that m has no square factor. Then, the square-free numbers that are divisible by m are the numbers of the form mn, where n is square-free and prime to m. The number of those which are at most equal to x is therefore equal to the number of square-free n's such that $n \leqslant \frac{x}{m}$ and $(n,m) = 1$.

Taking the set of prime divisors of m for E, Question I(c) gives:

$$\left|Q_m(x) - \frac{6}{\pi^2}\frac{x}{m} \prod_{p/m} \left(1 + \frac{1}{p}\right)^{-1}\right| \leqslant 2^{\nu(m)}\left(1 + 2\sqrt{\frac{x}{m}}\right),$$

or, since $m = \prod_{p/m} p$:

$$\left|Q_m(x) - \frac{6}{\pi^2} \gamma(m)x\right| \leqslant 2^{\nu(m)}\left(1 + 2\sqrt{\frac{x}{m}}\right).$$

If $x \geqslant m$ one has $\sqrt{\frac{x}{m}} \geqslant 1$, and the second member is equal at most to:

$$3.2^{\nu(m)}\sqrt{\frac{x}{m}} = 3\sqrt{x}\,\frac{2^{\nu(m)}}{\sqrt{m}}.$$

But:

$$\frac{2^{\nu(m)}}{\sqrt{m}} = \prod_{p/m} \frac{2}{\sqrt{p}} ,$$

and this is at most equal to $\frac{2\sqrt{2}}{\sqrt{3}}$, as $\frac{2}{\sqrt{p}} < 1$ for all p except 2 and 3, and $\frac{2}{\sqrt{2}}.\frac{2}{\sqrt{3}} = \frac{2\sqrt{2}}{\sqrt{3}}$. Therefore if $x \geqslant m$:

$$\left|Q_m(x) - \frac{6}{\pi^2}\gamma(m)x\right| \leqslant 2\sqrt{6}\sqrt{x}.$$

If $x < m$ one has $Q_m(x) = 0$, and consequently:

$$\left|Q_m(x) - \frac{6}{\pi^2}\gamma(m)x\right| = \frac{6}{\pi^2}\gamma(m)x \leqslant \frac{6}{\pi^2}\frac{x}{m} < \frac{x}{m} ,$$

since $\gamma(m) \leqslant \frac{1}{m}$. But:

$$\frac{x}{m} = \sqrt{x}\,\frac{1}{\sqrt{m}}\sqrt{\frac{x}{m}} < \sqrt{x}.$$

Finally, we see that for any m and any $x > 0$:

$$\left|Q_m(x) - \frac{6}{\pi^2}\gamma(m)x\right| \leqslant 2\sqrt{6}\sqrt{x}.$$

SOLUTION 1·19: (1): If $n = 1$ it is clear that there is the single solution $m = q = 1$. If $n = \prod_{j=1}^{r} p_j^{\alpha_j}$, where $p_1,p_2,\ldots,p_r$ are distinct prime numbers and $\alpha_1,\alpha_2,\ldots,\alpha_r \geqslant 1$ are integers, then m and q, divisors of n, must have the form:

$$m = \prod_{j=1}^{r} p_j^{\beta_j} \quad \text{and} \quad q = \prod_{j=1}^{r} p_j^{\gamma_j}, \quad \text{with } \beta_j,\gamma_j \geqslant 0.$$

With m and q given by these formulae, then $n = m^2q$, with q square-free if and only if, for each j, $2\beta_j + \gamma_j = \alpha$ and $\gamma\ = 0$ or 1. Therefore, there is an unique solution, obtained by taking β_j and γ_j equal respectively to the quotient and to the remainder when

a_j is divided by 2.

(2): We see that "d^2/n" is equivalent to "d/m". Indeed, if $n = 1$, the only possible d is $d = 1$, and $m = 1$. If $n = \prod_{j=1}^{r} p_j^{\alpha_j}$, d must be of the form $\prod_{j=1} p_j^{\delta_j}$, with $\delta_j \geq 0$. But the square of this number divides n if and only if $2\delta_j \leq \alpha_j$, that is to say $\delta_j \leq \beta_j$ (as indicated above). This says that:

$$\sum_{d^2/n} \mu(d) = \sum_{d/m} \mu(d) = \begin{cases} 1 & \text{if } m = 1, \\ 0 & \text{if } m > 1. \end{cases}$$

But it is clear that $m = 1$ if and only if n is square-free.

(3): By the preceding, we have:

$$Q(x:k,\ell) = \sum_{\substack{n \leq x \\ n \equiv \ell (\mathrm{mod}\, k)}} \left(\sum_{d^2/n} \mu(d) \right).$$

Since, for $d^2|n$ with $n \leq x$, it is necessary that $d^2 \leq x$, this only occurs for d's at most equal to $\sqrt{x}$.

On interchanging the order of summation, we obtain:

$$Q(x:k,\ell) = \sum_{d \leq \sqrt{x}} \mu(d) \left(\sum_{\substack{n \leq x \\ n \equiv \ell (\mathrm{mod}\, k) \\ d^2/n}} 1 \right).$$

Since "d^2/n" is equivalent to "$n = md^2$" with $m \in \mathbb{N}^*$, then:

$$\sum_{\substack{n \leq x \\ n \equiv \ell (\mathrm{mod}\, k) \\ d^2/n}} 1 = \sum_{\substack{m \leq x/d^2 \\ md^2 \equiv \ell (\mathrm{mod}\, k)}} 1.$$

If $(d^2,k) \nmid \ell$, the congruence $md^2 \equiv \ell (\mathrm{mod}\, k)$ has no solution in m, and the sum is zero. Hence one has only to consider the d's for which $(d^2,k)/\ell$. For such a d the congruence has exactly one solution in each sequence of $\frac{k}{(d^2,k)}$ consecutive integers. Consequently for any $X > 0$:

$$\sum_{\substack{m\leqslant X\\ md^2\equiv\ell(\mathrm{mod}\,k)}} 1 = \frac{(d^2,k)}{k} X + \rho(X), \quad \text{with } |\rho(X)| \leqslant 1,$$

since, if:

$$r \frac{k}{(d^2,k)} \leqslant X < (r+1) \frac{k}{(d^2,k)},$$

the sum is $\geqslant r$ and $\leqslant r+1$, and consequently the two numbers:

$$\frac{(d^2,k)}{k} X \quad \text{and} \quad \sum_{\substack{m\leqslant X\\ md^2\equiv\ell(\mathrm{mod}\,k)}} 1$$

belong to the interval $[r, r+1]$, and so differ by at most one. Hence:

$$Q(x:k,\ell) = \sum_{\substack{d\leqslant\sqrt{x}\\ (d^2,k)/\ell}} \mu(d)\left\{\frac{(d^2,k)}{k} \cdot \frac{x}{d^2} + \rho\left(\frac{x}{d^2}\right)\right\}$$

$$= \frac{x}{k} \sum_{\substack{d\leqslant\sqrt{x}\\ (d^2,k)/\ell}} \frac{\mu(d)(d^2,k)}{d^2} + \sum_{\substack{d\leqslant\sqrt{x}\\ (d^2,k)/\ell}} \mu(d)\rho\left(\frac{x}{d^2}\right).$$

Since $\left|\frac{\mu(d)(d^2,k)}{d^2}\right| \leqslant \frac{k}{d^2}$, the series $\sum_{(d^2,k)/\ell} \frac{\mu(d)(d^2,k)}{d^2}$ is absolutely convergent, and we can write:

$$Q(x:k,\ell) = A_{k,\ell}x + R_{k,\ell}(x),$$

with:

$$A_{k,\ell} = \frac{1}{k} \sum_{(d^2,k)/\ell} \frac{\mu(d)(d^2,k)}{d^2}$$

and:

$$R_{k,\ell}(x) = -\frac{x}{k} \sum_{\substack{d\leqslant\sqrt{x}\\ (d^2,k)/\ell}} \frac{\mu(d)(d^2,k)}{d^2} + \sum_{\substack{d\leqslant\sqrt{x}\\ (d^2,k)/\ell}} \mu(d)\rho\left(\frac{x}{d^2}\right).$$

We have:

$$\left|\sum_{\substack{d\leqslant\sqrt{x}\\ (d^2,k)/\ell}} \frac{\mu(d)(d^2,k)}{d^2}\right| \leqslant k \sum_{d\leqslant\sqrt{x}} \frac{1}{d^2}$$

$$\leqslant k\left(\frac{1}{x} + \int_{\sqrt{x}}^{+\infty} \frac{1}{t^2}\,dt\right) = k\left(\frac{1}{x} + \frac{1}{\sqrt{x}}\right),$$

and:

$$\left|\sum_{\substack{d\leqslant\sqrt{x}\\ (d^2,k)/\ell}} \mu(d)\rho\left(\frac{x}{d^2}\right)\right| \leqslant \sum_{d\leqslant\sqrt{x}} 1 \leqslant \sqrt{x},$$

and consequently:

$$|R_{k,\ell}(x)| \leqslant 2\sqrt{x} + 1.$$

The formula which defines $A_{k,\ell}$ can be written:

$$A_{k,\ell} = \frac{1}{k} \sum_{d=1}^{+\infty} h(d) \frac{\mu(d)(d^2,k)}{d^2} \quad \text{where } h(d) = \begin{cases} 1 & \text{if } (d^2,k)/\ell, \\ 0 & \text{if } (d^2,k)\nmid\ell. \end{cases}$$

It is immediately verified that the function $d \mapsto (d^2,k)$ and the function h are multiplicative. Thus:

$$A_{k,\ell} = \frac{1}{k} \sum_{d=1}^{+\infty} g(d),$$

where the function g, defined by:

$$g(d) = h(d)\,\frac{\mu(d)(d^2,k)}{d^2},$$

is multiplicative. As the series is absolutely convergent, its sum is equal to:

$$\prod_p \left(1 + \sum_{r=1}^{+\infty} g(p^r)\right).$$

We see that $g(p^r) = 0$ for $r > 1$, and:

$$g(p) = \begin{cases} 0 & \text{if } (p^2,k) \nmid \ell. \\ -\dfrac{(p^2,k)}{p} & \text{if } (p^2,k)/\ell. \end{cases}$$

Consequently:

$$A_{k,\ell} = \frac{1}{k} \prod_{(p^2,k)/\ell} \left(1 - \frac{(p^2,k)}{p^2}\right).$$

If $p \nmid k$, $(p^2,k) = 1$, and therefore $(p^2,k)/\ell$. If p/k and $p^2 \nmid k$, $(p^2,k) = p$ and therefore $(p^2,k)/\ell$ provided that p/ℓ. If p^2/k, $(p^2,k) = p^2$ and $(p^2,k) \nmid \ell$, since (k,ℓ) is square-free. Therefore:

$$\prod_{(p^2,k)/\ell} \left(1 - \frac{(p^2,k)}{p^2}\right)$$

$$= \prod_{p \nmid k} \left(1 - \frac{1}{p^2}\right) \prod_{\substack{p/(k,\ell) \\ p^2 \nmid k}} \left(1 - \frac{1}{p}\right)$$

$$= \left(\prod_p \left(1 - \frac{1}{p^2}\right)\right)\left(\prod_{p/k} \left(1 - \frac{1}{p^2}\right)^{-1}\right)\left(\prod_{\substack{p/(k,\ell) \\ p^2 \nmid k}} \left(1 - \frac{1}{p}\right)\right) =$$

(Contd)

(Contd) $$= \frac{6}{\pi^2}\left(\prod_{p/k}\left(1-\frac{1}{p^2}\right)^{-1}\right)\left(\prod_{\substack{p/(k,\ell)\\ p^2\nmid k}}\left(1-\frac{1}{p}\right)\right).$$

By separating, amongst the p which divide k, those which do not divide ℓ and those which divide (k,ℓ), we obtain:

$$\prod_{(p^2,k)/\ell}\left(1-\frac{(p^2,k)}{p^2}\right)$$

$$= \frac{6}{\pi^2}\left(\prod_{\substack{p/k\\ p\nmid\ell}}\left(1-\frac{1}{p^2}\right)^{-1}\right)\left(\prod_{\substack{p/(k,\ell)\\ p^2/k}}\left(1-\frac{1}{p^2}\right)^{-1}\right)\left(\prod_{\substack{p/(k,\ell)\\ p^2\nmid k}}\left(1+\frac{1}{p}\right)^{-1}\right).$$

SOLUTION 1·20:(1): The values that (m,n) assumes for $1 < m < n$ are the divisors of n. If d is a divisor of n and $n = dn'$, the number of m's such that $1 \leqslant m \leqslant n$ and $(m,n) = d$ is equal to $\varphi(n')$, for the condition $(m,n) = d$ is equivalent to $m = dm'$ with $(m',n') = 1$. Thus by grouping together the m's such that $1 \leqslant m \leqslant n$ according to the values of (m,n), we see that:

$$n = \sum_{d/n}\varphi\left(\frac{n}{d}\right),$$

which says that $i = z*\varphi$.

The First Möbius Inversion Formula then shows that $\varphi = \mu*i$. Since the functions μ and i are multiplicative, so is φ. Then, $\frac{\varphi}{i}$ is also multiplicative, and as it takes the value $1-\frac{1}{p}$ for $n = p^r$, we have:

$$\frac{\varphi(n)}{n} = \frac{\varphi(n)}{i(n)} = \prod_{p/n}\left(1-\frac{1}{p}\right).$$

(2): As $\varphi = \mu*i$, for $x \geqslant 1$ we have:

$$\sum_{n\leqslant x}\varphi(n) = \sum_{n\leqslant x}\mu(n)I\left(\frac{x}{n}\right), \quad \text{where } I(X) = \sum_{n\leqslant X} n. \tag{1}$$

But for $X \geqslant 1$, $I(X) = \frac{1}{2}[X]([X] + 1)$, and consequently:

$$\tfrac{1}{2}(X - 1)X \leqslant I(X) \leqslant \tfrac{1}{2}X(X + 1).$$

Therefore, for $X \geqslant 1$ we have:

$$I(X) = \tfrac{1}{2}X^2 + R(X), \quad \text{with } |R(X)| \leqslant \tfrac{1}{2}X.$$

Thus Equation (1) gives:

$$\sum_{n\leqslant x} \varphi(n) = \frac{x^2}{2} \sum_{n\leqslant x} \frac{\mu(n)}{n^2} + \sum_{n\leqslant x} \mu(n)R\left(\frac{x}{n}\right)$$

$$= \frac{x^2}{2} \sum_{n=1}^{+\infty} \frac{\mu(n)}{n^2} - \frac{x^2}{2} \sum_{n>x} \frac{\mu(n)}{n^2} + \sum_{n\leqslant x} \mu(n)R\left(\frac{x}{n}\right).$$

Since:

$$\left|\sum_{n>x} \frac{\mu(n)}{n^2}\right| \leqslant \sum_{n>x} \frac{1}{n^2} = O\left(\frac{1}{x}\right)$$

and:

$$\left|\sum_{n\leqslant x} \mu(n)R\left(\frac{x}{n}\right)\right| \leqslant \frac{x}{2} \sum_{n\leqslant x} \frac{1}{n} = O(x\log x),$$

and since

$$\sum_{n=1}^{+\infty} \frac{\mu(n)}{n^2} = \frac{\pi^2}{6},$$

we certainly have the result indicated in the problem.

(3): As "d/m and d/n" is equivalent to "$d/(m,n)$", we have:

$$\sum_{\substack{d/m\\ d/n}} \mu(d) = \sum_{d/(m,n)} \mu(d) = \begin{cases} 1 & \text{if } (m,n) = 1,\\ 0 & \text{if } (m,n) > 1.\end{cases}$$

(4): This allows us to write for $x,y \geqslant 1$:

$$\Phi(x,y) = \sum_{\substack{m\leqslant x\\ n\leqslant y}} \left(\sum_{\substack{d/m\\ d/n}} \mu(d) \right) = \sum_{d\leqslant \inf(x,y)} \mu(d) \sum_{\substack{m\leqslant x\\ n\leqslant y\\ d/m\\ d/n}} 1$$

$$= \sum_{d\leqslant \inf(x,y)} \mu(d) \left[\frac{x}{d}\right]\left[\frac{y}{d}\right] .$$

But if $X,Y \geqslant 1$, we have:

$$XY - X - Y + 1 = (X - 1)(Y - 1) \leqslant [X][Y] \leqslant XY,$$

and consequently:

$$[X][Y] = XY - R_1(X,Y) \quad \text{with } 0 \leqslant R_1(X,Y) < X + Y.$$

Thus, by setting $\inf(x,y) = u$, for $x,y \geqslant 1$ one obtains:

$$\Phi(x,y) = \sum_{d\leqslant u} \mu(d) \frac{xy}{d^2} - R_1\left(\frac{x}{d}, \frac{y}{d}\right)$$

$$= xy \sum_{d\leqslant u} \frac{\mu(d)}{d^2} - \sum_{d\leqslant u} \mu(d) R_1\left(\frac{x}{d}, \frac{y}{d}\right)$$

$$= \frac{6}{\pi^2} xy - xy \sum_{d>u} \frac{\mu(d)}{d^2} - \sum_{d\leqslant u} \mu(d) R_1\left(\frac{x}{d}, \frac{y}{d}\right) ,$$

whence:

$$\left|\frac{1}{xy} \Phi(x,y) - \frac{6}{\pi^2}\right| \leqslant \left|\sum_{d>u} \frac{\mu(d)}{d^2}\right| + \frac{1}{xy}\left|\sum_{d\leqslant u} \mu(d) R_1\left(\frac{x}{d}, \frac{y}{d}\right)\right|$$

$$\leqslant \sum_{d>u} \frac{1}{d^2} + \frac{1}{xy} \sum_{d\leqslant u} \left(\frac{x}{d} + \frac{y}{d}\right)$$

$$\leqslant \sum_{d>u} \frac{1}{d^2} + \left(\frac{1}{x} + \frac{1}{y}\right) \sum_{d\leqslant u} \frac{1}{d}$$

$$< \sum_{d>u} \frac{1}{d^2} + \frac{2}{u} \sum_{d\leqslant u} \frac{1}{d} = O\left(\frac{\log u}{u}\right) .$$

The result of (2) can be deduced from this by noticing that for $x \geqslant 1$:

$$\Phi(x,x) = 2 \sum_{n\leqslant x} \varphi(n) - 1.$$

(In order to count the pairs $[m,n]$ such that $m \leqslant x$, $n \leqslant x$ and $(m,n) = 1$, we can take those for which $m \leqslant n$, then those for which $n \leqslant m$. Thus we have counted the pair [1,1] twice, which is the only one for which $m = n$).

(5):(a): For all $n \in \mathbb{N}^*$ and $t > 0$:

$$N(n,t) = \sum_{m \leqslant tn} \left(\sum_{\substack{d/m \\ d/n}} \mu(d) \right) = \sum_{d/n} \mu(d) \sum_{\substack{m \leqslant tn \\ d/m}} 1 = \sum_{d/n} \mu(d) \left[\frac{tn}{d}\right] .$$

In particular, for all $n \in \mathbb{N}^*$:

$$N(n,1) = \varphi(n) = \sum_{d/n} \mu(d) \frac{n}{d} .$$

Consequently, for all $n \in \mathbb{N}^*$ and all $t > 0$:

$$N(n,t) - t\varphi(n) = \sum_{d/n} \mu(d) \left(\left[\frac{tn}{d}\right] - \frac{tn}{d} \right) ,$$

and if $n > 1$:

$$N(n,t) - t\varphi(n) = \sum_{d/n} \mu(d) \left(\frac{1}{2} + \left[\frac{tn}{d}\right] - \frac{tn}{d} \right) = - \sum_{d/n} \mu(d) B\left(\frac{tn}{d}\right) ,$$

since then $\sum_{d/n} \mu(d) = 0$. As $|B(x)| \leqslant \frac{1}{2}$ for all x, this gives:

$$|N(n,t) - t\varphi(n)| \leqslant \frac{1}{2} \sum_{d/n} |\mu(d)|.$$

But the number of square-free divisors of n is equal to the number of subsets of the set of prime divisors of n, that is to say $2^{\nu(n)}$. Hence, for all $n > 1$ and all $t > 0$ we have:

$$|N(n,t) - t\varphi(n)| \leqslant 2^{\nu(n)-1}.$$

(5):(b): By the preceding, for all $n > 1$ and all $t > 0$:

$$\left|\frac{1}{\varphi(n)} N(n,t) - t\right| \leqslant \frac{2^{\nu(n)-1}}{\varphi(n)} .$$

In order to establish the desired result, it suffices to show that for all $0 < \varepsilon < 1$ the expression $\frac{2^{\nu(n)} n^{1-\varepsilon}}{\varphi(n)}$ is bounded for $n > 1$.

Let $g_\varepsilon(n) = \frac{2^{\nu(n)} n^{1-\varepsilon}}{\varphi(n)}$. We see that g_ε is multiplicative and:

$$g_\varepsilon(p^r) = \frac{2p^{r(1-\varepsilon)}}{p^r - p^{r-1}} = \frac{2p^{-r\varepsilon}}{1 - \frac{1}{p}} \leqslant \frac{2p^{-\varepsilon}}{1 - \frac{1}{p}} .$$

Consequently, for all $n > 1$:

$$g_\varepsilon(n) \leqslant \prod_{p/n} \frac{2p^{-\varepsilon}}{1 - \frac{1}{p}} .$$

But when $p \to +\infty$, $\frac{2p^{-\varepsilon}}{1 - \frac{1}{p}} \to 0$. Therefore, there are only a finite number of p's for which $\frac{2p^{-\varepsilon}}{1 - \frac{1}{p}} > 1$. If E_ε is the set of these p's, then for all $n > 1$:

$$g_\varepsilon(n) \leqslant \prod_{p \in E_\varepsilon} \frac{2p^{-\varepsilon}}{1 - \frac{1}{p}} .$$

SOLUTION 1·21: (1): Let $q = q_1^{a_1} q_2^{a_2} \cdots q_j^{a_j}$ be the prime decomposition of n, with $q_1 \leqslant q_2 \leqslant \cdots \leqslant q_j$. Then, $2 \leqslant q_1$, $3 \leqslant q_2$, ..., $p_i \leqslant q_i$ for $1 \leqslant i \leqslant j$, which implies:

$$N_j = 2.3. \ \ldots \ .p_j \leqslant n.$$

Since, by hypothesis, $n < N_k$ and the sequence N_k is strictly increasing, we deduce from this $j \leqslant k - 1$. As $\nu(n) = j$, that yields $\nu(n) < k$.

Next:

$$\frac{\varphi(n)}{n} = \prod_{i=1}^{j} \left(1 - \frac{1}{q_i}\right)$$

$$\geqslant \prod_{i=1}^{j}\left(1 - \frac{1}{p_i}\right) \geqslant \prod_{i=1}^{k-1}\left(1 - \frac{1}{p_i}\right) = \frac{\varphi(N_{k-1})}{N_{k-1}},$$

and since we have:

$$\frac{\varphi(N_{k-1})}{N_{k-1}} = \frac{1}{1 - \frac{1}{p_k}} \cdot \frac{\varphi(N_k)}{N_k} > \frac{\varphi(N_k)}{N_k},$$

we obtain:

$$\frac{\varphi(n)}{n} > \frac{\varphi(N_k)}{N_k}.$$

(2): Let N be a number different from all N_k's. There exists k such that:

$$N_{k-1} < N < N_k.$$

Since $N < N_k$, Question (1) above shows that:

$$\frac{\varphi(N)}{N} \geqslant \frac{\varphi(N_{k-1})}{N_{k-1}}.$$

The number $m = N_{k-1}$ gives a counter-example to the implication:

$$m < N \Rightarrow \frac{\varphi(m)}{m} > \frac{\varphi(N)}{N}.$$

(3): We have:

$$\frac{\varphi(N_k)}{N_k} = \prod_{i=1}^{k}\left(1-\frac{1}{p_i}\right) = \prod_{p\leqslant p_k}\left(1-\frac{1}{p}\right) \sim \frac{e^{-\gamma}}{\log p_k}.$$

By the Prime Number Theorem:

$$\log N_k = \vartheta(p_k) \sim p_k.$$

That gives:

$$\frac{\varphi(N_k)}{N_k} \sim \frac{e^{-\gamma}}{\log\log N_k},$$

That is to say:

$$\lim_{k\to\infty} \frac{\varphi(N_k)\log\log N_k}{N_k} = e^{-\gamma}.$$

Now let n be an arbitrary integer, and let k be such that:

$$N_k \leqslant n < N_{k+1}$$

By part (1) above:

$$\frac{\varphi(n)}{n} \geqslant \frac{\varphi(N_k)}{N_k},$$

and:

$$\frac{\varphi(n)}{n}\log\log n \geqslant \frac{\varphi(N_k)}{N_k}\log\log N_k.$$

When $n \to +\infty$, $k \to +\infty$ also, and by taking the lower limit of the two we obtain:

$$\underline{\lim}\, \frac{\varphi(n)}{n}\log\log n \geqslant e^{-\gamma}.$$

(4): The function $\frac{\sigma(n)\varphi(n)}{n^2}$ is multiplicative, and for $n = \prod_{p/n} p^a$ with $a = a(p,n) \geq 1$:

$$\sigma(n) = \prod_{p/n} \frac{p^{a+1} - 1}{p - 1} = n \prod_{p/n} \left(\frac{1 - \frac{1}{p^{a+1}}}{1 - \frac{1}{p}} \right),$$

$$\varphi(n) = n \prod_{p/n} \left(1 - \frac{1}{p}\right),$$

from which we have:

$$\frac{\sigma(n)\varphi(n)}{n^2} = \prod_{p/n} \left(1 - \frac{1}{p^{a+1}}\right).$$

We now have:

$$1 \geq \prod_{p/n} \left(1 - \frac{1}{p^{a+1}}\right) \geq \prod_{p} \left(1 - \frac{1}{p^2}\right) = \frac{1}{\zeta(2)} = \frac{6}{\pi^2},$$

whence, for all $n \geq 1$, we deduce:

$$\frac{6}{\pi^2} \leq \frac{\sigma(n)\varphi(n)}{n^2} \leq 1.$$

Next, for all $\varepsilon > 0$:

$$\frac{\sigma(n)}{n\log\log n} \leq \frac{1}{\frac{\varphi(n)\log\log n}{n}} \leq e^\gamma + \varepsilon \quad \text{for } n \text{ large enough,}$$

which shows that:

$$\overline{\lim} \frac{\sigma(n)}{n\log\log n} \leq e^\gamma.$$

Let us now calculate:

$$\frac{\sigma(N_k^r)}{N_k^r \log\log(N_k^r)} = \frac{\prod_{i=1}^{k}\left(1 - \frac{1}{p_i^{r+1}}\right)}{\left(\prod_{i=1}^{k}\left(1 - \frac{1}{p_i}\right)\right)\log\log(N_k^r)} .$$

When $k \to +\infty$ the numerator tends to $\frac{1}{\zeta(r+1)}$. As for the denominator, by Mertens' Formula:

$$\left(\prod_{p \leq p_k}\left(1 - \frac{1}{p}\right)\right)\log\log(N_k^r) \sim \frac{e^{-\gamma}}{\log\log N_k}\log\log N_k^r$$

$$= e^{-\gamma}\left(1 + \frac{\log r}{\log\log N_k}\right) .$$

Therefore we have:

$$\lim_{k\to\infty} \frac{\sigma(N_k^r)}{N_k^r \log\log(N_k^r)} = \frac{e^{\gamma}}{\zeta(r+1)} .$$

This shows:

$$\overline{\lim}\ \frac{\sigma(n)}{n\log\log n} \geq \frac{e^{\gamma}}{\zeta(r+1)} .$$

As this is valid for all r, and as $\lim_{r\to\infty}\zeta(r) = 1$, we have:

$$\overline{\lim}\ \frac{\sigma(n)}{n\log\log n} = e^{\gamma}.$$

SOLUTION 1·22: (1): For all $x > 1$ we have:

$$\sum_{n\leq x} a(n)\log n = \sum_{1<n\leq x} a(n)\log(n)$$

$$= A(x)\log x - A(1)\log 1 - \int_1^x \frac{A(t)}{t}\,dt = A(x)\log x - \int_1^x \frac{A(t)}{t}\,dt.$$

If $|a(n)| \leqslant M$ for all $n \in \mathbb{N}^*$, we have for all $t \geqslant 1$:

$$|A(t)| \leqslant M[t] \leqslant Mt,$$

and from this it follows for all $x > 1$ that:

$$\left|\int_1^x \frac{A(t)}{t}\,dt\right| \leqslant \int_1^x M dt = M(x-1) = O(x).$$

(2): We know that $z * \Lambda = \log$. From this it follows that $\Lambda = \mu * \log$. In other words, for all $n \in \mathbb{N}^*$:

$$\Lambda(n) = \sum_{d/n} \mu(d) \log \frac{n}{d} = \sum_{d/n} \mu(d) \log n - \sum_{d/n} \mu(d) \log d$$

$$= \left(\sum_{d/n} \mu(d)\right) \log n - \sum_{d/n} \mu(d) \log d = -\sum_{d/n} \mu(d) \log d,$$

since $\sum_{d/n} \mu(d) = 0$ for $n > 1$, and $\log 1 = 0$. This says that $\Lambda = -z * \mu\log$, and from this it follows that $\mu\log = -\mu * \Lambda$.

(3): Since $\mu\log = -\mu * \Lambda$, we have for all $x \geqslant 1$:

$$\sum_{n \leqslant x} \mu(n) \log n = -\sum_{n \leqslant x} \mu(n) \psi\left(\frac{x}{n}\right) = -\sum_{n \leqslant x} \mu(n) \left(g\left(\frac{x}{n}\right) + \left[\frac{x}{n}\right]\right)$$

$$= -\sum_{n \leqslant x} \mu(n) g\left(\frac{x}{n}\right) - 1,$$

because $\sum_{n \leqslant x} \mu(n) \left[\frac{x}{n}\right] = 1$.

(4): The Prime Number Theorem implies that:

$$g(x) = o(x) \qquad (x \to +\infty).$$

Given $\varepsilon > 0$, there exists an $X > 1$ such that:

$$|g(x)| \leqslant \varepsilon x \quad \text{for } x > X.$$

On the other hand, there exists $K > 0$ such that:

$$|g(x)| \leqslant K \quad \text{for } 1 \leqslant x \leqslant X.$$

One can, for example, take $K = \sup(\psi(X),[X])$.

For $x > X$ we can write:

$$\left|\sum_{n\leqslant x} \mu(n) g\left(\frac{x}{n}\right)\right| \leqslant \sum_{n\leqslant x} \left|g\left(\frac{x}{n}\right)\right| \leqslant \sum_{n\leqslant \frac{x}{X}} \left|g\left(\frac{x}{n}\right)\right| + \sum_{\frac{x}{X}<n\leqslant x} \left|g\left(\frac{x}{n}\right)\right|$$

$$\leqslant \varepsilon x \sum_{n\leqslant \frac{x}{X}} \frac{1}{n} + K \sum_{\frac{x}{X}<n\leqslant x} 1 \leqslant \varepsilon x \sum_{n\leqslant x} \frac{1}{n} + Kx,$$

because $\left|g\left(\frac{x}{n}\right)\right| \leqslant \varepsilon \frac{x}{n}$ for $n \leqslant \frac{x}{X}$, and $\left|g\left(\frac{x}{n}\right)\right| \leqslant K$ for $\frac{x}{X} < n \leqslant x$.

From that it follows that:

$$\overline{\lim_{x\to+\infty}} \frac{1}{x\log x}\left|\sum_{n\leqslant x} \mu(n) g(\frac{x}{n})\right| \leqslant \varepsilon,$$

since $\sum_{n\leqslant x} \frac{1}{n} \sim \log x$ when $x \to +\infty$.

As ε is arbitrarily small, we have:

$$\overline{\lim_{x\to+\infty}} \frac{1}{x\log x}\left|\sum_{n\leqslant x} \mu(n) g\left(\frac{x}{n}\right)\right| = 0.$$

In other words:

$$\sum_{n\leqslant x} \mu(n) g\left(\frac{x}{n}\right) = o(x\log x) \quad \text{when } x \to +\infty.$$

The relation established in part (3) above then gives:

$$\sum_{n\leqslant x} \mu(n)\log n = o(x\log x).$$

However, by the result of Question (1) above, we have:

$$\sum_{n\leqslant x} \mu(n)\log n = M(x)\log x + O(x) = M(x)\log x + o(x\log x).$$

Hence we have:

$$M(x)\log x = o(x\log x),$$

whence:

$$M(x) = o(x).$$

SOLUTION 1·23: (1): The function $\dfrac{\mu^2(n)}{\varphi(n)}$ is multiplicative. In order to study the convergence look at the double sum:

$$\sum_{p,r} \frac{\mu^2(p^r)}{\varphi(p^r)p^{r\sigma}} .$$

As $\mu^2(p^r) = 0$ for $r \geqslant 2$, this sum reduces to:

$$\sum_p \frac{1}{(p-1)p^\sigma} ,$$

and it is convergent for all positive real σ. The abscissa of convergence is therefore less than or equal to zero.

For $s = 0$ the series $\sum_{n=1}^{\infty} \dfrac{\mu^2(n)}{\varphi(n)}$ diverges, since the sub-series $\sum_p \dfrac{1}{(p-1)}$ diverges. The abscissa of convergence of $F(s)$ is therefore exactly zero.

For $\operatorname{Re} s > 0$ we have the expansion of $F(s)$ into an Euler product:

$$F(s) = \sum_{n=1} \frac{\mu^2(n)}{\varphi(n)n^s} = \prod_p \left(1 + \frac{\mu^2(p)}{\varphi(p)p^s} + \cdots + \frac{\mu^2(p^r)}{\varphi(p^r)p^{rs}} + \cdots\right)$$

$$= \prod_p \left(1 + \frac{1}{(p-1)p^s}\right) .$$

(2): Since:

$$\zeta(s) = \prod_p \frac{1}{1 - \frac{1}{p^s}},$$

it follows from this that for Re$s > 0$ we have:

$$G(s) = \frac{F(s)}{\zeta(s+1)} = \prod_p \left(1 - \frac{1}{(p-1)p^s}\right)\left(1 - \frac{1}{p^{s+1}}\right)$$

$$= \prod_p \left(1 + \frac{1}{(p-1)p^{s+1}} - \frac{1}{(p-1)p^{2s+1}}\right).$$

Let us show that this product converges uniformly on K. This is the same as showing that the series:

$$\sum_p \left(\frac{1}{(p-1)p^{s+1}} - \frac{1}{(p-1)p^{2s+1}}\right)$$

converges normally on K. This series is extracted from the series:

$$\sum_{n\geqslant 2} \left(\frac{1}{(n-1)n^{s+1}} - \frac{1}{(n-1)n^{2s+1}}\right).$$

Since Res is minorised by $\sigma_0 > -\frac{1}{2}$ on K, we have the inequalities:

$$\left|\frac{1}{(n-1)n^{s+1}}\right| \leqslant \frac{1}{(n-1)^{\sigma_0+1}}$$

and:

$$\left|\frac{1}{(n-1)n^{2s+1}}\right| \leqslant \left|\frac{1}{(n-1)^{2\sigma_0+1}}\right|,$$

and this latter series is normally convergent on K. Now, we have:

$$G(0) = \prod_p \left(1 + \frac{1}{(p-1)p} - \frac{1}{(p-1)p}\right) = 1.$$

Since the infinite product:

$$G(s) = \prod \left(1 + \frac{1}{(p-1)p^{s+1}} - \frac{1}{(p-1)p^{2s+1}}\right)$$

converges normally on every compact set K contained in the region $\mathrm{Re}s > -\frac{1}{2}$, its logarithmic derivative is given, for $\mathrm{Re}s > \frac{1}{2}$, by:

$$\frac{G'(s)}{G(s)} = \sum_p \left[\frac{2(p-1)p^{2s+1} + p^s}{(p-1)p^{2s+1} + p^s - 1} - 2\right]\log p$$

$$= \sum_p \left(\frac{2 - p^s}{(p-1)p^{2s+1} + p^s - 1}\right)\log p.$$

In particular, for $s = 0$:

$$G'(0) = \sum_p \frac{\log p}{p(p-1)} G(0) = \sum_p \frac{\log p}{p(p-1)} .$$

(3): For $\mathrm{Re}s > 0$ the Dirichlet series:

$$F(s) = \sum_{n=1}^{\infty} \frac{\mu^2(n)}{\varphi(n)} \frac{1}{n^s} \quad \text{and} \quad \frac{1}{\zeta(s+1)} = \sum_{n=1}^{\infty} \frac{\mu(n)}{nn^s}$$

are absolutely convergent, and their product has the value:

$$G(s) = \sum_{n=1}^{\infty} \frac{g(n)}{n^s}, \quad \text{with } g = \frac{\mu^2}{\varphi} * \frac{\mu}{i} .$$

We have:

$$g(p) = \frac{1}{p(p-1)}, \quad g(p^2) = -\frac{1}{p(p-1)},$$

and:

$g(p^k) = 0$ for $k \geqslant 3$.

For real $s > -\frac{1}{2}$ we have:

$$\sum_{n\leqslant x} \frac{|g(n)|}{n^s} \leqslant \prod_{p\leqslant x} \left(1 + \frac{1}{(p-1)p^{s+1}} + \frac{1}{(p-1)p^{2s+1}}\right)$$

$$\leqslant \prod_{p} \left(1 + \frac{1}{(p-1)p^{s+1}} + \frac{1}{(p-1)p^{2s+1}}\right) < +\infty,$$

which shows that the abscissa of absolute convergence of the series $\sum \frac{g(n)}{n^s}$ is less than or equal to $-\frac{1}{2}$.

The series $\sum_p \frac{g(p^2)}{p^{2s}}$, for $s = -\frac{1}{2}$, has the value $\sum_p \left(-\frac{1}{p-1}\right)$ and diverges; therefore the abscissa of convergence of the series $\sum_p \frac{g(n)}{n^s}$ is exactly $-\frac{1}{2}$.

(4): Since $g = \frac{\mu^2}{\varphi} * \frac{\mu}{i}$, we have:

$$\frac{\mu^2}{\varphi} = g * \frac{1}{i},$$

which yields:

$$\sum_{n\leqslant x} \frac{\mu^2(n)}{\varphi(n)} = \sum_{n\leqslant x} g(n) L\left(\frac{x}{n}\right).$$

Next we have:

$$\sum_{n\leqslant x} g(n) L\left(\frac{x}{n}\right)$$

$$= \sum_{n\leqslant x} g(n)(\log x + \gamma) - \sum_{n\leqslant x} g(n)\log(n) + \sum_{n\leqslant x} g(n) \frac{x}{n}$$

(Contd)

$$= (\log x + \gamma) \sum_{n=1} g(n) - \sum_{n=1} g(n)\log(n) + R_1 + R_2 + R_3 =$$

(Contd) $= (\log x + \gamma)G(0) + G'(0) + R_1 + R_2 + R_3,$

with:

$$R_1 = -(\gamma + \log x)\sum_{n>x} g(n),$$

$$R_2 = \sum_{n>x} g(n)\log(n),$$

$$R_3 = \sum_{n\leqslant x} g(n)\eta\left(\frac{x}{n}\right).$$

Let c be a number such that $-\frac{1}{2} < c < 0$. Then:

$$\sum_{n>x} |g(n)| = \sum_{n>x} \frac{|g(n)|}{n^c} n^c \leqslant x^c \sum_{n>x} \frac{|g(n)|}{n^c} = O(x^c).$$

Similarly:

$$|R_2| \leqslant \sum_{n>x} |g(n)|\log n = \sum_{n>x} \frac{|g(n)|}{n^c}(\log n)n^c$$

$$\leqslant x^c \sum_{n>x} \frac{|g(n)|\log n}{n^c} = O(x^c);$$

finally:

$$|R_3| = \left|\sum_{n\leqslant x} g(n)\eta\left(\frac{x}{n}\right)\right| \leqslant \sum_{n\leqslant x} \left|g(n)\frac{n}{x}\right| = \frac{1}{x}\sum_{n\leqslant x} \frac{|g(n)|}{n^c} n^{1+c}$$

$$\leqslant x^c \sum_{n\leqslant x} \frac{|g(n)|}{n^c} = O(x^c).$$

for the series $\sum \frac{|g(n)|}{n^c}$ is convergent. Therefore we have:

$$\sum_{n\leqslant x} \frac{\mu^2(n)}{\varphi(n)} = \log x + \gamma - \sum_p \frac{\log p}{p(p-1)} + o(1).$$

SOLUTION 1·24: (1): With z being fixed, the arithmetic function f defined by:

$$f(n) = z^{\Omega(n)-\nu(n)}$$

is multiplicative, since Ω and ν are additive, and for p prime and $r \geqslant 1$ integral we have:

$$f(p^r) = z^{r-1}.$$

If $|z| < 2$ the double series:

$$\sum_{\substack{p,r\\ r\geqslant 1}} \frac{f(p^r)}{p^{r\sigma}}, \text{ that is to say, } \sum_{\substack{p,r\\ r\geqslant 1}} \frac{|z|^{r-1}}{p^{r\sigma}},$$

is convergent for $\sigma > 1$, since, for each p, the series $\sum_{r=1}^{+\infty} \frac{|z|^{r-1}}{p^{r\sigma}}$, a geometric series with ratio $\frac{|z|}{p^\sigma} < 1$, is convergent with sum $\frac{1}{p^\sigma - |z|}$, and the series $\sum_p \frac{1}{p^\sigma - |z|}$ is convergent.

Consequently, for $\operatorname{Re} s > 1$ and $|z| < 2$, the series $\sum_{n=1}^{+\infty} \frac{z^{\Omega(n)-\nu(n)}}{n^s}$ is absolutely convergent and has as sum the value of the absolutely convergent infinite product:

$$\prod_p \left(1 + \sum_{r=1}^{+\infty} \frac{z^{r-1}}{p^{rs}}\right), \quad \text{that is to say,} \quad \prod_p \left(1 + \frac{1}{p^s - z}\right).$$

But we have:

$$1 + \frac{1}{p^s - z} = \frac{p^s + 1 - z}{p^s - z} = \left(1 + \frac{1}{p^s}\right)\left(\frac{1 - \dfrac{z}{(p^s + 1)}}{1 - \dfrac{z}{p^s}}\right)$$

$$= \left(1 - \frac{1}{p^{2s}}\right)\left(1 - \frac{1}{p^s}\right)^{-1}\left(1 - \frac{z}{(p^s + 1)}\right)\left(1 - \frac{z}{p^s}\right)^{-1}.$$

It is easily seen that the infinite products

$$\prod \left(1 - \frac{1}{p^2}\right), \quad \prod \left(1 - \frac{1}{p^s}\right)^{-1} \quad \text{and} \quad \prod \left(\frac{1 - \dfrac{z}{(p^s + 1)}}{1 - \dfrac{z}{p^s}}\right)$$

are absolutely convergent. The first two are, moreover, equal respectively to $\frac{1}{\zeta(2s)}$ and $\zeta(s)$. Hence for $\operatorname{Re} s > 1$ and $|z| < 2$ we have:

$$\sum_{n=1}^{+\infty} \frac{z^{\Omega(n)-\nu(n)}}{n^s} = \frac{\zeta(s)}{\zeta(2s)} \prod \left(\frac{1 - \dfrac{z}{(p^s + 1}}{1 - \dfrac{z}{p^s}}\right). \tag{1}$$

(2): We see that the infinite product $\prod \left(\frac{1 - \frac{z}{(p^s + 1)}}{1 - \frac{z}{p^s}}\right)$ is uniformly convergent on every compact set of $\mathbb{C}^2$ contained in the set determined by pairs (s,z) such that $\operatorname{Re} s > \frac{1}{2}$ and $|z| < \sqrt{2}$.

In fact, if K is such a compact set there exists $\sigma_0 > \frac{1}{2}$ and satisfying $0 < R < \sqrt{2}$ such that for $(s,z) \in K$, $\operatorname{Re} s \geqslant \sigma_0$, and $|z| \leqslant R$. Then for $(s,z) \in K$:

$$\left|\frac{1 - \dfrac{z}{(p^s + 1)}}{1 - \dfrac{z}{p^s}} - 1\right| = \left|\frac{z}{(p^s + 1)(p^s - z)}\right|$$

$$\leqslant \frac{R}{(p^{\sigma_0} - 1)(p^{\sigma_0} - R)}.$$

But the series $\sum_p \frac{R}{(p^{\sigma_0} - 1)(p^{\sigma_0} - R)}$ is convergent.

If $F(s,z)$ is the value of the infinite product, the function F is holomorphic in s and z for $\operatorname{Re} s > \frac{1}{2}$ and $|z| < \sqrt{2}$, because each factor is holomorphic in this domain.

(3): For $\mathrm{Re}\, s > \frac{1}{2}$ and $|z| < \sqrt{2}$ we can write:

$$F(s,z) = \sum_{q=0}^{+\infty} A_q(s)z^q \tag{2}$$

where the functions A_q are holomorphic in the half-plane $\mathrm{Re}\, s > \frac{1}{2}$. For $s = 1$ this gives:

$$\sum_{q=0}^{+\infty} A_q(1)z^q = \prod \left(\frac{1 - \frac{z}{(p+1)}}{1 - \frac{z}{p}}\right) .$$

Moreover the equality in fact holds for $|z| < 2$, for it is easily seen that the infinite product converges uniformly on every compact set contained in the disc $|z| < 2$, and therefore represents a function holomorphic in this disc.

Notice that the numbers $A_q(1)$ are all real and positive. In fact, for each p, for $|z| < 2$:

$$\frac{1 - \frac{z}{p+1}}{1 - \frac{z}{p}} = \exp\left\{\sum_{q=1}^{+\infty} \frac{z^q}{q}\left(\frac{1}{p^q} - \frac{1}{(p+1)^q}\right)\right\} .$$

From this it follows, for $|z| < 2$, that:

$$\prod \left(\frac{1 - \frac{z}{p+1}}{1 - \frac{z}{p}}\right) = \exp\left\{\sum_p \left(\sum_{q=1}^{+\infty} \frac{z^q}{q}\left(\frac{1}{p^q} - \frac{1}{(p+1)^q}\right)\right)\right\}$$

$$= \exp\left\{\sum_{q=1}^{+\infty} \alpha_q z^q\right\} ,$$

where:

$$\alpha_q = \frac{1}{q}\sum \left(\frac{1}{p^q} - \frac{1}{(p+1)^q}\right) ,$$

because the double series $\sum_{p,q} \frac{z^q}{q}\left(\frac{1}{p^q} - \frac{1}{(p+1)^q}\right)$ is absolutely convergent.

Each α_q is clearly real and greater than zero. Now, the coefficients of the expansion as a power series of $\exp\left\{\sum_{q=1} \alpha_q z^q\right\}$ are polynomials, with positive coefficients, in $a_1, a_2, \ldots, \alpha_q, \ldots$. They are therefore real and greater than zero.

Whenever the series $\sum_{n=1}^{+\infty} \frac{z^{\Omega(n)-\nu(n)}}{n^s}$ is absolutely convergent, we can calculate its sum by grouping together the terms for which $\Omega(n) - \nu(n)$ has the same value, which gives:

$$\sum_{q=0}^{+\infty} \left(\sum_{\Omega(n)-\nu(n)=q} \frac{1}{n^s}\right) z^q.$$

Thus equalities (1) and (2) show that for $\mathrm{Re}\, s > \frac{1}{2}$:

$$\sum_{\Omega(n)-\nu(n)=q} \frac{1}{n^s} = \frac{\zeta(s)}{\zeta(2s + 1)} A_q(s).$$

The function on the right is holomorphic in the half-plane $\mathrm{Re}\, s > \frac{1}{2}$ except at the point 1, which is a simple pole with residue $\frac{A_q(1)}{\zeta(2)} = \frac{6}{\pi^2} A_q(1)$.

Ikehara's Theorem of Section 1.12 of the Introduction shows that for $x \to +\infty$:

$$N_q(x) \sim \frac{6}{\pi^2} A_q(1) x.$$

If we set $\frac{6}{\pi^2} A_q(1) = d_q$, this may be written $N_q(x) \sim d_q x$, or:

$$N_q(x) = d_q x + o(x).$$

Clearly $d_q > 0$.

On the other hand, Equation (3) shows that for $|z| < 2$:

$$\sum_{q=0}^{\infty} d_q z^q = \frac{6}{\pi^2} \prod \left(1 - \frac{z}{(p + 1)}\right)\left(1 - \frac{z}{p}\right)^{-1}.$$

By setting $z = 1$ we obtain:

$$\sum_{q=0}^{+\infty} d_q = \frac{6}{\pi^2} \prod \left(\frac{1 - \dfrac{1}{(p+1)}}{1 - \dfrac{1}{p}} \right) = \frac{6}{\pi^2} \prod \left(1 - \frac{1}{p^2}\right)^{-1} = 1.$$

SOLUTION 1·25: (1): Let D_m be the set of divisors of m. It is a finite set whose cardinality is $d(m)$.

Now let $n = \prod p^a$ be an element of E_m. We have:

$$\phi(n) = \prod p^{a-1}(p - 1) = m.$$

Therefore $(p - 1) \in D_m$. There are therefore only a finite number of possible prime factors for n. For each of these factors one must have $p^{a-1} \leqslant m$, so:

$$a \leqslant 1 + \frac{\log m}{\log p} \leqslant 1 + \frac{\log m}{\log 2},$$

and there are only a finite number of possible exponents. Finally, we have:

$$\mathrm{card} E_m \leqslant \left(1 + \frac{\log m}{\log 2}\right) d(m).$$

When m is odd, $p - 1$ must be a divisor of m, and therefore is odd. If $n \in E_m$, one must therefore have $n = 2^a$. For $a \geqslant 2$, $\varphi(2^a)$ is even. Hence we have:

$$E_1 = \{2\} \quad \text{and} \quad E_m = \emptyset \quad \text{for odd } m > 3.$$

(2): We have:

$$\frac{n}{\varphi(n)} = \frac{1}{\prod_{p/n} \left(1 - \frac{1}{p}\right)} = \prod_{p/n} \frac{1 + \frac{1}{p}}{1 - \frac{1}{p^2}} \leqslant$$

(Contd)

(Contd) $$\leqslant \prod_{p\text{ prime}} \frac{1}{1-\frac{1}{p^2}} \prod_{p/n}\left(1+\frac{1}{p}\right) = \zeta(2)\prod_{p/n}\left(1+\frac{1}{p}\right),$$

and:

$$\prod_{p/n}\left(1+\frac{1}{p}\right) = \sum_{\substack{d/n\\ \mu(d)\neq 0}} \frac{1}{d} \leqslant \sum_{d/n}\frac{1}{d}.$$

Whence:

$$\frac{n}{\varphi(n)} \leqslant C\sum_{d/n}\frac{1}{d},$$

with $C = \zeta(2) = \frac{\pi^2}{6}$.

Next:

$$\frac{n}{\varphi(n)} \leqslant C\sum_{d/n}\frac{1}{d} \leqslant C\sum_{d\leqslant n}\frac{1}{d} \leqslant C\left(1+\int_1^n\frac{\mathrm{d}t}{t}\right) = C(1+\log n),$$

that is to say:

$$\frac{n}{\varphi(n)} = O(\log n).$$

Now,

$$A(x) = \operatorname{card}\{n : \varphi(n) \leqslant x\}.$$

If $n \leqslant x$ then $\varphi(n) \leqslant n \leqslant x$, so we conclude $A(x) \geqslant [x]$. If $\varphi(n) \leqslant x$ the relation $\frac{n}{\varphi(n)} = O(\log n)$ yields:

$$\frac{n}{\log n} = O(x).$$

On taking logarithms this becomes:

$$\log n - \log\log n = O(\log x),$$

and since $\log\log n$ is small compared with $\log n$:

$$\log n = O(\log x).$$

We have then:

$$n = O(x)\log(n) = O(x\log x),$$

and from this it follows that

$$A(x) = O(x\log x).$$

(3): Let us show that the series $\sum_{n=1}^{\infty} \frac{1}{\varphi(n)^s}$ is absolutely convergent for Re$s > 1$. There holds:

$$\left|\frac{1}{\varphi(n)^s}\right| = \frac{1}{\varphi(n)^{\text{Re}s}} \leqslant \left(\frac{C(1 + \log n)}{n}\right)^{\text{Re}s}.$$

This series is convergent for Re$s > 1$.

Since the series $\sum \frac{1}{\varphi(n)^s}$ is absolutely convergent, all the terms with the same denominator can be grouped. For each m there will be exactly a_m terms with denominator $m^s = \varphi(n)^s$, and we have:

$$\sum_{m=1}^{\infty} \frac{a_m}{m^s} = \sum_{n=1}^{\infty} \frac{1}{\varphi(n)^s}.$$

This shows that the series $\sum \frac{a_m}{m^s}$ converges for Re$s > 1$. To decompose it into an Euler product the Theorem in Section 1.10 of the Introduction is used.

(4): We have:

$$\left|\frac{1}{(p-1)^s} - \frac{1}{p^s}\right| = \left|\int_{p-1}^{p} \frac{s}{x^{s+1}}\,dx\right|$$

$$\leqslant \int_{p-1}^{p} \frac{|s|}{x^{\text{Re}s+1}}\,dx \leqslant \frac{|s|}{(p-1)^{\text{Re}s+1}}.$$

Let K be a compact set contained in the region Re$s > 0$. To show that f is holomorphic we will show that the product defining $f(s)$ is normally convergent on K, or, which amounts to the same thing, that the series:

$$\sum_p \left(\frac{1}{(p-1)^s} - \frac{1}{p^s}\right)$$

is normally convergent on K. We have:

$$\sum_p \left|\frac{1}{(p-1)^s} - \frac{1}{p^s}\right| \leqslant |s| \sum \frac{1}{(p-1)^{\mathrm{Re}s+1}} \leqslant M \sum_p \frac{1}{(p-1)^{\sigma_0+1}}$$

(on K, $|s|$ is majorised by M, and Res is minorised by $\sigma_0 > 0$).

We have:

$$\hat{f}(1) = \prod_p \left(1 + \frac{1}{p-1} - \frac{1}{p}\right) = \prod_p \frac{p^2 - p + 1}{p(p-1)}$$

$$= \prod \frac{p^3 + 1}{p(p-1)(p+1)} = \prod_p \frac{p^6 - 1}{p(p^2 - 1)(p^3 - 1)}$$

$$= \prod_p \left(\frac{p^6 - 1}{p^6}\right)\left(\frac{p^3}{p^3 - 1}\right)\left(\frac{p^2}{p^2 - 1}\right).$$

Since:

$$\zeta(k) = \prod_p \frac{1}{1 - \frac{1}{p^k}} = \prod_p \frac{p^k}{p^k - 1},$$

we have:

$$\hat{f}(1) = \frac{\zeta(3)\zeta(2)}{\zeta(6)} = 1.94\cdots\cdots .$$

(5): Ikehara's Theorem, of Section 1.12 of the Introduction, is applied to the Dirichlet series $\sum_{m=1}^{\infty} \frac{a_m}{m^s} = \zeta(s)\hat{f}(s)$. The function $\zeta(s)\hat{f}(s)$ is meromorphic for $\mathrm{Re}\, s > 0$ with a single pole, the pole of $\zeta(s)$ for $s = 1$, whose residue is $\hat{f}(1)$. We then have:

$$A(x) = \sum_{m \leqslant x} a_m \sim \hat{f}(1)x = \frac{\zeta(3)\zeta(2)}{\zeta(6)}\, x.$$

SOLUTION 1·26: (1): If $A' \subset A$ every partition whose summands are in A' is a partition whose summands are in A. Hence:

$$p_A(n) \geqslant p_{A'}(n)$$

and:

$$P(A') \Rightarrow P(A).$$

(2): By Bezout's Theorem† there exists $x', y' \in \mathbb{Z}$ such that:

$$n = x'a + y'b.$$

Let us assume $x' < 0$. There exists a unique $\lambda \in \mathbb{N}^*$ such that:

$$0 \leqslant x' + \lambda b < b$$

Then:

$$n = (x' + \lambda b)a + (y' - \lambda a)b = xa + yb,$$

on setting $x = x' + \lambda b$, $y = y' - \lambda a$. By the choice of λ, we have $x \geqslant 0$. We also have $y > 0$, for if y were negative, we would have:

$$n = xa + yb < xa < ab,$$

which is contrary to the hypothesis.

If Bezout's Theorem gives $n = x'a + y'b$ with $x' \geqslant 0$ and $y' < 0$, one carries out a similar argument by constructing $y = y' + \mu a$.

For $A = \{a,b\}$ the number of partitions of n whose summands are

† (*Editor's Note*): Bezout's Theorem is the French name for the proposition that the g.c.d. of a and b is a linear combination of a and b.

in A is the number of solutions (x,y) of $n = xa + yb$ with $x \geqslant 0$ and $y \geqslant 0$. For $n \geqslant ab$ one has $p_A(n) > 0$ and $P(A)$ holds.

If A contains two relatively prime elements a and b, we apply Question (1) above with $A' = \{a,b\}$, and we see that $P(A)$ is true.

(3): Let us decompose into partial fractions:

$$F_A(x) = \frac{1}{(1 - x^{a_1})(1 - x^{a_2})\cdots(1 - x^{a_r})} = \sum_{\zeta \text{pole}} E_\zeta(x),$$

where $E_\zeta(x)$ is the sum of those corresponding to the pole ζ. For $\zeta = 1$, which is a pole of order r, we have:

$$E_1(x) = \frac{\lambda_r}{(1 - x)^r} + \frac{\lambda_{r-1}}{(1 - x)^{r-1}} + \cdots + \frac{\lambda_1}{(1 - x)},$$

and $\lambda_r = (a_1 a_2 \cdots a_r)^{-1}$. The expansion of $E_1(x)$ into a power series i

$$E_1(x) = \sum_n g(n) x^n,$$

with:

$$g(n) = \lambda_1 + (n + 1)\lambda_2 + \cdots + \frac{(n + r - 1) \quad (n + 1)}{(r - 1)!} \lambda_r.$$

We see that $g(n)$ is a polynomial whose term of highest degree is:

$$\frac{(a_1 a_2 \cdots a_r)^{-1}}{(r - 1)!} n^{r-1}.$$

Now let ζ be a pole different from 1. It is a root of unity, and let us assume that it is a primitive n-th root of unity. ζ will be a pole of $(1 - x^{a_i})$ if n divides a . The order of the pole ζ will therefore be the number s of elements of A that are multiples of n. These s elements are not relatively prime. Therefore we must have $s \leqslant q$.

We have:

$$E_\zeta(x) = \frac{\alpha_s}{(\zeta - x)^s} + \frac{\alpha_{s-1}}{(\zeta - x)^{s-1}} + \cdots + \frac{\alpha_1}{(\zeta - x)} .$$

The expansion of $E_\zeta(x)$ as a power series will be:

$$E_\zeta(x) = \sum_{n=0}^{\infty} h_\zeta(n)x ,$$

with:

$$h_\zeta(n) = \frac{1}{\zeta}\left[\frac{\alpha_1}{\zeta^n} + \frac{(n+1)\alpha_2}{\zeta^{n+1}} + \cdots + \frac{(n+1)(n+2) \quad (n+s-1)}{(s-1)^{n+s-1}}\alpha_s\right] .$$

As $|\zeta| = 1$,

$$h_\zeta(n) = O(n^{s-1}) = O(n^{q-1})$$

holds. Finally, we have:

$$p_A(n) = g(n) + \sum_{\zeta \neq 1} h_\zeta(n) = g(n) + O(n^{q-1}).$$

(4): We have the following equivalences:

Not $P_2(A)$ <=> there exists a prime number p such that $p|a$ for all $a \in A$

<=> there exists a prime number p such that $p|$(g.c.d. of the elements of A)

<=> the g.c.d. of the elements of A is greater than one

<=> Not $P_1(A)$.

This shows: $P_1(A)$ <=> $P_2(A)$.

In order to prove $P(A)$ => $P_1(A)$ let us show that:

Not $P_1(A) \Rightarrow$ Not $P(A)$.

Now, we have:

$P_1(A) \Rightarrow$ the g.c.d. of the elements of A is $d > 1$.

If n is not a multiple of d it is impossible to write n as the sum of elements of A, and $p_A(n) = 0$. Property $P(A)$ is therefore not satisfied.

It is now necessary to show that for a finite set $A = \{a_1 \cdots a_r\}$ $P_1(A) \Leftrightarrow P(A)$. We use (3) above:

$$p_A(n) = g(n) + O(n^{q-1}).$$

If $P_1(A)$ is true, then $q \leqslant r - 1$, and:

$$p_A(n) \sim g(n) \sim \frac{(a_1 a_2 \quad a_r)^{-1}}{(r-1)!} n^{r-1},$$

which shows that $p_A(n)$ is positive for n large enough. The Propertv $P(A)$ is therefore satisfied.

(5): Let $a_0 \in A$ and let $p_1, p_2, \ldots, p_k$ be prime factors of a_0. Since $P_2(A)$ is true, for every prime number p there exists $a \in A$ such that $p \nmid a$. In particular, there exists $a_1, \ldots, a_k$ such that p_i does not divide a_1. (It may be that certain of the numbers a_i are the same). We set $A' = \{a_0, a_1, \ldots, a_k\}$.

Property $P_2(A')$ is satisfied. In fact let p be a prime number:

If $p \neq p_i$ for $1 \leqslant i \leqslant k$, there exists $a = a_0 \in A$ such that p does not divide a;

If $p = p_i$, there exists $a = a_i$ such that p does not divide a.

Let A be an infinite set of $\mathbb{N}^*$ satisfying $P_1(A)$. It also satisfies $P_2(A)$. We construct a finite subset $A' \subset A$ satisfying $P_2(A')$. As A' is finite, $P(A')$ is true and $P(A)$ is true by Question (1) above.

Thus $P_1(A) \Rightarrow P(A)$ has been shown, and as it was already known that $P(A) \Rightarrow P_1(A)$, we conclude that $P(A) \Leftrightarrow P_1(A)$.

SOLUTION 1·27: (a): When Q is the set of prime numbers less than or equal to N, denoted Q_N, let us define B_N by the relation:

$$\sum_{m\leqslant N} \frac{1}{m} = \sum_{m\in A(N,Q_N)} \frac{1}{m} = B_N \prod_{p\leqslant N} \left(1 - \frac{1}{p}\right)^{-1}.$$

By Mertens' Theorem (cf., the Introduction) and a classical asymptotic estimate, we have:

$$\log N = (1 + o(1))B_N e^{+\gamma} \log N.$$

From this it follows that:

$$\lim_{N\to\infty} B_N = e^{-\gamma} > 0,$$

and the sequence B_N has a lower bound $B > 0$. We will show that B answers our Question.

From this point on, N is to be a fixed integer greater than two. Relation (*):

$$\sum_{m\in A(N,Q)} \frac{1}{m} \geqslant B \prod_{p\in Q} \left(1 - \frac{1}{p}\right)^{-1} \qquad (*)$$

of course holds when $Q = Q_N$. Let us assume that it holds for a non-empty set Q of prime numbers less than N; let q be an element of Q; we will show that Relation (*) still holds for $Q' = Q - \{q\}$.

From the relations:

$$\sum_{m\in A(N,Q')} \frac{1}{m} = \sum_{m\in A(N,Q)} \frac{1}{m} - \sum_{\substack{m\in A(N,Q)\\ q|m}} \frac{1}{m},$$

$$\sum_{\substack{m\in A(N,Q)\\ q|m}} \frac{1}{m} \leq \sum_{m'\in A(N,Q)} \frac{1}{qm'} = \frac{1}{q} \sum_{m\in A(N,Q)} \frac{1}{m},$$

we deduce:

$$\sum_{m\in A(N,Q)} \frac{1}{m} \geq \left(1 - \frac{1}{q}\right) \sum_{m\in A(N,Q)} \frac{1}{m}.$$

By now using the induction hypothesis (*) we obtain:

$$\sum_{m\in A(N,Q')} \frac{1}{m} \geq B\left(1 - \frac{1}{q}\right) \prod_{p\in Q} \left(1 - \frac{1}{p}\right)^{-1} = B \prod_{p\in Q'} \left(1 - \frac{1}{p}\right)^{-1}.$$

REMARK: This result gives a good lower bound, to the extent that one has the trivial (or almost trivial) upper bound:

$$\sum_{m\in A(N,Q)} \frac{1}{m} \leq \lim_{N\to\infty} \sum_{m\in A(N,Q)} \frac{1}{m} = \prod_{p\in Q} \left(1 - \frac{1}{p}\right)^{-1}.$$

(b): Let us start by showing the first inequality. Let m be an integer less than or equal to z. We can write:

$$m = \prod_{s=1}^{\sigma} p_s^{\sigma_s} \prod_{t=1}^{\tau} q_t^{\beta_t} \text{ with } p_s \in P_1,\ q_t \in P_2.$$

If one writes $m_2 = \prod_{t=1}^{\tau} q_t^{\beta_t}$, we have:

$$d(m_2) = \prod_{t=1}^{\tau} (\beta_t + 1) \quad\text{and}\quad \omega(m) = \prod_{t=1}^{\tau} 2^{\beta_t},$$

whence:

$$d(m_2) \leq \omega(m).$$

We also have:

$$\sum_{k\leqslant\sqrt{z}}\frac{1}{k}\sum_{\substack{\ell\leqslant\sqrt{z}\\(\ell,P_1)=1}}\frac{1}{\ell}=\sum_{k\leqslant\sqrt{z}}\sum_{\substack{\ell\leqslant\sqrt{z}\\(\ell,P_1)=1}}\frac{1}{k\ell}$$

$$\leqslant\sum_{m\leqslant z}\frac{1}{m}\sum_{\substack{\ell\leqslant\sqrt{z}\\ \ell|m\\(\ell,P_1)=1}}1$$

$$\leqslant\sum_{m\leqslant z}\frac{1}{m}\sum_{\substack{\ell|m\\(\ell,P_1)=1}}1=\sum_{m\leqslant z}\frac{1}{m}d(m_2)$$

$$\leqslant\sum_{m\leqslant z}\frac{\omega(m)}{m}\,.$$

Let us now prove the second inequality. Because $\sqrt{z}\geqslant 2$ there exists an absolute constant C_1 such that:

$$\sum_{k\leqslant\sqrt{z}}\frac{1}{k}\geqslant 2C_1\log\sqrt{z}=C_1\log z.$$

By Exercise 1·27(a) we have:

$$\sum_{\substack{\ell\leqslant\sqrt{z}\\(\ell,P_1)=1}}\frac{1}{\ell}\geqslant B\prod_{p\in P_2}\left(1-\frac{1}{p}\right)^{-1}=B\prod_{p\in P_1}\left(1-\frac{1}{p}\right)\prod_{p\leqslant\sqrt{z}}\left(1-\frac{1}{p}\right)^{-1}.$$

By Mertens' Theorem (cf., Introduction) there exists a constant C_2 such that:

$$\prod_{p\leqslant\sqrt{z}}\left(1-\frac{1}{p}\right)^{-1}\geqslant 2C_2\log\sqrt{z}=C_2\log z.$$

The required lower bound is clearly obtained (with $C=BC_1C_2$).

SOLUTION 1·28: Here it will be assumed that m is positive. With the notations of Section 1.13 of the Introduction, let us choose:

$$n = [x], \quad A = \{1,2,\dots,n\}, \quad z = n^{1/3}, \quad P = \{p \leqslant z\}.$$

If $p|m$, set $\omega(p) = 1$ and $R_1^{(p)} = \{\ell | \ell \equiv 0[p]\}$;

If $p\nmid m$, set $\omega(p) = 2$ and $\begin{cases} R_1^{(p)} = \{\ell|\ell \equiv 0[p]\}; \\ R_2^{(p)} = \{\ell|\ell \equiv -m[p]\}. \end{cases}$

In order to verify relation (H) of Section 1.13 of the Introduction, we will verify that:

$$0 \leqslant \operatorname{card}A_d - \omega(d)\left[\frac{n}{d}\right] \leqslant \omega(d).$$

Let us now show the majorisation:

$$\operatorname{card}\{p|p \leqslant x,\ p + m \text{ a prime number}\} \leqslant S(A,P) + z. \qquad (*)$$

Apart from the prime numbers of the first set which satisfy $p \leqslant z$ (there are at most z) all the other p's satisfy $z < p \leqslant n$; for every element $p_i \in P$ we have:

$$p \not\equiv 0[p_i] \quad \text{and} \quad p + m \not\equiv 0[p_i],$$

which certainly shows that p is counted in $S(A,P)$.

From Selberg's Theorem and the upper bound (*) we deduce the upper bound:

$$\operatorname{card}\{p|p \leqslant x,\ p + m \text{ a prime number}\}$$

$$\leqslant x\left(\sum_{k\leqslant z} \frac{\omega(k)}{k}\right)^{-1} + 5x^{2/3} \prod_{2<p\leqslant z} \left(1 - \frac{2}{p}\right)^{-2} + z. \qquad (**)$$

From Exercise 1·27 we have:

$$\sum_{k \leqslant z} \frac{\omega(k)}{k} \geqslant C\left(\prod_{p|m}\left(1-\frac{1}{p}\right)\right)\log^2 z \geqslant C_3\left(\prod_{p|m}\left(1-\frac{1}{p}\right)\right)\log^2 x.$$

It is easy to see that:

$$\prod_{2<p\leqslant z}\left(1-\frac{2}{p}\right)^{-2} = \prod_{2<p\leqslant z}\left(1-\frac{1}{p}\right)^{-4}\prod_{2<p\leqslant z}\frac{\left(1-\frac{1}{p}\right)^4}{\left(1-\frac{2}{p}\right)^2}.$$

The second product is majorised by the convergent product:

$$\prod_p\left(1+\frac{1}{p^2-2p}\right)^2,$$

and the product $\prod_{2<p\leqslant z}\left(1-\frac{1}{p}\right)^{-4}$ is majorised (because of Mertens' Theorem) by a term in $(\log z)^4$. From (**) we deduce then the majorisation:

$$\operatorname{card}\{p|p \leqslant x,\ p+m \text{ a prime number}\} \leqslant D\prod_{p|m}\left(1-\frac{1}{p}\right)^{-1}\frac{x}{\log^2 x}.$$

which is the first sought.

The second is obtained by noticing that the product:

$$\prod_{p|m}\frac{1}{\left(1-\frac{1}{p}\right)\left(1+\frac{1}{p}\right)}$$

is majorised by the convergent infinite product $\prod_p \frac{1}{1-\frac{1}{p^2}}$.

SOLUTION 1·29: By Exercise 1·28 above there exists an absolute constant C such that:

$$\operatorname{card} I_x \leqslant C\frac{x}{\log^2 x},$$

setting:

$$I_x = \{p \mid p \leqslant x,\ p + 2 \text{ a prime number}\}.$$

By the formula of Section 1.2 of the Introduction:

$$\sum_{p \in I_x} \frac{1}{p} \leqslant \frac{C}{\log^2 x} + \int_2^x \frac{C}{t \log^2 t}\, dt = O(1).$$

(This formulation is due to Viggo Brun (1919) who deduced it from the more coarse upper bound:

$$\operatorname{card} I_x \leqslant C \frac{x(\log\log x)^2}{\log^2 x}$$

which he had obtained by the first version of his sieve).

SOLUTION 1·30: (a): If N is odd or if $N < 7$ it is clear that $r(N) \leqslant 2$, and the upper bound stated is certainly satisfied. If N is even and $N > 7$ we use Exercise 1·28 with $x = N$ and $m = -N$, from which we deduce:

$$r(N) \leqslant E \frac{N}{\log^2 N} \prod_{p \mid N} \left(1 + \frac{1}{p}\right) < E \frac{N}{\log^2 N} \sum_{d \mid N} \frac{1}{d}.$$

(b): From the preceding Question (a) we deduce:

$$\sum_{N \leqslant x} r^2(N) \leqslant F^2 \frac{x^2}{\log^4 x} \sum_{0 < N \leqslant x} \left(\sum_{d \mid N} \frac{1}{d}\right)^2.$$

Thus:

$$\sum_{0 < N \leqslant x} \left(\sum_{d \mid N} \frac{1}{d}\right)^2 = \sum_{0 < N \leqslant x} \sum_{d_1 \mid N} \sum_{d_2 \mid N} \frac{1}{d_1 d_2}$$

$$= \sum_{d_1 \leqslant x} \sum_{d_2 \leqslant x} \frac{1}{d_1 d_2} \sum_{\substack{0 < N \leqslant x \\ [d_1, d_2] \mid N}} 1.$$

From this follows:

$$\sum_{0<N\leqslant x}\left(\sum_{d|N}\frac{1}{d}\right)^2 \leqslant x\sum_{d_1\leqslant x}\sum_{d_2\leqslant x}\frac{(d_1,d_2)}{(d_1\ d_2)^2}\ .$$

Lastly, we have:

$$\sum_{d_1\leqslant x}\sum_{d_2\leqslant x}\frac{(d_1,d_2)}{(d_1\ d_2)^2} \leqslant \sum_{\delta<x}\frac{1}{\delta^3}\sum_{\delta_1\leqslant x}\sum_{\substack{\delta_2\leqslant x\\(\delta_1,\delta_2)=1}}\frac{1}{(\delta_1\ \delta_2)^2}$$

$$\leqslant \left(\sum\frac{1}{\delta^3}\right)\left(\sum_d\frac{1}{d^2}\right)^2,$$

whence the upper bound sought.

(c): We have:

$$\sum_{N\leqslant x} r(N) = \sum_{N\leqslant x}\sum_{\substack{p_1,p_2\\p_1+p_2=N}} 1 = \sum_{p_1\leqslant x}\pi(x-p_1)$$

$$\geqslant \sum_{p_1\leqslant x/2}\pi(\tfrac{1}{2}x) \geqslant \pi^2(\tfrac{1}{2}x) \geqslant F''\frac{x^2}{\log^2 x} \quad \text{if } x\geqslant 4.$$

(d): By the Cauchy-Schwarz Inequality we have:

$$\left(\sum_{N\leqslant x} r(N)\right)^2 \leqslant \left(\sum_{\substack{N\leqslant x\\r(N)\neq 0}} 1^2\right)\left(\sum_{N\leqslant x} r^2(N)\right)\ .$$

With the help of the upper and lower bounds given in Questions (b) and (c) above, we have:

$$M(x) = \sum_{\substack{N\leqslant x\\r(N)\neq 0}} 1 \geqslant \frac{F''^2}{F'}\,x = Cx \text{ if } x>4.$$

(e): Let S be the sequence formed of 0,1, and the prime numbers. By Question (d) above the sequence S has a positive Šnirelman density (greater than or equal to C because it contains 0,1,2,3, and 4); therefore the sequence S is an additive basis, and it follows that S is also a basis (cf., Chapter 2).

REMARK: It was to obtain this result that Šnirelman introduced the notion of density in 1930.

SOLUTION 1·31: (a): Let us set $E = \{i \mid p_{i+1} \leqslant x, d_i \geqslant A\log x\}$ and let i_0 be the largest element of E; we have:

$$x \geqslant \sum_{i \leqslant i_0} d_i \geqslant \sum_{i \in E} d_i \geqslant A\log x \sum_{i \in E} 1 = A\log x \operatorname{card} E.$$

Therefore we have:

$$\operatorname{card} E \leqslant \frac{x}{A\log x} .$$

(b): By the definition of $M(x)$:

$$M(x) = \sum_{a\log x < m \leqslant A\log x} \operatorname{card}\{i \mid \tfrac{1}{2}x < p_i \leqslant x \text{ and } d_i = m .$$

If $d_i = m$, the numbers p_i and $p_i + m$ are prime numbers, and by Exercise 1·28:

$$\operatorname{card}\{i \mid \tfrac{1}{2}x < p_i \leqslant x \text{ and } d_i = m\} \leqslant E \frac{x}{\log^2 x} \prod_{p \mid m} \left(1 + \frac{1}{p}\right) ,$$

and so the desired upper bound is deduced.

(c): We have:

$$\prod_{p \mid m} \left(1 + \frac{1}{p}\right) = \sum_{d \mid m} \frac{|\mu(d)|}{d} ,$$

where μ denotes the Möbius function. From this we deduce:

$$\sum_{a\log x<m\leqslant A\log x}\prod_{p|m}\left(1+\frac{1}{p}\right)=\sum_{a\log x<m\leqslant A\log x}\sum_{d|m}\frac{|\mu(d)|}{d}$$

$$=\sum_{d\leqslant A\log x}\frac{|\mu(d)|}{d}\sum_{\substack{a\log x<m\leqslant A\log x\\ m\equiv 0(\mathrm{mod}\, d)}}1$$

$$=\sum_{d\leqslant A\log x}\frac{|\mu(d)|}{d}\frac{(A-a)\log x}{d}+O\left(\sum_{d\leqslant A\log x}\frac{1}{d}\right)$$

$$=(A-a)\left(\sum_{d=1}\frac{\mu(d)}{d^2}\right)(1+o(1))\log x.$$

From this latter estimate and from the result of Question (b) above, the desired upper bound certainly follows: (One can take $F' = F\sum_{d=1}^{\infty}\frac{|\mu(d)|}{d^2}$).

(d): Let us assume for fixed a that $d_i > a\log x$ whenever $i \leqslant \pi(x)$ (i.e., $p_i \leqslant x$) for all sufficiently large x. For large enough, then:

$$S(x)=\sum_{a\log x<d_i\leqslant A\log x}d_i+\sum_{d_i>A\log x}d_i \quad \text{for } \pi(\tfrac{1}{2}x)<i\leqslant\pi(x).$$

The number of terms in the first sum is $M(x)$, the number in the second is therefore:

$$\pi(x)-\pi(\tfrac{1}{2}x)-M(x).$$

Hence:

$$S(x)\geqslant(\pi(x)-\pi(\tfrac{1}{2}x))A\log x-M(x)(A-a)\log x.$$

By the Prime Number Theorem and the preceding Question (c) we obtain:

$$\liminf_{x\to\infty}\frac{S(x)}{x}\geqslant\tfrac{1}{2}A-F'(A-a)^2.$$

(e): Let us choose $A = 1 + t$, $a = 1 - t$, with $t = \frac{1}{16F'}$. We obtain:

$$\tfrac{1}{2}A - F'(A - a)^2 = \tfrac{1}{2} + \tfrac{1}{2}t - F'4.t.\frac{1}{16F'} = \tfrac{1}{2} + \tfrac{1}{4}t.$$

By the Prime Number Theorem:

$$\sum_{i \leqslant \pi(x)} d_i = p_{\pi(x)+1} - 2 \sim x,$$

from which one deduces $S(x) \sim \frac{1}{2}x$, which contradicts:

$$\liminf_{x\to\infty} \frac{S(x)}{x} \geqslant \tfrac{1}{2}A - F'(A - a)^2 = \tfrac{1}{2} + \tfrac{1}{4}t \geqslant \tfrac{1}{2}.$$

The hypothesis of Question (d) above is therefore false. Therefore, there exist arbitrarily large x's for which:

$$d_i \leqslant a\log x$$

for at least one i satisfying $\frac{1}{2}x < p_i \leqslant x$. For this i we have:

$$\frac{d_i}{\log p_i} \leqslant \frac{d_i}{\log\frac{1}{2}x} \leqslant \frac{a\log x}{\log\frac{1}{2}x},$$

which implies:

$$\liminf_{i\to\infty} \frac{d_i}{\log p_i} \leqslant a < 1.$$

(By this method one can obtain $a = \frac{15}{16}$).

CHAPTER 2

Additive Theory

INTRODUCTION

2.1 DEFINITIONS

DEFINITIONS: Let A be a strictly increasing sequence of non-negative integers; the strictly positive elements of A arranged in increasing order will be denoted $a_1, a_2, \ldots$ (if 0 is an element of A it will be written $a_0 = 0$). For every positive real number X, we *denote* by $A(X)$ the *NUMBER OF POSITIVE ELEMENTS OF* A which are *LESS THAN OR EQUAL TO* X, and one defines the *ŠNIRELMAN DENSITY OF THE SEQUENCE* A by the relation:

$$\sigma A = \inf_{N \in \mathbb{N}^*} \left(\frac{A(N)}{N} \right) .$$

If A and B are two sequences of integers, we *denote* by A + B the set of all the integers which are the sum of an element of A and an element of B. If B = A we write 2A = A + A, and by induction we define $hA = (h - 1)A + A$.

We say that the sequence A is a *BASIS* if there exists an integer h such that $hA = \mathbb{N}$. We call the *ORDER OF A BASIS* A the smallest integer h such that $hA = \mathbb{N}$. (Notice that every basis necessarily contains 0 and 1).

2.2 RESULTS

THEOREM: (Šnirelman-Mann): *If* 0 *belongs to both* A *and* B *then:*

$$\sigma(A + B) \geqslant \min(1, \sigma A + \sigma B).$$

By noticing that the only sequence containing 0 with Šnirelman density one is the set of integers, it follows from the preceding Theorem that *every sequence with strictly positive Šnirelman density containing* 0 *is a basis.*

2.3 BIBLIOGRAPHY

The basic work in this field, rich in results and references, and pleasant to read, is [Hal].

PROBLEMS

EXERCISE 2·1:(1): Show that if A is a basis there exist two positive real numbers β and γ such that for all n:

$$a_n < \beta n^\gamma.$$

(2): Show that if a sequence A is such that:

$$A(n) = O(n^\varepsilon) \text{ for any real number } \varepsilon,$$

then A is not a basis.

EXERCISE 2·2: Let A and B be two increasing sequences of integers; we *denote* by $A(B)$ the *SET OF ELEMENTS OF* A which are *INDEXED BY ELEMENTS OF* B, that is to say, the elements of the form a_n with n an element of B.

Prove the inequalities:

$$\min(\sigma(A),\sigma(B)) \geqslant \sigma(A(B)) \geqslant \sigma A.\sigma B.$$

Show by examples that these inequalities are best possible.

EXERCISE 2·3: Let f be a mapping of $\mathbb{N}^*$ into $[0,1]$ such that $f(n) \to 0$ as $n \to \infty$.

Show that there exists a sequence A possessing the following properties:

(i): A contains 0 and 1;

(ii): For every integer N, $A(N) > f(N).N$;

(iii): The sequence A is not a basis.

(We will be able to construct a sequence A satisfying (i), (ii) and the following Condition: For every integer $j > 1$ there exists an integer n_j such that A contains no element in the interval $[n_j, j.n_j)$).

EXERCISE 2·4: Let A be a basis and $\mathcal{D}$ a bounded sequence of integers $\geqslant 0$ such that $d_0 = d_1 = 0$.

Show that the sequence $\mathcal{B} = \{a_i + d_i | i \in \mathbb{N}\}$ is a basis.

EXERCISE 2·5: Let A be a sequence of integers containing 0 and 1, and let α be a positive real number; let $\mathcal{B} = \{1\} \cup \{[\alpha a] | a \in A\}$.

Show that $\mathcal{B}$ is a basis if and only if A is. Begin by establishing the relation:

$$[x_1] + \cdots + [x_\ell] \leqslant [x_1 + x_2 + \cdots + x_\ell]$$

$$\leqslant [x_1] + \cdots + [x_\ell] + \ell - 1).$$

EXERCISE 2·6: Show that for every pair of positive rational numbers a and b, $a + b < 1$, one can construct two sequences A and $\mathcal{B}$ such that:

$$\sigma A = a, \qquad \sigma\mathcal{B} = b, \qquad \sigma(A + \mathcal{B}) = a + b.$$

EXERCISE 2·7: (a): Let A be a sequence of integers; assume that there exists an irrational number α such that:

$$\lim_{n\to\infty}(\{\alpha a_n\}) = 0. \tag{H}$$

Show that A is not a basis.

(One should prove by induction on h that the density of the sequence hA is zero by using the equipartition of the sequence $(\{\alpha n\})_{n\in\mathbb{N}}$, cf., [Rau]).

(b): The question as above, replacing (H) by (H'):

The set of fractional parts of the elements αa_n has only a finite number of limit points. (H')

EXERCISE 2·8: Use Exercise 2·7(a) to give another proof of the result of Exercise 2·3.

(An irrational number α should be chosen, and for every integer $k \leqslant 2$ set:

$$A_k = \{n \in \mathbb{N} \mid \{\alpha n\} \leqslant \frac{2}{k}\}.$$

By a diagonal process construct a sequence:

$$A = \{0 = a_0 < a_1 = 1 < a_2 < \dots\}$$

satisfying Conditions 2·3(i),(ii), and such that $\{\alpha a_i\} \to 0$).

EXERCISE 2·9: Assume that every integer divisible by a prime number q congruent to 3 modulo 4, but not divisible by q^2, is not the sum of two squares of integers ([Har], Theorem 355).

Deduce from this that one can find arbitrarily long chains of consecutive integers which are not the sums of two squares; show that for no positive real number α is the sequence $([\alpha n^2])$ a basis of order two.

SOLUTIONS

SOLUTION 2·1:(1): Let us assume that A is a basis of order h. Every element x in the set $E = \{0,1,\dots,a_n - 1\}$ may be written in the form:

$$x = a_{i_1} + \cdots + a_{i_h} \quad \text{with} \quad 0 \leqslant i_1 \leqslant i_2 \leqslant \cdots \leqslant n - 1.$$

The number of such forms is the number of combinations with repetitions of the n objects $a_0, a_1, \dots, a_{n-1}$ taken h by h, say $\binom{n+h-1}{h}$. We must therefore have:

$$\operatorname{card} E = a_n \leqslant \binom{n+h-1}{h} \leqslant \frac{(n + h)^h}{h!}.$$

Now we have:

$$\frac{a_n}{n^h} \leqslant \frac{\left(1 + \frac{h}{n}\right)^h}{h!}.$$

Since we have:

$$\lim_{n\to\infty} \frac{\left(1 + \frac{h}{n}\right)^h}{h!} = \frac{1}{h!},$$

the sequence $\frac{a_n}{n^h}$ is bounded above; let β be an upper bound of this sequence. We have $a_n \leqslant \beta n^h$.

(2): From the above, if A is a basis there exist two positive real numbers β and γ such that: $a_n < \beta n^\gamma$ $(n = 1,2,\ldots)$, which can be written $A(\beta n^\gamma) \geqslant n$. By setting $x = \beta n^\gamma$ we obtain:

$$A(x) \geqslant \left(\frac{x}{\beta}\right)^{1/\gamma},$$

so

$$\overline{\lim}\, \frac{A(x)}{x^{1/\gamma}} > 0,$$

which contradicts the hypothesis $A(n) = O(n^\varepsilon)$ for $\varepsilon < \frac{1}{\gamma}$.

SOLUTION 2·2: To say that an element $a_i \leqslant N$ is equivalent to say that $i < A(N)$. Hence:

$$a_{b_j} \leqslant N \iff b_j \leqslant A(N) \iff j \leqslant B(A(N))$$

Let us set $C = A(B)$; hence $C(N) = B(A(N))$. By the definition of the Šnirelman density we have:

$$C(N) = B(A(N)) \geqslant A(B).\sigma B \geqslant N.\sigma A.\sigma B.$$

From which we deduce:

$$\sigma(A(B)) \geqslant \sigma A.\sigma B.$$

On the other hand, $A(N) \leqslant N$ and $B(M) \leqslant M$. Since B is an increasing function, we have that $C(N) \leqslant B(N)$ and $C(N) \leqslant A(N)$, whence:

$$\sigma(A(B)) \leqslant \min(\sigma A,\sigma B).$$

EXAMPLE 1: $A = \{1,4,5,6,7,8,\ldots\}$, $\sigma A = \frac{1}{3}$;

$B = \{1,2,4,5,6,7,8,\ldots\}$, $\sigma B = \frac{2}{3}$;

$A(B) = \{1,4,6,7,8,\ldots\}$, $\sigma(A(B)) = \frac{1}{3} = \min(\sigma A, \sigma B)$.

EXAMPLE 2: $A = \{1,4,7,10,11,12,13,\ldots\}$, $\sigma A = \frac{1}{3}$;

$B = \{1,2,4,5,6,7,8,\ldots\}$, $\sigma B = \frac{2}{3}$;

$A(B) = \{1,4,10,11,12,13,\ldots\}$, $\sigma(A(B)) = \frac{2}{9} = \sigma A, \sigma B$.

SOLUTION 2·3: Set $\psi(m) = \sup_{n \geqslant m} f(n)$. The function ψ is a decreasing mapping of $\mathbb{N}^*$ into $[0,1]$, with limit zero at infinity, which bounds f. By induction we define the sequence n_i in the following way: $n_0 = 0$, n_i $(i \geqslant 1)$ is the smallest integer satisfying $n_i > 2(i-1)n_{i-1} + 1$ such that $\psi(n_i) < \frac{1}{2i}$. One then sets:

$$A = \{n \in \mathbb{N} \mid \exists i \geqslant 0 : in_i \leqslant n < n_{i+1}\}.$$

(i) is satisfied, as $n_1 > 1$.

(ii) Let N be an integer; let us consider three cases:

(a): $1 \leqslant N < n_1$, $\frac{A(N)}{N} = 1 \geqslant \psi(N)$.

(b): There exists a positive integer i such that $n_i \leqslant N < in_i$:

$$\frac{A(N)}{N} = \frac{A(n_i)}{N} > \frac{1}{i}\frac{A(n_i)}{n_i} \geqslant \frac{n_i - (i-1)n_{i-1}}{in_i}$$

$$\geqslant \frac{n_i}{2in_i} = \frac{1}{2i} \geqslant \psi(n_i) \geqslant \psi(N).$$

(c): There exists a positive integer i such that $in_i \leqslant N < n_{i+1}$:

$$\frac{A(N)}{N} = \frac{N - in_i + 1 + A(in_i - 1)}{N} = 1 - \frac{in_i - 1 - A(in_i - 1)}{N}$$

$$\geqslant 1 - \frac{in_i - 1 - A(in_i - 1)}{in_{ii} - 1} = \frac{A(in_i - 1)}{in_i - 1}$$

$$> \psi(in_i - 1),$$

by using (a) or (b), whence:

$$\frac{A(N)}{N} \geqslant \psi(N).$$

(iii) Let $h \geqslant 2$ be an integer. The element $(h + 1)n_{h+1} - 1$ is not a sum of less than h elements of A; in fact the elements of A which are less than it are bounded by n_{h+1}; therefore A is not a basis.

SOLUTION 2·4: Let h be the order of the basis A; for every integer N, at least one of the numbers $N, N + 1, \ldots, N + hd$ is an element of hB, where the maximum of the elements of D has been denoted by d. In fact we can write $N = a_{i_1} + \cdots + a_{i_h}$; the element $a_{i_1} + d_{i_1} + \cdots + a_{i_h} + d_{i_h}$ is an element of hB, and it certainly lies between N and $N + hd$. Moreover, the sequence hB contains 0 and 1, and it therefore has positive density (greater than or equal to $\frac{1}{hd + 1}$); it is therefore a basis, and the sequence B is also (of order $h(hd + 1)$ at most).

SOLUTION 2·5: *First Step:* By writing $x_i = [x_i] + \{x_i\}$, with $0 \leqslant \{x_i\} < 1$, we have:

$$[x_1] + \cdots + [x_\ell] \leqslant x_1 + x_2 + \cdots + x_\ell < [x_1] + \cdots + [x_\ell] + \ell.$$

Since $[x_1 + \cdots + x_\ell]$ is the largest integer less than or equal to $x_1 + x_2 + \cdots + x_\ell$, and since $[x_1] + \cdots + [x_\ell]$ is an integer less than or equal to $x_1 + x_2 + \cdots + x_\ell$, we have:

$$[x_1] + \cdots + [x_\ell] \leqslant [x_1 + x_2 + \cdots + x_\ell].$$

Since $[x_1 + \cdots + x_\ell] \leqslant x_1 + x_2 + \cdots + x_\ell$,

$$[x_1 + \cdots + x_\ell] < [x_1] + \cdots + [x_\ell] + \ell,$$

and since both members of this relation are integers, we have:

$$[x_1 + \cdots + x_\ell] \leqslant [x_1] + \cdots + [x_\ell] + \ell - 1.$$

Second Step: Let us first assume that A is a basis of order h. To every integer N, we can associate an (unique) integer n such that:

$$[\alpha n] \leqslant N < [\alpha(n + 1)].$$

One can then write:

$$0 \leqslant N - [\alpha n] \leqslant 1 + [\alpha],$$

and:

$$\alpha n = \alpha(a^{(1)} + \cdots + a^{(h)}),$$

where the $a^{(i)}$ are elements of A. By the inequalities proved in the First Step, we have:

$$[\alpha a^{(1)}] + \cdots + [\alpha a^{(h)}] \leqslant [\alpha a^{(1)} + \cdots + \alpha a^{(h)}] = [\alpha n]$$

$$\leqslant [\alpha a^{(1)}] + \cdots + [\alpha a^{(h)}] + h - 1.$$

N can therefore be written as the sum of h elements $[\alpha a^{(i)}]$ and at most $h + [\alpha]$ elements equal to one. By hypothesis, $1 \in B$, hence B is a basis of order at most $2h + [\alpha]$.

Third Step: Now assume that B is a basis of order h. Let $N > \frac{h}{\alpha}$ be an integer. The integer $[\alpha N] - h$ is positve or zero, hence it is an element of hB; hence can be written as the sum of at most h elements $[\alpha a^{(i)}]$, i.e., there exists a real number ϑ (with $0 \leqslant \vartheta \leqslant 1$) and an integer ℓ (with $0 \leqslant \ell \leqslant h$) such that:

$$[\alpha N] - h = [\alpha a^{(1)}] + \cdots + [\alpha a^{(h-\ell)}] + \vartheta\ell \quad (\text{where } a^{(i)} \in A).$$

From this one deduces:

$$\begin{aligned} \alpha a^{(1)} + \cdots + \alpha a^{(h-\ell)} - h &< [\alpha N] - h \\ &\leqslant \alpha a^{(1)} + \cdots + \alpha a^{(h-\ell)} + h, \end{aligned}$$

or again:

$$\begin{aligned} \alpha a^{(1)} + \cdots + \alpha a^{(h-\ell)} &< \alpha N \\ &\leqslant \alpha a^{(1)} + \cdots + \alpha a^{(h-\ell)} + 2h + 1, \end{aligned}$$

hence one has:

$$\begin{aligned} a^{(1)} + \cdots + a^{(h-\ell)} &< N \\ &\leqslant a^{(1)} + \cdots + a^{(h-\ell)} + \left[\frac{2h+1}{\alpha}\right] \end{aligned}$$

N is therefore written as the sum of at most h elements of A and $\left[\frac{2h+1}{\alpha}\right]$ ones; now $1 \in A$, hence N is the sum of at most $h + \left[\frac{2h+1}{\alpha}\right]$ elements of A. It is easily verified that the same holds when $N \leqslant \frac{h}{a}$. A is therefore a basis of order equal at most

$h + \left[\frac{2h+1}{\alpha}\right]$. (The same method also leads to the upper bound:

bound:

$$\sup_{0 \leq \ell \leq h} \left(h - \ell + \left[\frac{h + \ell + 1}{\alpha}\right] \right),$$

or,

$$1 + \left[\frac{2h+1}{\alpha}\right] \text{ if } \alpha > 1, \quad \text{and} \quad h + 1 + \left[\frac{h+1}{\alpha}\right] \text{ if } \alpha < 1).$$

SOLUTION 2·6: Set $a = \frac{u}{p}$ and $b = \frac{v}{q}$; to say that $a + b < 1$ reduces to saying that $uq + vp < pq$. Set:

$$A = \{0,1,2,\ldots,uq,pq+1,pq+2,\ldots\}, \qquad \sigma A = \frac{u}{p},$$

and

$$\mathcal{B} = \{0,1,2,\ldots,vp,pq+1,pq+2,\ldots\}, \qquad \sigma\mathcal{B} = \frac{v}{q}.$$

Then:

$$A + \mathcal{B} = \{0,1,2,\ldots,uq+vp,pq+1,pq+2,\ldots\}$$

and

$$\sigma(A + \mathcal{B}) = \frac{uq + vp}{pq} = \frac{u}{p} + \frac{v}{q} = \sigma A + \sigma\mathcal{B}.$$

SOLUTION 2·7:(a): For every positive real number ε we shall set:

$$A'_\varepsilon = \{a_i \in A \mid \{\alpha a_i\} > \varepsilon\} \quad \text{and} \quad A_\varepsilon = \complement_A A'_\varepsilon.$$

Notice that A'_ε is a finite set; let A'_ε be an upper bound for its cardinality. For every integer N we have:

$$A(N) = A'_\varepsilon(N) + A_\varepsilon(N) \leqslant A'_\varepsilon + A_\varepsilon(N)$$

$$\leqslant A'_\varepsilon + \text{card}\{n \leqslant N \mid \{\alpha n\} \leqslant \varepsilon\}$$

It is known that the sequence $\{\alpha n\}$ is equidistributed mod 1 (cf., Exercise 5·8), and we have:

$$\limsup_{N\to\infty} \frac{A(N)}{N} \leqslant \limsup_{N\to\infty} \frac{1}{N}\,\text{card}\{n \leqslant N \mid \{\alpha n\} \leqslant \varepsilon\} \leqslant \varepsilon.$$

This proves that the sequence A has a zero Šnirelman density. A useful convention is to call the quantity $\limsup_{N\to\infty} \frac{A(N)}{N}$ the *UPPER ASYMPTOTIC DENSITY of A*.

Now let $h \geqslant 1$ be an integer. Assume that the sequence hA has upper asymptotic density zero. We can write:

$$(h + 1)\mathrm{A} = (h + 1)(\mathrm{A}_\varepsilon \cup \mathrm{A}'_\varepsilon) = (\mathrm{A}'_\varepsilon + h\mathrm{A}) \cup (h + 1)\mathrm{A}_\varepsilon.$$

(In fact every element of $(h + 1)$A is either the sum of $(h + 1)$ elements of A_ε, or the sum of an element of A'_ε and h elements of A).

The sequence $\mathrm{A}'_\varepsilon + h\mathrm{A}$, which is the union of a finite number of translates of the sequence hA, has upper asymptotic density zero.

Now, every integer m which is an element of the sequence $(h + 1)\mathrm{A}_\varepsilon$ satisfies the relation $\{\alpha m\} \leqslant (h + 1)\varepsilon$, hence we have:

$$\limsup_{N\to\infty} \frac{1}{N}\,\text{card}\{n \leqslant N \mid n \in (h + 1)\mathrm{A}_\varepsilon\}$$

$$\leqslant \limsup_{N\to\infty} \frac{1}{N}\,\text{card}\{n \leqslant N \mid \{\alpha N\} \leqslant (h + 1)\varepsilon\}$$

$$\leqslant (h + 1)\varepsilon.$$

The sequence $(h + 1)A_\varepsilon$ therefore has upper asymptotic density zero, and the same holds for the sequence $(h + 1)A$, since this is included in the union of two sequences with zero upper asymptotic density.

(b): Let $x_1,\dots,x_k$ be the limit points of the sequence A. Set:

$$A'_\varepsilon = \{a \in A \mid \forall j \in [1,k], \|\alpha a - x_j\| > \varepsilon\}$$

if $1 \leqslant j \leqslant k$, and:

$$A_\varepsilon^{(j)} = \{a \in A \mid \|\alpha a - x_j\| \leqslant \varepsilon\},$$

$$A_\varepsilon = A_\varepsilon^{(1)} \cup \dots \cup A_\varepsilon^{(k)}.$$

The sequence A'_ε is finite, and each of the sequences $A_\varepsilon^{(j)}$ has upper asymptotic desnity less than 2ε; from this we deduce that the sequence A has upper asymptotic density zero.

Let h be an integer such that the sequence hA has upper asymptotic density zero; as before, we have the equality:

$$(h + 1)A = (A'_\varepsilon + hA) \cup (h + 1)A_\varepsilon.$$

Denote by E the set of points $x_{i_1} + \dots + x_{i_{h+1}}$ taken modulo 1. E has only a finite number (at most $k^{(h+1)}$) of elements, and $(h + 1)A_\varepsilon$ is included in the set of integers m for which there exists a $y \in E$ such that $\|\alpha m - y\| < (h + 1)\varepsilon$. From this one deduces the upper asymptotic density of the sequence $(h + 1)A_\varepsilon$ (and therefore that of $(h + 1)A$) is less than $2k^{(h+1)}(h + 1)\varepsilon$, the sequence $(h + 1)A$ therefore has zero density.

SOLUTION 2·8: Let α be a fixed irrational number, and let $k \geqslant 2$ be an integer. Since $f(n) \to 0$ there exists an integer M_k such that:

$$n \geqslant M_k \Rightarrow f(n) < \frac{1}{k}.$$

Set $A_k = \{n \in \mathbb{N} \mid \{\alpha n\} \leqslant 2/k\}$; since the sequence $(\alpha n)_{n \in \mathbb{N}}$ is equidistributed modulo 1 (cf., [Rau] or Exercise 5·8), we can find an integer M_k' such that:

$$n \geqslant M_k' \Rightarrow \frac{A_k(n)}{n} \geqslant \frac{1}{k}.$$

The sequence $(N_i)_{i \in \mathbb{N}}$ is defined by the relations:

$$N_1 = 0, \qquad N_k = \sup(M_k, M_k', N_{k-1} + 1) \quad \text{for } k = 2,3,\ldots,$$

and the sequence A in the following way:

$$A \cap [N_1, N_2] = \mathbb{N} \cap [N_1, N_2],$$

$$A \cap [N_k, N_{k+1}] = A_k \cap [N_k, N_{k+1}] \quad \text{for } k = 2,3,\ldots .$$

(i): The sequence A contains 0 and 1 by construction.

(ii): Let N be an integer.

— If $N \leqslant N_2$, $A(N) = N \geqslant Nf(N)$.

— If $N_k \leqslant N < N_{k+1}$ we observe that the sequences A_k are included in each other, and we have:

$$A_k \cap [1, N] \subset A \cap [1, N],$$

therefore

$$A(N) \geqslant \frac{N}{k} > Nf(N).$$

(iii): The sequence A is not a basis, for the sequence $\{\alpha a_i\}$ $\to 0$ by construction; one then uses Exercise 2·7(a).

SOLUTION 2·9:(i): Let us show that there is an infinite number of prime numbers congruent to 3 modulo 4. Let $q_1,\dots,q_k$ be the first k such prime numbers; the number $4(q_1\cdots q_k) - 1$ is congruent to 3 modulo 4, hence at least one of its prime factors is congruent to 3 modulo 4, and it is necessarily distinct from $q_1,\dots,q_k$.

(ii): Let us consider the system of congruences:

$$n \equiv q_1 - 1[q_1^2], \quad \dots, \quad n \equiv q_k - k[q_k^2],$$

where q_i is the i-th prime number congruent to 3 modulo 4; then numbers q_j^2 and q_j^2 being relatively prime if i is distinct from j, this system has a solution ([Har] Theorem 121). Each of the numbers $n+1,\dots,n+k$ is divisible by a prime number congruent to 3 modulo 4, without being divisible by the square of this number; hence it is not the sum of two squares.

(iii): Let α be a fixed positive real number. We are going to assume that the sequence $([\alpha n^2])$ is a basis of order two; from this we are going to deduce that in every interval of the form $[Y, Y + \frac{3}{\alpha}]$ there is a sum of two squares, which contradicts Part (ii).

Set $X = Y + \frac{1}{\alpha}$; there exist two integers n_1 and n_2 such that

$$[\alpha X] = [\alpha n_1^2] + [\alpha n_2^2].$$

We then have:

$$\alpha X = \alpha n_1^2 + \alpha n_2^2 - (\{\alpha n_1^2\} + \{\alpha n_2^2\} - \{\alpha X\}),$$

$$\alpha n_1^2 + \alpha n_2^2 - 2 < \alpha X < \alpha n_1^2 + \alpha n_2^2 + 1,$$

$$n_1^2 + n_2^2 - \frac{2}{\alpha} < X < n_1^2 + n_2^2 + \frac{1}{\alpha} ,$$

$$Y = X - \frac{1}{\alpha} < n_1^2 + n_2^2 < X + \frac{2}{\alpha} = Y + \frac{3}{\alpha} .$$

CHAPTER 3
Rational Series

INTRODUCTION

3.0 INTRODUCTION

Let K be a field of characteristic zero. The *algebraic closure of* K will be *denoted* $\bar{K}$, $K[X]$ will *denote* the *polynomial algebra* of K, and $K[[X]]$ the *algebra of formal power series with coefficients in* K.

If P and Q are polynomials with coefficients in K, we can ex-expand the rational function P/Q as a power series if $Q(0) \neq 0$. Conversely, we will say that a power series is a *RATIONAL SERIES*, or *RATIONAL*, if it is the development of a rational function.

3.1 FIRST CHARACTERISATION

Let $f(X) = \sum_n u_n X^n$ be a rational series with coefficients in K; on decomposing it into partial fractions one shows that the coefficients u_n can be written, starting from a certain index n_0, as:

$$u_n = \sum_{1 \leqslant i \leqslant r} P_i(n)\alpha_i^n,$$

where $\frac{1}{\alpha_i}$, $1 \leqslant i \leqslant r$, is a pole of f, and P_i is a polynomial with coefficients in the algebraic closure $\bar{K}$ of K of degree equal to the order of the multiplicity of $\frac{1}{\alpha_i}$ less 1.

— *Conversely:* If $f(X) = \sum_{n\geq 0} u_n X^n$ and if starting from a certain index n_0, for all $n \geq n_0$ we have:

$$u_n = \sum_{1\leqslant i\leqslant r} P_i(n)\alpha_i^n,$$

where the α_i, $1 \leqslant i \leqslant r$, are algebraic on K and distinct, and the P_i, $1 \leqslant i \leqslant r$, are polynomials with coefficients in $\bar{K}$, then $f = \sum_{n\geq 0} u_n X^n$ represents in its disc of convergence a rational series which has r poles $\frac{1}{\alpha_1}, \ldots, \frac{1}{\alpha_r}$; the order of each of these poles $\frac{1}{\alpha_i}$ is equal to the degree of the polynomial P_i plus 1.

Proof: By expanding $\frac{1}{\left(X - \frac{1}{\alpha}\right)^s}$, where $s \in \mathbb{N}$, as a power series we see that the series $\sum_{n\geq 0} P(n)\alpha^n X^n$, where P is a polynomial of degree d, is a rational fraction with pole $\frac{1}{\alpha}$ with multiplicity $(d + 1)$.

If, for $n \geqslant n_0$, we have:

$$u_n = \sum_{1\leqslant i\leqslant r} P_i(n)\alpha_i^n,$$

it follows that:

$$f(X) = \sum_{n\geq 0} u_n X^n$$

is a rational series — with coefficients in K. When the $u_n \in K$ it remains to be shown that $\frac{u}{v}$ has its coefficients in K.

When $K = \mathbb{R}$, $\bar{K} = \mathbb{C}$, we have:

$$\frac{\bar{U}}{\bar{V}} = \sum_{n\geq 0} u_n X^n = \sum_{n\geq 0} u_n X^n = \frac{U}{V},$$

whence:

$$\bar{U}V = U\bar{V} = \overline{U\bar{V}}.$$

The fraction $\frac{UV}{V\bar{V}}$ is equal to $\frac{U}{V}$ and its coefficients are real.

In the general case let K' be a Galois extension of K containing the coefficients of U and V. The fraction $\frac{U}{V}$ is invariant under the elements σ of the Galois group G. The fraction:

$$U = \prod_{\sigma \in G \setminus \{id\}} \frac{\sigma(V)}{\prod_{\sigma \in G} \sigma(V)}$$

is a quotient of polynomials with coefficients in K and is equal to $\frac{U}{V}$.

3.2 SECOND CHARACTERISATION [Sal]

A necessary and sufficient condition in order that $f(X) = \sum u_n X^n$, where $u_n \in K$, be rational is that the coefficients u_n satisfy a recurrence relation starting from a certain index n_0:

For all $n \geqslant n_0$ the u_n satisfy:

$$a_s u_n + a_{s-1} u_{n+1} + \cdots + a_1 u_{n+s-1} + u_{n+s} = 0, \quad \text{where } a_i \in K.$$

The proof uses the formal series:

$$\text{If} \quad f(X) = \sum_n u_n X^n \quad \text{and} \quad Q(X) = 1 + a_1 X + \cdots + a_s X^s,$$

then: $f(X)Q(X) \in K[X]$.

3.3 THIRD CHARACTERISATION: HANKEL DETERMINANTS (cf., [Sal] and Exercise 3·1)

Let $f(X) = \sum_n u_n X^n$, we call the *HANKEL DETERMINANT* $D_n^{(s)}$ *of the series* $f(X)$ the following determinant:

$$D_n^{(s)} = \begin{vmatrix} u_n & u_{n+1} & \cdots & u_{n+s} \\ u_{n+1} & u_{n+2} & \cdots & u_{n+s+1} \\ \cdot & \cdot & \cdot & \cdot \\ \cdot & \cdot & \cdot & \cdot \\ u_{n+s} & u_{n+s+1} & \cdots & u_{n+2s} \end{vmatrix}.$$

In order that $f(X)$ be a rational fraction it is necessary and sufficient that there exist two integers s and n_0 such that for all $n \geqslant n_0$ $D_n^{(s)} = 0$.

3.4 FOURTH CHARACTERISATION: KRONECKER DETERMINANTS

We call the *KRONECKER DETERMINANT of the series* $\sum_n u_n X^n$ the determinant:

$$\Delta_t = D_0^{(t)} = \begin{vmatrix} u_0 & u_1 & \cdots & u_t \\ u_1 & u_2 & \cdots & u_{t+1} \\ \cdot & \cdot & \cdot & \cdot \\ \cdot & \cdot & \cdot & \cdot \\ u_t & u_{t+1} & \cdots & u_{2t} \end{vmatrix}.$$

In order that the series $\sum_n u_n X^n$ be rational, it is necessary and sufficient that there exists t_0 such that for all $t \geqslant t_0$ we have $\Delta_t = 0$.

BIBLIOGRAPHY

[Sal], [Ami], [Pis].

PROBLEMS

EXERCISE 3·1: HANKEL AND KRONECKER DETERMINANTS

(1): Let $f(X) = \sum_{n\geq 0} u_n X^n$ be a rational series with coefficients in K; show that there exist two integers s and n_0 such that for $n \geq n_0$:

$$D_n^{(s)} = \begin{vmatrix} u_n & u_{n+1} & \cdots & u_{n+s} \\ u_{n+1} & u_{n+2} & \cdots & u_{n+s+1} \\ \cdot & \cdot & \cdot & \cdot \\ \cdot & \cdot & \cdot & \cdot \\ u_{n+s} & u_{n+s+1} & \cdots & u_{n+2s} \end{vmatrix} = 0.$$

(2): Let $D = [a_{i,j}]_{\substack{1\leq i\leq n \\ 1\leq j\leq n}}$ be a determinant with coefficients in K, and let:

$$d = \begin{vmatrix} a_{2,2} & \cdots & a_{2,n-1} \\ \cdot & \cdot & \cdot \\ \cdot & \cdot & \cdot \\ a_{n-1,2} & \cdots & a_{n-1,n-2} \end{vmatrix}.$$

(d is obtained by starting from D and deleting in D the first and the last line and the first and the last column).

Denote the cofactor of $a_{i,j}$ in D by $A_{i,j}$. Prove *SYLVESTER'S IDENTITY*

$$Dd = A_{1,1}A_{n,n} - A_{n,1}A_{1,n}.$$

(3): Assume that for fixed s $D_n^{(s)} = 0$ for $n \geqslant n_0$. Assume that there exists $n_1 \geqslant n_0$ such that $D_n^{(s-1)} = 0$ for $n = n_1$. Show that $D_n^{(s-1)} = 0$ for $n \geqslant n_1$.

(4): Show that if one can associate with the series $\sum_{n\geqslant 0} u_n X^n$ two integers s and n_0 such that for $n \geqslant n_0$ we have $D_n^{(s)} = 0$, then this series represents a rational function.

(5): Show that if the series $\sum_{n\geqslant 0} u_n X^n$ is rational, there exists t_0 such that for all $t \geqslant t_0$, $\Delta_t = D_0^{(t)} = 0$. Conversely, if $\Delta_t = 0$ for $t \geqslant t_0$, then the series $\sum_{n\geqslant 0} u_n X^n$ is rational.

EXERCISE 3·2: FATOU'S LEMMA

(1): A formal series in $\mathbb{Z}[[X]]$ is called a *PRIMITIVE SERIES* if the greatest common divisor of its coefficients is one. Show that the product of two primitive series of $\mathbb{Z}[[X]]$ is primitive.

(2): Let $\sum_{n\geqslant 0} u_n X^n \in \mathbb{Z}[[X]]$ be a formal series which is a rational function $\frac{P(X)}{Q(X)}$ with $P, Q \in \mathbb{Q}[X]$. Show that one can choose P and Q in $\mathbb{Z}[X]$ with $Q(0) = 1$.

EXERCISE 3·3: ANALYTIC CHARACTERISATION OF POLYNOMIALS AND RATIONAL FUNCTIONS

Let $f(z) = \sum_{n\geqslant 0} u_n z^n$ be a power series with coefficients in $\mathbb{Z}$ and let $R > 0$ be its radius of convergence.

(1): Assume $R > 1$. Show that $\sum_{n\geq 0} u_n z^n$ reduces to a polynomial.

(2): Assume $R < 1$ and that f is extended to a meromorphic function in a disc $|z| < \rho$, $\rho > 1$. Show that $f(z)$ is a rational function.

(3): Give $\mathbb{Q}$ the p-adic valuations, where p is a prime number. For this valuation the completion of $\mathbb{Q}$ will be denoted $\mathbb{Q}_p$, and the algebraic closure of $\mathbb{Q}_p$ by Ω_p.

Assume that f is extended into a meromorphic function in $\mathbb{C}$ in the disc $|z| < \rho$ and into a meromorphic function in Ω_p in the disc $|z|_p < \rho_p$. Show that if $\rho\rho_p > 1$, then $f(z)$ is a rational function.

EXERCISE 3·4: HADAMARD'S PRODUCT

Let $f(z) = \sum_n u_n z^n$ and $g(z) = \sum_n v_n z^n$ be two rational series with coefficients in K. Then the series $h(z) = \sum_n (u_n \times v_n) z^n$ is rational.

EXERCISE 3·5: HADAMARD'S QUOTIENT

Let $f(z) = \sum_n u_n z^n$ be a rational function, and $g = \sum_n v_n z^n$ a rational function having a simple pole of absolute value strictly less than the others, with coefficients in $\mathbb{Q}$.

Let a_n be the coefficient of a series defined in the following way:

If: $v_n \neq 0$ set $a_n = u_n v_n^{-1}$,

if: $v_n = 0$ set $a_n = 0$.

If the series $\sum_{n\geq 0} a_n z^n$ has coefficients in $\mathbb{Z}$, then it is rational.

EXERCISE 3·6: FORMAL SERIES AND MATRICES

Given a series $f(X) = \sum_{n\geq 0} u_n X^n \in K[[X]]$. We say that the *square matrix* $A = M_N(K)$ *is associated with the series* f if for all $n \geq 1$:

$$u_n = [A^n]_{1,N},$$

where $[A^n]_{1,N}$ denotes the coefficient in the first row and the last column of the matrix A^n.

(1): Show that if there exists a matrix A associated with the series $\sum_{n\geq 0} u_n X^n$, then this series is rational.

(2): We want to prove the converse of Question (1) above. Several steps are needed:

(a): Show that to the polynomial:

$$P(X) = u_0 + u_1 X + \cdots + u_{N-1} X^{N-1}$$

we can associate the matrix:

$$A = \begin{bmatrix} 0 & 1 & 0 & \cdots & 0 & u_1 \\ 0 & 0 & 1 & \cdots & 0 & u_2 \\ \cdot & \cdot & \cdot & \cdot & \cdot & \cdot \\ \cdot & \cdot & \cdot & \cdot & \cdot & \cdot \\ 0 & 0 & 0 & \cdots & 1 & u_{N-2} \\ 0 & 0 & 0 & \cdots & 0 & u_{N-1} \\ 0 & 0 & 0 & & 0 & 0 \end{bmatrix}.$$

(b): Assume that the N-dimensional matrix A is associated with the series $\sum_{n\geq 0} u_n X^n$, and the M-dimensional matrix B with the series $\sum_{n\geq 0} v_n X^n$. Assume, furthermore, that $u_0 = 0$.

Show that the product series:

$$\sum_{n\geq 0} w_n X^n = \left(\sum_{n\geq 0} u_n X^n\right)\left(\sum_{n\geq 0} v_n X^n\right)$$

is associated the block matrix:

$$C = \begin{bmatrix} A & \tilde{A} \\ 0 & B \end{bmatrix},$$

where $\tilde{A}$ is the $N \times M$ matrix all the columns of which are zero except the first, which is equal to the last column of A.

(c): Let $f(X) = \sum_{n\geq 0} u_n X^n$ be a rational series and $u_0 = 0$. Let A be a matrix, of order $N > 1$, associated with f. Show that to the rational series $\frac{1}{1 - f(X)}$ is associated the matrix $\hat{A} = A + \tilde{A}$, where $\tilde{A}$ is the $N \times N$ matrix defined, in Question (b), starting from A.

(By induction on n prove the relation:

For all j, $1 \leqslant j \leqslant N$:

$$(\hat{A}^n)_{1,j} = (A^n)_{1,j} + \sum_{0\leqslant i\leqslant n} (A^i)_{1,N}(\hat{A}^{n-i})_{1,j}.)$$

(d): Show that every rational function $\frac{P}{Q}$ with $Q(0) \neq 0$ can be put into the form:

$$\frac{P}{Q} = a + \frac{P_1}{1 - Q_1},$$

where $a \in K$ and P_1 and Q_1 are polynomials with no constant terms. From this deduce that there exists a matrix associated to $\frac{P}{Q}$. Show that if the fraction $\frac{P}{Q} \in \mathbb{Z}[[X]]$ there exists a matrix, with coefficients in $\mathbb{Z}$, associated to $\frac{P}{Q}$.

SOLUTIONS

SOLUTION 3·1: (1): If $f(X)$ is a rational series, for $n \geqslant n_0$ there exists a recurrence relation of order s on the u_n by the Second Characterisation. This gives a relation on the rows of the determinant $D_n^{(s)}$, which is therefore zero for $n \geqslant n_0$.

(2): We have:

$$\begin{bmatrix} a_{1,1} & a_{1,2} & \cdots & a_{1,n-1} & a_{2,n} \\ a_{2,1} & a_{2,2} & \cdots & a_{2,n-1} & a_{2,n} \\ \cdots & \cdots & \cdots & \cdots & \cdots \\ \cdots & \cdots & \cdots & \cdots & \cdots \\ a_{n-1,1} & a_{n-1,2} & \cdots & a_{n-1,n-1} & a_{n-1,n} \\ a_{n,1} & a_{n,2} & \cdots & a_{n,n-1} & a_{n,n} \end{bmatrix}$$

$$\times \begin{bmatrix} A_{1,1} & 0 & & 0 & A_{n,1} \\ A_{1,2} & 1 & & & A_{n,2} \\ & & 1 & & \\ & & & \ddots & \\ A_{1,n-1} & & & 1 & A_{n,n-1} \\ A_{1,n} & 0 & & 0 & A_{n,n} \end{bmatrix} = \qquad \text{(Contd)}$$

$$\text{(Contd)} \quad = \begin{bmatrix} D & a_{1,2} & a_{1,3} & \cdots & a_{1,n-1} & 0 \\ 0 & a_{2,2} & \cdot & \cdots & \cdot & 0 \\ \cdot & \cdot & \cdot & \cdots & \cdot & \cdot \\ \cdot & \cdot & \cdot & \cdots & \cdot & \cdot \\ 0 & a_{n-1,2} & \cdot & \cdots & \cdot & 0 \\ 0 & a_{n,2} & \cdot & \cdots & a_{n,n-1} & D \end{bmatrix} = D^2 d,$$

whence:

$$D^2 d = D(A_{1,1}A_{n,n} - A_{n,1}A_{1,n}),$$

therefore:

If $D \not\equiv 0$: we have:

$$Dd = A_{1,1}A_{n,n} - A_{n,1}A_{1,n}.$$

If $D = 0$: the determinants $A_{1,1}, A_{n,n}, A_{n,1}, A_{1,n}$, d and D are polynomials with coefficients $(a_{i,j})_{\substack{1\leq i\leq n \\ 1\leq j\leq n}}$.

We have:

$$D[Dd - A_{1,1}A_{n,n} - A_{n,1}A_{1,n}] = 0,$$

D being a polynomial in $(a_{i,j})_{\substack{1\leq j\leq n \\ 1<i<n}}$ one considers the polynomial P in n^2 variables that gives the expression for D, this polynomial P is not identically zero.

Similarly, consider the polynomial Q which corresponds to:

$$Dd - A_{1,1}A_{n,n} + A_{n,1}A_{1,n};$$

then we have:

$$P(a_{i,j})Q(a_{i,j}) = 0.$$

At all the points of K^{n^2} where $P \neq 0$, we have $Q \equiv 0$.

Now, the ring $K[X_1, X_2, \ldots, X_{n^2}]$ is an integral domain, whence $Q \equiv 0$. Hence we have:

$$Dd = A_{1,1}A_{n,n} - A_{n,1}A_{1,n}$$

in all cases.

(3): Let us apply Sylve ter's relation to the Hankel determinant. We obtain:

$$D_n^{(s)} D_{n+2}^{(s-2)} = D_{n+2}^{(s-1)} D_n^{(s-1)} - (D_{n+1}^{(s-1)})^2,$$

which lets us prove by induction that, if $D_n^{(s)} = 0$ for $n \geqslant n_0$, and if there exists $n_1 \geqslant n_0$ with $D_{n_1}^{(s-1)} = 0$, then $D_n^{(s-1)} = 0$ for all $n \geqslant n_1$, in fact:

$$(D_{n+1}^{(s-1)})^2 = D_{n+2}^{(s-1)} D_n^{(s-1)} - D_n^{(s)} D_{n+2}^{(s-2)} = 0,$$

as soon as $D_n^{(s-1)} = 0$ (this by induction) and $D_n^{(s)} = 0$.

(4): Let s be the smallest integer such that there exists n_0 with $D_n^{(s)} = 0$ for all $n \geqslant n_0$. The preceding allows us to assume that $D_n^{(s-1)} \neq 0$ for all $n \geqslant n_0$.

Consider the following system of equations for $n \geqslant n_0$:

$$\begin{cases} a_s u_n + a_{s-1} u_{n+1} + \cdots + a_1 u_{n+s-1} = -u_{n+s}, \\ \cdots\cdots\cdots\cdots\cdots\cdots \\ \cdots\cdots\cdots\cdots\cdots\cdots \\ a_s u_{n+s+1} + a_{s-1} u_{n+s} + \cdots + a\, u_{n+2s-2} = -u_{n+2s-1}, \end{cases}$$

where the unknowns are $(a_1, a_2, \ldots, a_s)$ and the coefficients are $(u_n, \ldots, u_{n+2s-2})$; the determinants of the coefficients is not zero, this system possesses one and only one solution $(a_1, \ldots, a_s)$ in K^s.

Now, $D_n^{(s)} = 0$, hence this solution is compatible with taking $s + 1$ equations. Therefore there exists a recurrence relation between the (u_n), and by the Second Characterisation of rational series the series $\sum_n u_n X^n$ is rational.

(5): By the Second Characterisation for $n \geqslant n_0$ the u_n are connected by the recurrence relation:

$$a_s u_n + a_{s-1} u_{n+1} + \cdots + a_1 u_{n+s-1} + u_{n+s} = 0.$$

From this one deduces for $t \geqslant n_0 + s$ a linear dependence relation between the last $s + 1$ columns of the Kronecker determinant which is zero.

Conversely, assume $\Delta_t = D_0^{(t)} = 0$ for $t \geqslant t_0$. The Sylvester's relation gives:

$$(D_1^{(t)})^2 = - D_0^{(t+1)} D_2^{(t-1)} + D_2^{(t)} D_0^{(t)},$$

and one has $D_1^{(t)} = 0$ for all $t \geqslant t_0$. Using Sylvester's relation:

$$(D_{p+1}^{(t)})^2 = - D_p^{(t+1)} D_{p+2}^{(t-1)} + D_{p+2}^{(t)} D_p^{(t)},$$

by induction on p one shows that for all $p \geqslant 0$, for all $t \geqslant t_0$, $D_p^{(t)} = 0$. Then using the characterisation of rational series by the vanishing of the Hankel determinants gives the rationality of the series $\sum_{n \geqslant 0} u_n X^n$.

SOLUTION 3·2: (1): Let $\sum_{n \geqslant 0} a_n X^n$ and $\sum_{n \geqslant 0} b_n X^n$ be two primitive series, and let us assume that the product:

$$\sum_{n \geqslant 0} c_n X^n = \left(\sum_{n \geqslant 0} a_n X^n\right)\left(\sum_{n \geqslant 0} b_n X^n\right)$$

is not. There then exists a prime number p that divides all the c_n. This number does not divide all the a_n. Let h be the smallest index such that p does not divide a_h. Similarly, let k be the smallest index such that p does not divide b_k. We then have:

$$c_{h+k} = (a\, b_{h+k} + \cdots + a_{h-1}b_{k+1}) + a_h b_k$$

$$+ (a_{h+1}b_{k-1} + \cdots + a_{h+k}b_0).$$

The terms in parentheses are divisible by p, but $a_h b_k$, and therefore c_{h+k}, is not divisible by p, which contradicts the hypothesis.

(2): After possibly multiplying by a common denominator, we can assume that P and Q are in $\mathbb{Z}[X]$.

Let us assume P and Q are relatively prime, there then exist $A,B \in \mathbb{Z}[X]$ and $a \in \mathbb{Z}$, $a \neq 0$, such that:

$$PA + QB = a,$$

whence:

$$\frac{a}{Q} = \frac{PA}{Q} + B;$$

as $\frac{P}{Q}$, $A,B \in \mathbb{Z}[[X]]$, from this is deduced:

$$\frac{a}{Q} \in \mathbb{Z}[[X]];$$

by expanding $\frac{a}{Q}$ in $\mathbb{Z}[[X]]$ we obtain:

$$\frac{a}{Q} = C = \sum_{n \geqslant 0} c_n X^n, \quad \text{where } c_n \in \mathbb{Z} \text{ for all } n \geqslant 0.$$

Let q (*resp.* c) be the g.c.d. of the coefficients of Q (*resp.* C). Hence:

$Q = qQ^*$ where Q^* is primitive, $Q^* \in \mathbb{Z}[X]$,

$C = cC^*$ where $C^* \in \mathbb{Z}[[X]]$ and C^* is primitive.

From the relation $\frac{a}{Q} = C$ we conclude $\frac{a}{qc} = C^*Q^*$. By Question (1) above C^*Q^* is primitive, thus:

$$\frac{a}{qc} = \pm 1.$$

If we write:

$$Q^* = q_0 + q_1X + \cdots + q_nX^n,$$

then $c_0q_0 = \frac{a}{qc} = \pm 1$, which implies $q_0 = \pm 1$. This now gives:

$$P(X) = Q(X) \sum_{n\geq 0} u_nX^n$$

$$= qQ^* \sum_{n\geq 0} u_nX^n,$$

whence $P(X) \in \mathbb{Z}[X]$ and q divides the coefficients of P, that is to say, that $\frac{P(X)}{q} \in \mathbb{Z}[X]$. We then have:

$$\sum_{n\geq 0} u_nX^n = \frac{\frac{P(X)}{q}}{Q^*(X)} \text{ with } Q^*(0) = 1.$$

Which ends the proof.

REMARK: In this Exercise one can replace $\mathbb{Z}$ and $\mathbb{Q}$ by an arbitrary Dedeking ring A and its quotient field K. In particular, K can be a number field and A its ring of integers.

SOLUTION 3·3: (1): The radius of convergence of a series is given by the formula:

$$\frac{1}{R} = \overline{\lim}\sqrt[n]{|u_n|},$$

now if $R > 1$, $\frac{1}{R} < 1$, whence $\overline{\lim}\sqrt[n]{|u_n|} < 1$, that is to say $|u_n| < 1$ (now, $u_n \in \mathbb{Z}$, whence $u_n = 0$) starting from a certain rank n_0; the series $\sum_{n\geq 0} u_n z^n$ reduces to a polynomial.

(2): Let $1 < \rho' < \rho$. In the disc $|z| \leqslant \rho'$ the function f has a finite number of poles and there exists a polynomial:

$$q(z) = a_0 + a_1 z + \cdots + a_k z^k,$$

such that $g(z) = f(z)q(z)$ is holomorphic for $|z| < \rho'$. We will show that the Hankel determinants of the series $f(z)$ tend to zero; since these determinants are integers it follows that they are zero starting from a certain point. We have:

$$f(z)q(z) = \left(\sum_{n\geq 0} u_n z^n\right)\times(a_0 + a_1 z + \cdots + a_k z^k) = \sum_{n\geq 0} v_n z^n,$$

where:

$$v_n = u_n a_0 + u_{n-1}a_1 + \cdots + u_{n-k}a_k,$$

whence for $s \geqslant k$:

$$D_n^{(s)} = \begin{bmatrix} u_n & u_{n+1} & \cdots & u_{n+k-1} & \cdots & u_{n+s} \\ u_{n+1} & u_{n+2} & \cdots & u_{n+k} & \cdots & u_{n+s+1} \\ \cdot & \cdot & \cdot & \cdot & \cdot & \cdot \\ \cdot & \cdot & \cdot & \cdot & \cdot & \cdot \\ u_{n+s} & u_{n+s+1} & \cdots & u_{n+s+k-1} & \cdots & u_{n+2s} \end{bmatrix}$$

$$= \begin{bmatrix} u_n & \cdots & u_{n+k-1} & v_{n+k} & \cdots & v_{n+s} \\ u_{n+1} & \cdots & u_{n+k} & v_{n+k+1} & \cdots & v_{n+s+1} \\ \cdot & \cdot & \cdot & \cdot & \cdot & \cdot \\ \cdot & \cdot & \cdot & \cdot & \cdot & \cdot \\ u_{n+s} & \cdots & u_{n+k-1+s} & v_{n+k+s} & \cdots & v_{n+2s} \end{bmatrix} \times (a_0)^{s-k+1}.$$

After writing out the $(s+1)!$ terms of this determinant we obtain:

$$|a_0|^{-s+k+1} D_n^{(s)} \leqslant \prod_{h=0}^{k-1} (|u_{n+h}| + \cdots + |u_{n+h+s}|)$$

$$\times \prod_{h=k}^{s} (|v_{n+h}| + \cdots + |v_{n+h+s}|).$$

Let A be a real number such that $0 < A < R$ and let B satisfy $1 < B < \rho'$. Set:

$$M_1 = \sup_{|x|\leqslant A} |f(x)|, \qquad M_2 = \sup_{|x|\leqslant B} |g(x)|,$$

and:

$$M = \max(M_1, M_2).$$

Apply the Cauchy inequalities:

$$|u_n| \leqslant \frac{M}{A^n}, \qquad |v_n| \leqslant \frac{M}{B^n}.$$

The following upper bound for $D_n^{(s)}$ is obtained:

$$|D_n^{(s)}| \leqslant C\left(\frac{1}{A^k B^{s+1-k}}\right)^n$$

where C is a constant depending upon M, s, k and a_0 (for example, one can take:

$$C = \frac{(M(s+1))^{s+1}}{A^{2ks}} (a_0)^{s-k+1}).$$

As $B > 1$, s can be fixed large enough so that $A^k B^{s+1-k} > 1$, whence:

$$\lim_{n\to\infty} |D_n^{(s)}| = 0.$$

With $D_n^{(s)}$ being an integer, therefore $D_n^{(s)} = 0$ starting from a certain index n_0.

(3): The Cauchy inequalities are valid in Ω_p for A_p, B_p satisfying $0 < A_p < R_p$, $0 < B_p < \rho_p$ (where R_p is the radius of convergence of the series f in Ω_p. The determinants $D_n^{(s)}$ are majorised exactly as in Question (2) preceding:

$$|D_n^{(s)}|_p \leqslant C_p A_p^{-nk_p} B_p^{-n(s+1-k_p)}.$$

We always have:

$$|D_n^{(s)}| \leqslant CA^{-nk} B^{-n(s+1-k)}.$$

Set:

$$\Delta_p(s) = A_p^{-k_p} B_p^{-(s+1-k_p)},$$

$$\Delta_0(s) = A^{-k} B^{-(s+1-k)},$$

$$\Delta(s) = \Delta_p(s) \times \Delta_0(s),$$

$$C' = C.C_p.$$

Since $\rho\rho_p > 1$ there exists an integer s such that:

$$|D_n^{(s)}|\,|D_n^{(s)}|_p \leqslant C'\Delta^n, \quad \text{where } \Delta = \Delta(s) < 1.$$

One can therefore choose s such that:

$$|D_n^{(s)}| \times |D_n^{(s)}|_p < 1 \quad \text{for } n \geqslant n_0.$$

As the coefficients $u_n \in \mathbb{Z}$, the determinant $D_n^{(s)} \in \mathbb{Z}$ is as well. By using the relation that for all $x \in \mathbb{Q}^*$:

$$|x| \prod_{p \text{ prime}} |x|_p = 1,$$

we see that $|x|\,|x|_p \geq 1$ if x is a non-zero integer.

It follows that $D_n^{(s)} = 0$ for s fixed and n large enough, and the series $\sum_n u_n z^n$ is rational.

SOLUTION 3·4: By the First Characterisation, as $f(z)$ is rational, there exist k algebraic numbers $\alpha_1,\ldots,\alpha_k$, and k polynomials $P_1,\ldots,P_k$, such that:

$$u_n = \sum_{1\leq i\leq k} P_i(n)\alpha_i^n \quad \text{for } n \geq n_1.$$

As $g(z)$ is rational, there exist ℓ numbers $\alpha_1,\ldots,\alpha_\ell$, and ℓ polynomials $Q_1,\ldots,Q_\ell$, such that:

$$v_n = \sum_{1\leq i\leq \ell} Q_i(n)\beta_i^n \quad \text{for } n \geq n_2,$$

whence for $n \geq \sup(n_1,n_2)$ we have the relation:

$$u_n \times v_n = \sum_{1\leq i\leq k} P_i(n)\alpha_i^n \times \sum_{1\leq j\leq \ell} Q_j(n)\beta_j^n$$

$$= \sum_{\substack{1\leq i\leq k\\ 1\leq j\leq \ell}} P_i(n)Q_j(n)(\alpha_i\beta_j)^n.$$

By the First Characterisation of rational series the series $h(x)$ is rational.

SOLUTION 3·5: We will show that the series $\sum_n a_n z^n$ is meromorphic on arbitrary large discs, in particular on discs containing the unit disc. Exercise 3·3 gives us that:

— Starting from a certain index n_0 the coefficients of $g(z)$ can be written:

$$v_n = a(\alpha_1^n - \sum_{2\leqslant i\leqslant r} P_i(n)\alpha_i^n),$$

where the P_i $(2 \leqslant i \leqslant r)$ are polynomials and the α_i are algebraic numbers satisfying the condition: $\left|\frac{\alpha_i}{\alpha_1}\right| < 1.$

Starting from a certain index n_1, the term $\sum_{2\leqslant i\leqslant r} P_i(n)\left(\frac{\alpha_i}{\alpha_1}\right)^n$ is smaller in absolute value than $\lambda < 1$. Whence for $n \geqslant n_1$ one has $v_n \neq 0$ and:

$$u_n v_n^{-1} = u_n a^{-1}\alpha_1^{-n}\left[1 - \sum_{2\leqslant i\leqslant r} P_i(n)\left(\frac{\alpha_i}{\alpha_1}\right)^n\right]^{-1}$$

$$= u_n a^{-1}\alpha_1^{-n}\left[1 + \sum_{2\leqslant i\leqslant r} P_i(n)\left(\frac{\alpha_i}{\alpha_1}\right)^n + \cdots\right.$$

$$\left. + \left[\sum_{2\leqslant i\leqslant r} P_i(n)\left(\frac{\alpha_i}{\alpha_1}\right)^n\right]^p\right] + u_n a^{-1}\alpha_1^{-n}\frac{\left[\sum_{2\leqslant i\leqslant r} P_i(n)\left(\frac{\alpha_i}{\alpha_1}\right)^n\right]^{p+1}}{1 - \sum_{2\leqslant i\leqslant r} P_i(n)\left(\frac{\alpha_i}{\alpha_1}\right)^n}$$

this expansion is valid for all $p \geqslant 0$.

For all $p \geqslant 1$ the series $\sum_{n\geqslant n_1} t_n z^n$, with:

$$t_n = u_n a^{-1}\alpha_1^{-n}[1 + \sum_{2\leqslant i\leqslant r} P_i(n)(\frac{\alpha_i}{\alpha_1})^n + \ldots + [\sum_{2\leqslant i\leqslant r} P_i(n)(\frac{\alpha_i}{\alpha_1})^n]^p]$$

is rational because it is a linear combination of Hadamard products of rational series (see 4).

We first have to see the series $\sum_{n>n_1} w_n z^n$ with:

$$w_n = u_n a^{-1}\alpha_1^{-n}\left[\sum_{2\leqslant i\leqslant r} P_i(n)\frac{\alpha_i}{\alpha_1}^n\right]^{p+1}\frac{1}{1 - \sum_{2\leqslant i\leqslant r} P_i(n)\left(\frac{\alpha_i}{\alpha_1}\right)^n} =$$

(Contd)

(Contd)
$$= u_n \frac{\left[\sum_{2\leqslant i\leqslant r} P_i(n)\left(\frac{\alpha_i}{\alpha_1}\right)^n\right]^{p+1}}{a\alpha_1\left[1 - \sum_{2\leqslant i\leqslant r} P_i(n)\left(\frac{\alpha_i}{\alpha_1}\right)^n\right]} .$$

p can be chosen large enough so that the radius of convergence of this series is greater than one.

The series $\sum_{n\geqslant n_1} a_n z^n$ is the sum of a rational function $\sum_{n\geqslant n_1} t_n z^n$ and of the series $\sum_{n\geqslant n_1} w_n z^n$ whose radius of convergence is greater than one. As $a_n \in \mathbb{Z}$ Exercise 3·3(1) says that this is a rational function

REMARK: If the rational series $\sum_{n\geqslant 0} v_n z^n$ has a pole of modulus strictly *greater* than the others, if the series $\sum_n v_n z^n$ is rational and if the series $a = \sum_n \frac{u_n}{v_n} z^n$ has integer coefficients (or algebraic integers) then the series a is rational (cf., D.G. Cantor: 'On Arithmetic Properties of the Taylor Series of Rational Functions. II', *Pacif. J. Math.*, **41**, No. 2, (1972), pp. 329-334).

SOLUTION 3·6: (1): Assume that for $n \geqslant 1$ we have:

$$u_n = (A^n)_{1,N} \quad \text{where } A \in M_N(K).$$

The matrix A satisfies its minimal polynomial, the coefficients u_n of the series $\sum u_n X^n$ satisfy a recurrence relation, and this series is therefore rational.

(2): (a): Let $\{e_1,e_2,\ldots,e_N\}$ be the canonical basis of K^N. In this basis the matrix A is the matrix of the linear mapping, denoted g, defined by:

$$g(e_1) = 0,$$

$$g(e_i) = e_{i-1} \quad \text{for } 2 \leq i \leq N - 1,$$

$$g(e_N) = u_1e_1 + u_2e_2 + \cdots + u_{N-1}e_{N-1}.$$

We have:

$$g^i(e_N) = u_ie_1 + u_{i+1}e_2 + \cdots + u_{N-1}e_{N-1},$$

whence we obtain:

$$(A^i)_{1,N} = u_i \quad \text{for } 1 \leq i \leq N - 1.$$

Finally $A^N = 0$, and for $i \geq N$, $(A^i)_{1,N} = 0$.

(2): (b): First of all we have:

$$C^n = \begin{bmatrix} A^n & \sum_{0 \leq k \leq n-1} A^{n-1-k}\tilde{A}B^k \\ 0 & B^n \end{bmatrix}.$$

The matrix $A^{n-1-k}\tilde{A}$ has all its columns zero except the first, and:

$$[A^{n-1-k}\tilde{A}]_{1,1} = \sum_{1 \leq i \leq N} [A^{n-1-k}]_{1,i}[\tilde{A}]_{i,1}$$

$$= \sum_{1 \leq i \leq N} [A^{n-1-k}]_{1,i}[A]_{i,N} = [A^{n-k}]_{1,N}.$$

Next,

$$[A^{n-1-k}\tilde{A}B^k]_{1,M} = [A^{n-1-k}\tilde{A}]_{1,1}[B^k]_{1,M} = [A^{n-k}]_{1,N}[B^k]_{1,M}$$

$$= u_{n-k}v_k.$$

Finally we have:

$$[C^n]_{1,N+M} = \sum_{0\leqslant k\leqslant n-1} [A^{n-1-k}\tilde{A}B^k]_{1,M} = \sum_{0\leqslant k\leqslant n-1} u_{n-k}v_k$$

$$= \sum_{0\leqslant k\leqslant n} u_{n-k}v_k = w_n,$$

as $u_0 = 0$.

(2): (c): Let us prove the relation by induction. For $n = 1$ this reduces to:

$$[A]_{1,j} = [A]_{1,j} + [A]_{1,N}[I]_{1,j},$$

where I is the identity matrix.

For $j \neq 1$: $[\hat{A}]_{1,j} = [A]_{1,j}$ certainly holds;

For $j = 1$: we have:

$$[\hat{A}]_{1,1} = [A]_{1,1} + [\tilde{A}]_{1,1} = [A]_{1,1} + [A]_{1,N}.$$

Assume the formula to be true for n, and let us prove it for $n + 1$. We have:

$$[\hat{A}^{n+1}]_{1,j} = \sum_{1\leqslant k\leqslant N} [\hat{A}^n]_{1,k}[\hat{A}]_{k,j}.$$

$[\hat{A}^n]_{1,k}$ is given by the induction hypothesis, therefore:

$$[\hat{A}^{n+1}]_{1,j} = \sum_{1\leqslant k\leqslant N} [A^n]_{1,k}[\hat{A}]_{k,j} + \sum_{\substack{0\leqslant i\leqslant n\\ 1\leqslant k\leqslant N}} [A^i]_{1,N}[\hat{A}^{n-i}]_{1,k}[\hat{A}]_{k,j}.$$

Let us calculate the first sum, which has the value:

$$S_1 = \sum_{1\leqslant k\leqslant N} [A^n]_{1,k}([A]_{k,j} + [\tilde{A}]_{k,j}) = \qquad \text{(Contd)}$$

(Contd) $= [A^{n+1}]_{1,j} + \sum_{1 \leqslant k \leqslant N} [A^n]_{1,k} [\tilde{A}]_{k,j}.$

For $j \neq 1$: $[\tilde{A}]_{k,j} = 0$ and $S_1 = [A^{n+1}]_{1,j}.$

For $j = 1$:

$$S_1 = [A^{n+1}]_{1,j} + \sum_{1<k<N} [A^n]_{1,k} [A]_{k,N}$$

$$= [A^{n+1}]_{1,j} + [A^{n+1}]_{1,N}.$$

In two cases we have:

$$S_1 = [A^{n+1}]_{1,j} + [A^{n+1}]_{1,N} [\hat{A}^0]_{1,j}.$$

The second sum has the value:

$$S_2 = \sum_{0 \leqslant i \leqslant n} [A^i]_{1,N} \sum_{1 \leqslant k \leqslant N} [\hat{A}^{n-i}]_{1,k} [\hat{A}]_{k,j}$$

$$= \sum_{0 \leqslant i \leqslant n} [A^i]_{1,N} [\hat{A}^{n-i+1}]_{1,j}.$$

Finally we obtain:

$$[\hat{A}^{n+1}]_{1,j} = S_1 + S_2 = [A^{n+1}]_{1,j} + \sum_{0 \leqslant i \leqslant n+1} [A^i]_{1,N} [A^{n-i+1}]_{1,j}.$$

For $j = N$, and setting $b_n = [A^n]_{1,N}$, and $b_0 = 0$, the relation we have just proved gives:

$$b_n = u_n + \sum_{0 \leqslant i \leqslant n} u_i b_{n-1},$$

which proves that the series $g(X) = \sum_{n \geqslant 0} b_n X^n$ satisfies the equation:

$$g(x) = f(x) + y(x) f(x),$$

that is to say:

$$g(X) = \frac{f(X)}{1 - f(X)} = \frac{1}{1 - f(X)} - 1.$$

The series $\frac{1}{1 - f(X)}$ differs from $g(x)$ only by the constant term; the matrix $\hat{A}$ associated with $g(x)$ is also associated with it.

(2): (d): Let $p_0 = P(0)$ and $q_0 = Q(0)$. We have:

$$\frac{P}{Q} = \frac{p_0}{q_0} + \frac{P_1}{1 - Q_1},$$

with:

$$Q_1(X) = \frac{1}{q_0}(q_0 - Q(X)) \quad \text{and} \quad P_1(X) = \frac{1}{q_0}\left(P(X) - \frac{p_0}{q_0}Q(X)\right).$$

A matrix associated with $\frac{P_1}{1 - Q_1}$ will also be associated with $\frac{P}{Q}$.

Associate with the polynomial Q_1 the matrix A of order $N \geqslant 2$ defined in Question (2)(a) above. To the fraction $\frac{1}{1 - Q_1}$ will, by Question (2)(c) above, be associated the matrix $A + \tilde{A}$. And Question (2)(b) above gives the matrix associated with the product $(P_1)\left(\frac{1}{1 - Q_1}\right)$.

If $\frac{P}{Q} \in \mathbb{Z}[[X]]$, by Exercise 3·2 we can choose another representation $\frac{P^*}{Q^*}$ of $\frac{P}{Q}$, with $P^*, Q^* \in \mathbb{Z}[X]$ and $Q^*(0) = 1$. All the preceding calculations show that the associated matrix that we have just constructed has integer coefficients.

REMARK: This relation between rational series and matrices allows us to generalise the notion of rational series. If X is a finite set, X^* the free monoid that it generates, a rational series on X^* with coefficients in $\mathbb{Z}$ or $\mathbb{Q}$ is related to the representation of the free monoid X^* on $M_N(\mathbb{Z})$ or $M_N(\mathbb{Q})$ for a certain $N \geqslant 1$. If

X has a single element, then $X^* = \mathbb{N}$, and one shows that the usual definition of rational series on $\mathbb{Z}$ or $\mathbb{Q}$ is obtained (cf., M.P. Schutzenberger: 'Properties of the Families of the Automata', *Inf. and Control*, **6**, (1963), pp. 246-264).

CHAPTER 4

Algebraic Theory

INTRODUCTION

4.1 FIELD EXTENSIONS

4.1·1 FINITE EXTENSIONS: ALGEBRAIC EXTENSIONS

Let L be a field; if K is a subfield of L we also say that L is an *EXTENSION FIELD OF* K. If L, as a vector space over K, has finite dimension n, we say that the extension $L|K$ is a *FINITE EXTENSION OF DEGREE* n, and this is written $n = [L:K]$; otherwise we say that $L|K$ is an *INFINITE EXTENSION*. If $M|L$ and $L|K$ are two finite extensions, $M|K$ is then finite, and $[M:K] = [M:L] \times [L:K]$.

Let K_1, K_2 be two extensions of K contained in the same field L. The *COMPOSITE OF* K_1, K_2, which is denoted by K_1K_2, is the subfield of L generated by K_1 and K_2.

Let $(\alpha_i)_{1 \leq i \leq k}$ be k elements of a field L that is an extension field of K. We denote by $K(\alpha_1, \ldots, \alpha_k)$ the subfield of L generated by K and by the α_i $(1 \leq i \leq k)$.

Let $L|K$ be an extension of fields and let α be an element of L. One says that α *IS ALGEBRAIC* (*resp. TRANSCENDENTAL*) *OVER* K if the extension $K(\alpha)|K$ is finite (*resp.* infinite).

An extension $L|K$ is called an *ALGEBRAIC EXTENSION* if every element of L is algebraic over K. Every finite extension is algebraic. By an *ALGEBRAIC NUMBER FIELD* we mean a finite extension of the field $\mathbb{Q}$ of rational numbers.

4.1·2 IRREDUCIBLE POLYNOMIAL OF AN ALGEBRAIC ELEMENT

Let $L|K$ be an extension of fields and let α be an element of L algebraic over K. The set of polynomials $P \in K[X]$ such that $P(\alpha) = 0$ is an ideal of $K[X]$ generated by an irreducible polynomial, assumed unitary, which is called the *IRREDUCIBLE POLYNOMIAL* of α over K (the notation used is $\mathrm{Irr}(\alpha,K)$). The field $K(\alpha)$ is isomorphic to the field $K[X]/(\mathrm{Irr}(\alpha,K))$. The *DEGREE OF AN ELEMENT* α is the degree of the polynomial $\mathrm{Irr}(\alpha,K)$; this is also the degree $[K(\alpha):K]$.

Conversely, if P is an irreducible polynomial of $K[X]$ there exists an extension of K in which P admits a root, and this extension contains a field K-isomorphic to $K[X]/(P)$.

4.1·3 SPLITTING FIELD: ALGEBRAIC CLOSURE

Let P be a polynomial of $K[X]$. There exists an extension L of K such that:

(1): P splits in $L[X]$ into a product of first degree polynomials;

(2): If $(\alpha_1,\ldots,\alpha_k)$ is the set of roots of P in L, $L = K(\alpha_1,\ldots,\alpha_k)$.

Such an extension L is unique up to K-isomorphism and is called a *SPLITTING FIELD OF THE POLYNOMIAL* P OVER K

A field Ω is said to be *ALGEBRAICALLY CLOSED* if every polynomial of $\Omega[X]$ of degree greater than or equal to one has a root in Ω.

Let K be a field. There exists an extension Ω of K such that:

(a): $\Omega|K$ is algebraic;

(b): Ω is algebraically closed.

The field K is unique up to K-isomorphism; Ω is called an *ALGEBRAIC CLOSURE of* K.

4.1·4 Conjugate Fields

In an algebraic closure Ω of K given two extensions L, L' of K; we say that L and L' are *CONJUGATE FIELDS over K* if there exists a K-isomorphism of L onto L'. Analogously, if α, α' are two elements of Ω we say that they are *CONJUGATE ELEMENTS over K* if there exists a K-isomorphism φ of $K(\alpha)$ onto $K(\alpha')$ such that $\varphi(\alpha) = \alpha'$. Two elements of Ω are conjugate on K if and only if they have the same irreducible polynomial over K.

If K is a field of characteristic zero or a finite field, an irreducible polynomial of $K[X]$ has all its roots distinct in an algebraic closure Ω of K; if L is an extension of K of finite degree n there exist n distinct K-isomorphisms of L into Ω.

4.1·5 Norm: Trace: Characteristic Polynomial

Let $L|K$ be a finite extension. To every element $\alpha \in L$ we associate the endomorphism φ_α of the K-vector space L defined by $\varphi_\alpha(x) = \alpha x$ for all $x \in L$.

The *NORM* (*resp. TRACE, CHARACTERISTIC POLYNOMIAL*) *OF AN ELEMENT* α *IN THE EXTENSION* $L|K$ is the determinant (*resp.* trace, characteristic polynomial) of the endomorphism φ_α. The norm is denoted $N_{L|K}$ and the trace is denoted $\mathrm{Tr}_{L|K}$.

If the field K is of characteristic zero or is finite, on writing $k = [K(\alpha):K]$ and $(\alpha_i)_{1\leqslant i\leqslant k}$ for the roots of the polynomial $\mathrm{Irr}(\alpha, K)$ in an algebraic closure Ω of L we have:

$$N_{L|K}(\alpha) = (\alpha_1\alpha_2\cdots\alpha_k)^{n/k},$$

$$\mathrm{Tr}_{L|K}(\alpha) = \frac{n}{k}(\alpha_1 + \alpha_2 + \cdots + \alpha_k).$$

The characteristic polynomial of α is the polynomial $\mathrm{Irr}(\alpha, K)^{n/k}$.

4.2 GALOIS THEORY

We assume the field K is either characteristic zero or a finite field.

4.2·1 GALOIS EXTENSION

Let $L|K$ be a finite extension. We say that $L|K$ is a *GALOIS EXTENSION* if for all $\alpha \in L$ the polynomial $\mathrm{Irr}(\alpha, K)$ has all its roots in L. The set of K-automorphisms of L has a group structure that is called the *GALOIS GROUP OF THE EXTENSION* $L|K$, and is denoted $\mathrm{Gal}(L|K)$; its order is equal to the degree $[L:K]$.

4.2·2 FUNDAMENTAL THEOREM OF GALOIS THEORY

Let $L|K$ be a finite Falois extension, and let $G = \mathrm{Gal}(L|K)$. If H is a subgroup of G, the set of elements of L invariant by every $\sigma \in H$ is a field intermediate between K and L that is denoted L^{H}. If F is a field intermediate between K and L the extension $L|F$ is Galois and $\mathrm{Gal}(L|F)$ is a subgroup of G.

There exists a bijection between the set of fields F between K and L and the set of subgroups H of G; this bijection is given by:

$$F \mapsto H = \mathrm{Gal}(L|F),$$

the inverse map being:

$$H \mapsto F = L^{H}.$$

The extension $F|K$ is Galois if and only if $\mathrm{Gal}(L|F)$ is a normal subgroup of G; in this case we have an isomorphism $\mathrm{Gal}(F|K) \simeq G/\mathrm{Gal}(L|F)$. We say that $L|K$ is an *ABELIAN* (*resp. CYCLIC*) *EXTENSION* if $L|K$ is Galois and if $\mathrm{Gal}(L|K)$ is abelian (*resp.* cyclic).

4.2·3 EXTENSIONS OF FINITE FIELDS

Let $K \simeq \mathbb{F}_q$ be a finite field with q elements and let $L \simeq \mathbb{F}_{q^n}$ be a finite extension of K of degree n. The extension $L|K$ is cyclic and its Galois group is generated by the automorphism $x \mapsto x^q$.

4.2·4 Norm and Trace

If $L|K$ is a finite Galois extension, then for all $\alpha \in L$:

$$N_{L|K}(\alpha) = \prod_{\sigma \in \mathrm{Gal}(L|K)} \sigma(\alpha),$$

$$\mathrm{Tr}_{L|K}(\alpha) = \sum_{\sigma \in \mathrm{Gal}(L|K)} \sigma(\alpha).$$

4.3 INTEGERS

4.3·1 Elements Integral over a Ring

Let B be a commutative ring and A a subring of B. An element $x \in B$ is called an *INTEGRAL ELEMENT OVER A* if there exists an integer n and elements $a_0, a_1, \ldots, a_{n-1} \in A$ such that:

$$x^n + a_{n-1}x^{n-1} + \cdots + a_1 x + a_0 = 0.$$

The set of elements of B integral over A is a subring of B which contains A and is called the *INTEGRAL CLOSURE OF A IN B*. If the integral closure of A in B is B itself, we say that B is *INTEGRAL OVER A*.

Let C be a commutative ring, B a subring of C, and A a subring of B. If C is integral over B and if B is integral over A, then C is integral over A.

If a ring A is an integral domain and K is quotient field, we say that A is *INTEGRALLY CLOSED* if A coincides with its integral closure in K. A principal ideal dom. in is integrally closed.

4.3·2 Algebraic Integers. Example of Quadratic Fields

Let K be an algebraic number field, that is to say, a finite extension of $\mathbb{Q}$. An element of K is called an *ALGEBRAIC INTEGER* if it is integral over the ring $\mathbb{Z}$ of rational integers. The set of integers of K is a free $\mathbb{Z}$-module with $[K:\mathbb{Q}]$ generators, which are called an *INTEGRAL BASIS* of K.

EXAMPLE: K is a quadratic field, that is to say, of degree two over $\mathbb{Q}$. There exists a square-free integer d, such that $K = \mathbb{Q}(\sqrt{d})$. If $d \equiv 2$ or 3 modulo 4, $\{1,\sqrt{d}\}$ is an integral basis of K. If $d \equiv 1$ modulo 4, $\{1,(1+\sqrt{d})/2\}$ is an integral basis of K.

4.3.3 DISCRIMINANT

Let A be an integrally closed domain, K the quotient field of A, L a finite extension of K, and B the integral closure of A in L. If $x \in B$, the polynomial $\mathrm{Irr}(x,K)$ has its coefficients in A; $N_{L|K}(x)$ and $T_{L|K}(x)$ belong to A.

The *DISCRIMINANT OF A BASIS* $(e_i)_{1<i<n}$ of L over K is the element of K defined by:

$$D(e_1,\ldots,e_n) = \det(T_{L|K}(e_ie_j))_{1\leqslant i,j\leqslant n}.$$

If K is a field of characteristic 0 or a finite field, the discriminant of a basis of L over K is not zero; if the n distinct K-isomorphisms of L into an algebraic closure Ω of K are denoted $\sigma_1,\ldots,\sigma_n$, then:

$$D(e_1,\ldots,e_n) = \det(\sigma_i(e_j))^2_{1\leqslant i,j\leqslant n}.$$

Example: If $L = K(\alpha)$, and if we write $n = [K(\alpha):K]$, $P = \mathrm{Irr}(\alpha,K)$, we have:

$$D(1,\alpha,\ldots,\alpha^{n-1}) = N_{L|K}(P'(\alpha)).$$

If $(e_i)_{1\leqslant i\leqslant N}$ is a basis of L over K formed of elements of B, the discriminant $D(e_1,\ldots,e_n) \in A$. The *DISCRIMINANT IDEAL of* B *over* A, denoted by $\Delta_{B|A}$, is the ideal of A generated by the discriminants of the bases of L over K which are contained in B. If B is a free A-module, $\Delta_{B|A}$ is the ideal generated by the discriminant of any basis of B on A. If $A = \mathbb{Z}$, $K = \mathbb{Q}$, B is a free $\mathbb{Z}$-module, and all $\mathbb{Z}$ bases of B have the same discriminant.

4.4 IDEALS

4.4·1 DEDEKIND DOMAINS

Let A be a commutative ring; we say that A is a *NOETHERIAN RING* if every ideal of A is an A-module of finite type. A *DEDEKIND DOMAIN* is a Noetherian integrally closed domain of which every non-zero prime ideal is maximal.

Let A be a Dedekind domain, K the quotient field of A, L a finite extension of K, B the integral closure of A in L. Assume K is of characteristic zero. Then B is a Dedekind domain.

A principal ideal domain is a Dedekind domain. The ring of integers of a number field is a Dedekind domain.

4.4·2 IDEALS OF A DEDEKIND DOMAIN

If A is an integral domain with quotient field K a *FRACTIONAL IDEAL* of A is a sub-A-module I of K such that there exists a non-zero element d of A satisfying $dI \subset A$. The product of two fractional ideals I,J is the set of finite sums $\sum x_i y_i$, where $x_i \in I$, $y_i \in J$; this is a fractional ideal denoted IJ.

Let A be a Dedekind domain, P the set of non-zero prime ideals of A. The set of non-zero fractional ideals of A has a group structure. Every non-zero fractional ideal A of A can be written uniquely in the form:

$$\mathrm{A} = \prod_{1 \leqslant i \leqslant k} p_i^{n_i}, \quad \text{where } p_i^{n_i} \in P \text{ and } n_i \in \mathbb{Z}.$$

The principal fractional ideals form a subgroup of the group of ideals of A; the quotient group is called the *CLASS GROUP OF A*.

4.4·3 DECOMPOSITION OF PRIME IDEALS IN AN EXTENSION

Let A be a Dedekind ring, K its quotient field is assumed to be of characteristic zero, L a finite extension of K of degree n, B the integral closure of A in L.

If p is a non-zero prime ideal of A one has the decomposition:

$$pB = \prod_{1 \leqslant i \leqslant q} P_i^{e_i},$$

where the P_i are distinct prime ideals of B, and where the $e_i \geqslant 1$ are integers. The ideals P_i are the prime ideals P of B over p, that is to say such that $P \cap A = p$. The ring B/P_i is a field which is a finite extension of the field A/p; we write $f_i = [B/P_i : A/P]$ and call:

f_i the *RESIDUAL DEGREE of* P_i *on* A,

e_i the *RAMIFICATION INDEX of* P_i *on* A.

We have the equality:

$$\sum_{1 \leqslant i \leqslant q} e_i f_i = n.$$

We say that p is *RAMIFIED IN* (*THE EXTENSION*) L if one of the ramification indices e_i is $\geqslant 2$. The prime ideals p ramified in L are finite in number; these are the prime divisors of the discriminant ideal $\Delta_{B|A}$.

4.4·4 DECOMPOSITION IN A GALOIS EXTENSION

The notations and hypotheses are the same as in Section 4.3; we assume, further, that $L|K$ is Galois, and we write $G = \mathrm{Gal}(L|K)$.

For every $\sigma \in G$ one has $\sigma B = \sigma$; if P is a prime ideal of B, σP is a prime ideal of B which is called the *CONJUGATE IDEAL OF THE PRIME IDEAL*) P.

The decomposition of a non-zero prime ideal p of A is written:

$$pB = \Big(\prod_{1 \leqslant i \leqslant g} P_i \Big)^e ,$$

where the ideals P_i are all conjugate, have the same residual degree f, and the same ramification index e. We have the equality:

$$efg = n.$$

The *DECOMPOSITION GROUP D OF* (*A PRIME IDEAL*) P of B is the set of $\sigma \in G$ such that $\sigma P = P$. D is a subgroup of G of order $\frac{n}{g}$ if P is one of the ideals P_i. The *DECOMPOSITION FIELD OF* (*THE PRIME IDEAL*) P is the field L^D of elements of L invariant under D.

Let σ be an element of D; by passage to the quotient field σ defines an A/p automorphism $\bar{\sigma}$ of B/P. The mapping $\sigma \mapsto \bar{\sigma}$ is a group homomorphism the kernel of which is called the *INERTIA GROUP I OF* (*THE PRIME IDEAL*). P. The field L^I is called the *INERTIA FIELD OF* (*THE PRIME IDEAL*) P. We have:

$$[L:L]^I = e, \qquad [L^I:L^D] = f, \qquad [L^D:K] = g.$$

4.4·5 EXAMPLE: DECOMPOSITION IN A QUADRATIC FIELD

Let $A = \mathbb{Z}$, $K = \mathbb{Q}$, $L = \mathbb{Q}(\sqrt{d})$ (where d is an integer of $\mathbb{Z}$ with no square factor), and B the integral closure of $\mathbb{Z}$ in L.

If p is a prime number, the splitting of p in B has one of the three following forms:

(1): $pB = P_1 P_2$, where P_1 and P_2 are two distinct prime ideals of residual degree one; it we say that p is decomposed in

(2): $pB = P$, where P is a prime ideal of residual degree two; we say that p is *INERT IN L*;

(3): $pB = P^2$, where P is a prime ideal of degree one; we say that p is *RAMIFIED IN* L.

Given an odd prime number p and an integer d prime to p, one says that d is a *QUADRATIC RESIDUE MODULO* p if the class of d modulo p is a square. The *LEGENDRE SYMBOL* $\left(\frac{d}{p}\right)$ is defined by:

$$\left(\frac{d}{p}\right) \equiv \begin{cases} +1 \text{ if } d \text{ is a quadratic residue modulo } p, \\ -1 \text{ otherwise.} \end{cases}$$

The decomposition of p in B is given by:

(i) p is decomposed in L if: $\begin{cases} \text{either: } p \text{ is odd and } \left(\frac{d}{p}\right) = +1, \\ \text{or: } p = 2 \text{ and } d \equiv 1 \bmod 8; \end{cases}$

(ii) p is inertial in L if: $\begin{cases} \text{either: } p \text{ is odd and } \left(\frac{d}{p}\right) = -1, \\ \text{or: } p = 2 \text{ and } d \equiv 5 \bmod 8; \end{cases}$

(iii) p is ramified in L if: $\begin{cases} \text{either: } p \text{ is odd and divides } d, \\ \text{or: } p = 2 \text{ and } d \equiv 2 \text{ or } 3 \bmod 4. \end{cases}$

In order to determine the value of the Legendre symbol $\left(\frac{d}{p}\right)$ we observe that this symbol is multiplicative, that is to say:

$$\left(\frac{ab}{p}\right) = \left(\frac{a}{p}\right)\left(\frac{b}{p}\right) \quad \text{(where } a,b \text{ are integers prime with } p\text{).}$$

One also uses the *QUADRATIC RECIPROCITY LAW*: *If p and q are two distinct odd prime numbers, then* :

$$\left(\frac{p}{q}\right)\left(\frac{q}{p}\right) = (-1)^{\frac{1}{4}(p-1)(q-1)}.$$

4.5 IDEAL CLASSES AND UNITS OF ALGEBRAIC NUMBER FIELDS

K denotes a number field and $n = [K:\mathbb{Q}]$ its degree.

4.5·1 THE INTEGERS r_1 AND r_2

There exist exactly n distinct $\mathbb{Q}$-isomorphisms σ_i $(1 \leqslant i \leqslant n)$ of K into the complex number field $\mathbb{C}$. We denote by r_1 the number of indices i such that $\sigma_i(K)$ is contained in the real number field $\mathbb{R}$. If i is an index such that $\sigma_i(K)$ is not contained in $\mathbb{R}$, and if τ is complex comjugation in $\mathbb{C}$, $\tau\sigma_i$ is also a $\mathbb{Q}$-isomorphism of K in $\mathbb{C}$ and it is distinct from σ_i. The number of indices i such that $\sigma_i(K)$ is not contained in $\mathbb{R}$ is therefore even, a number which will be denoted $2r_2$. We have:

$$n = r_1 + 2r_2.$$

4.5·2 IDEAL CLASSES OF B

Let Δ be the discriminant of B (that is to say, the discriminant of an integer basis of B). The *MINKOWSKI CONSTANT OF A FIELD* K is the number:

$$M = \left(\frac{4}{\pi}\right)^{r_2} \frac{n!}{n^n} |\Delta|^{\frac{1}{2}}.$$

One can show that every class of ideals of B contains an integral ideal A such that:

$$N(A) \leqslant M,$$

where the norm $N(A)$ of the integer ideal A is by definition the cardinality of B/A. As the set of integral ideals of B for which the norm is a given integer is finite, it follows that: *The class group of a number field is finite.*

4.5·3 THE UNITS OF K

Dirichlet's Theorem gives the structure of the group E of units of the ring B of integers of K:

Let $r = r_1 + r_2 - 1$, and let U be the group of roots of unity contained in K. The group E is the direct product of the group U and of r groups isomorphic to $\mathbb{Z}$. In other words, there exist r units $\varepsilon_1,\ldots,\varepsilon_r$, called *FUNDAMENTAL UNITS*, such that:

$$E = U \times \varepsilon_1^{\mathbb{Z}} \times \cdots \times \varepsilon_r^{\mathbb{Z}}.$$

REMARK: As every finite subgroup of the multiplicative group K^* of elements of an arbitrary field K is cyclic, in particular therefore, the group U is a cyclic group.

BIBLIOGRAPHY

[Bor], [Bou], [CaF], [Che], [Har], [Has], [Joly], [Lang 1], [Mac], [Mor], [Sam], [Ser 2], [Wei].

The book giving the quickest and most complete access to algebraic number theory is that of P. Samuel, cited above, and from which this Introduction has largely been drawn.

PROBLEMS

EXERCISE 4·1: EXTENSION OF A FINITE FIELD

Notations and revision:

$k = \mathbb{F}_q$ denotes the finite field with $q = p^r$ elements (p a prime number);

$K = \mathbb{F}_{q^n}$ denotes the extension, of degree n, of k.

We denote by σ the k-automorphism of K defined by $\sigma(x) = x^q$ and we recall that σ is a generator of $\mathrm{Gal}(K|k)$.

We denote by $T(x)$ and $N(x)$, respectively, the trace and norm, relative to the extension $K|k$ of an element $x \in K$.

I:(1): Show that the mapping $T:K \to k$ is surjective.

I:(2): Let $\vartheta \in K$ be an element of the form $\vartheta = x - \sigma(x)$ (where $x \in K$); show that $T(\vartheta) = 0$.

Conversely, let $\vartheta \in K$ be an element satisfying $T(\vartheta) = 0$; set:

$$x = \frac{1}{T(u)} (\vartheta\sigma(u) + (\vartheta + \sigma(\vartheta))\sigma^2(u) + \cdots + (\vartheta + \sigma(\vartheta) + \cdots + \sigma^{n-2}(\vartheta))\sigma^{n-1}(u)),$$

(where $u \in K$ is an element such that $T(u) \neq 0$). Verify that $\vartheta = x - \sigma(x)$, and that every $x_1 \in K$ such that $\vartheta = x_1 - \sigma(x_1)$ is of the form $x_1 = x + \lambda$, where $\lambda \in k$.

II:(1): Let $\vartheta \in K$ be an element of the form $\vartheta = \frac{x}{\sigma(x)}$ (where $x \in K$ is a non-zero element); show that $N(\vartheta) = 1$.

Conversely, let $\vartheta \in K$ be an element such that $N(\vartheta) = 1$; show that there exists an element $x \in K$ such that $\vartheta = \frac{x}{\sigma(x)}$. Verify that every element $x_1 \in K$ such that $\vartheta = \frac{x_1}{\sigma(x_1)}$ is of the form $x_1 = \lambda x$, where $\lambda \in k$ is a non-zero element.

II:(2): Show that the mapping $N:K \to k$ is surjective.

III: In this Part it is assumed that n divides $q - 1$. We denote by μ_n the subgroup of the n-th roots of unity of k^*.

III:(1): Let $A = \{\alpha \in K^*:\alpha^n \in k^*\}$; A is a multiplicative subgroup of K^* containing k^*. Denote by $\varphi:A \to \mu_n$ the mapping defined by $\varphi(\alpha) = \frac{\alpha}{\sigma(\alpha)}$. Verify that φ is a group homomorphism and induces an isomorphism:

$$\bar{\varphi}:A/k^* \xrightarrow{\sim} \mu_n.$$

III:(2): Let α be an element of A and let d be the order of $\varphi(\alpha)$ in μ_n. Show that $[k(\alpha):k] = d$. Give the factorisation of the polynomial $X^n - \alpha^n$ as a product of irreducible polynomials of $k[X]$.

III:(3): Show that the mapping $\psi:A \to k^*$ defined by $\psi(\alpha) = \alpha^n$ is a group homomorphism which induces an isomorphism:

$$\bar{\psi}:A/k^* \xrightarrow{\sim} k^*/k^{*n}.$$

Show that every element of k is an n-th power of an element of K.

EXERCISE 4·2: EXTENSIONS OF DEGREE p OF A FIELD OF CHARACTERISTIC p

Let k be a field of characteristic $p \neq 0$, and let π be its prime subfield. Let $P(X) = X^p - X - a$, where $a \in k$.

(1): Let α be a root of $P(X)$ in an algebraic closure of K. Express all the roots of $P(X)$ as a function of α.

(2): Assume that there exists an irreducible polynomial $Q(X)$ of $k[X]$ of degree $r \geq 2$ that is a factor of $P(X)$.

Show that $r = p$. (Consider the term of degree $r - 1$ of $Q(X)$). Give a necessary and sufficient condition for $P(X)$ to be irreducible over k.

(3): Assume $P(X)$ to be irreducible on k. Let α be a root of $P(X)$; show that $K = k(\alpha)$ is a cyclic extension of k of degree p.

(4): Let $P(X) = X^p - X - a$ and $Q(X) = X^p - X - b$ be two polynomials of $k[X]$ which are assumed to be irreducible.

Show that in an algebraic closure of k the splitting fields of these two polynomials coincide if and only if b is of the form $b = ta + c^p - c$ (where $c \in k$, $t \in \pi$, $t \neq 0$).

(5): Let K be a cyclic extension of k of degree p.

Show that there exists an element $\alpha \in K$ such that $K = k(\alpha)$, α being a root of a polynomial $P(X) = X^p - X - a$ (where $a \in k$).

EXERCISE 4·3: EMBEDDING OF A QUADRATIC FIELD IN A DIHEDRAL EXTENSION OF $\mathbb{Q}$

Notations: m denotes a rational integer different from 1 and -3 that is not divisible by any square p^2 of a prime number p. $\sqrt{m}$ (*resp.* $\sqrt{-3}$) denotes one of the square roots in $\mathbb{C}$ of m (*resp.* -3); $\sqrt{-3m}$ denotes the product $\sqrt{-3}\sqrt{m}$; j denotes the complex number $\frac{1 - \sqrt{-3}}{2}$.

PRELIMINARY QUESTION: Let L be the subfield of $\mathbb{C}$ defined as a splitting field over $\mathbb{Q}$ of the polynomial $X^3 - 2$. Recall why the extension $L/\mathbb{Q}$ is Galois and non-abelian. From this deduce that the quadratic field $\mathbb{Q}(\sqrt{-3})$ possesses an extension L of degree three that is Galois and non-abelian over $\mathbb{Q}$.

We propose to show that every quadratic field possesses this property.

(1): Let $k = \mathbb{Q}(\sqrt{-3m})$. Show that the extension $k/\mathbb{Q}$ is Galois of degree two; write f for the $\mathbb{Q}$-automorphism of k different from the identity, and for $\mu \in k$ set $\bar{\mu} = f(\mu)$. Show that there exist elements $\mu \in k$ such that $\mu^2\bar{\mu}$ is not the cube of an element of k.

In what follows μ denotes an element of k such that $\mu^2\bar{\mu}$ is not the cube of an element of k.

(2): Let K be the subfield of $\mathbb{C}$ defined as the splitting field on k of the polynomial $X^3 - \mu^2\bar{\mu}$. The roots of this polynomial in $\mathbb{C}$ will be denoted $\alpha, j\alpha, j^2\alpha$.

(a): Show that the extension K/k is Galois of degree six.

(b): Show that there exist two k-automorphisms σ, τ of K such that:

$$\sigma(\sqrt{m}) = \sqrt{m}, \qquad \sigma(\sqrt{-3}) = \sqrt{-3}, \qquad \sigma(\alpha) = j\alpha,$$

$$\tau(\sqrt{m}) = -\sqrt{m}, \qquad \tau(\sqrt{-3}) = -\sqrt{-3}, \qquad \tau(\alpha) = \alpha.$$

(c): Show that the group $\langle\sigma, \tau\rangle$ generated by σ and τ is exactly the group $G = \mathrm{Gal}(K|k)$. Is it commutative? Find all the intermediate fields between k and K.

(3): Find the polynomial $\mathrm{Irr}(\alpha, \mathbb{Q})$; what are the conjugates of α over $\mathbb{Q}$?

Show that the extension $K/\mathbb{Q}$ is Galois of degree twelve.

Show that there exists a $\mathbb{Q}$-automorphism s of K such that:

$$s(\sqrt{m}) = \sqrt{m}, \qquad s(\sqrt{-3}) = -\sqrt{-3}, \qquad s(\alpha) = \frac{\alpha^2}{\mu}.$$

Calculate s^2.

(4): Show that the group $G = \mathrm{Gal}(K/\mathbb{Q})$ is the direct product of the subgroup $\langle s\rangle$ (generated by s) and of the subgroup G. Let N be the subfield of K defined as the fixed field of the subgroup $\langle s\rangle$.

Verify that $\mathbb{Q}(\sqrt{m}) \subset N$ and that $[N:\mathbb{Q}(\sqrt{m})] = 3$. Show that the extension $N/\mathbb{Q}$ is Galois and non-abelian. From this deduce that every quadratic field possesses an extension of degree three that is Galois and non-abelian over $\mathbb{Q}$.

EXERCISE 4·4: BIQUADRATIC EXTENSIONS OF $\mathbb{Q}$

Let m_1, m_2 be two distinct square-free integers different from zero and one; k_1 (*resp.* k_2) denotes the extension obtained by adjoining to the rationals a root ϑ_1 (*resp.* ϑ_2) of the equation $x^2 - m_1 = 0$ (*resp.* $x^2 - m_2 = 0$). K denotes the field composite of k_1 and k_2. The ring of integers of a finite extension E of the rational numbers is denoted O_E and the discriminant of an extension F of E is denoted $\Delta_{(F/E)}$. Recall that if α_i, ($i = 1,2,\ldots,n = [F:E]$) are elements of O_F whose conjugates relative to E are denoted $\alpha_i^{(j)}$, $i,j = 1,2,\ldots,n$, then $\Delta_{(F/E)}$ divides $(\det(\alpha_i^{(j)}))^2$.

I:(1): Show that $K|\mathbb{Q}$ is Galois.

I:(2): What is the degree $[K:\mathbb{Q}]$ of K relative to $\mathbb{Q}$?

I:(3): Prove that there exists a $\mathbb{Q}$-automorphism σ of K satisfying:

$$\sigma(\vartheta_1) = -\vartheta_1, \qquad \sigma(\vartheta_2) = \vartheta_2,$$

and a $\mathbb{Q}$-automorphism τ of K such that:

$$\tau(\vartheta_1) = \vartheta_1, \qquad \tau(\vartheta_2) = -\vartheta_2;$$

deduce from this the Galois group of $K/\mathbb{Q}$.

I:(4): Show that K contains one and only one quadratic field k_3 distinct from k_1 and k_2, what is the discriminant as a function of m_1 and m_2?

I:(5): Can a prime ideal of $\mathbb{Q}$ different from (2) be ramified simultaneously in k_1, k_2, k_3? Give an example showing the same does not happen with the ideal (2).

I:(6): Show that Δ_{K/k_i} divides $\Delta_{k_j/\mathbb{Q}}$ if $j \neq i$. From this deduce that a prime ideal of $\mathbb{Q}$ which is ramified in K is ramified in at least two of the fields k_1, k_2, k_3.

II: Assume that $m_1 = -1$ and $m_2 \equiv 1 \bmod 4$.

II:(1): What is the number of ideal classes (class number) of O_{k_1}. Is the O_{k_1}-module O_k free?

II:(2): By using Question I(b) above, calculate the discriminant of K/k_3. What prime ideals of O_{k_3} ramify in K?

III: What are the possible different decompositions of a rational prime ideal in K?

EXERCISE 4·5 STUDY OF A CYCLIC CUBIC FIELD WITH PRIME CONDUCTOR

Let $\mathbb{Q}$ be the field of rational numbers and p a prime number congruent to 1 modulo 3.

I: Let $j \neq 1$ be a cube root of unity, k the field $\mathbb{Q}(j)$, A the ring of integers of k.

I:(1): Verify that A is a principal ideal domain and that in A, p becomes a product of two distinct prime ideals, therefore that there exist rational integers α_0, β_0 for which $p = N_{k/\mathbb{Q}}(\alpha_0 + j\beta_0)$

I:(2): Find, as functions of α_0, β_0, all the pairs (x,y) of rational integers satisfying $p = \frac{3x^2 + y^2}{4}$. Deduce from this that there exists one and only one pair (a,b) of rational integers for which:

$$a > 0, \tag{1}$$

$$b \equiv 1 \bmod 3, \tag{2}$$

$$p = \frac{27a^2 + b^2}{4}. \tag{3}$$

II: Let $\omega \neq 1$ be a p-th root of unity, and $\mathbb{Q}'$ the extension of $\mathbb{Q}$ obtained by the adjunction of ω. Denote by G the Galois group $\mathbb{Q}'/\mathbb{Q}$ and its elements by s_i (i mod p, $i \not\equiv 0$ mod p), s being defined by the condition $s_i\omega' = \omega'^i$ for every p-th root of unity ω'.

II:(1): Verify that $\mathbb{Q}'$ possesses one and only one subfield of degree three over $\mathbb{Q}$. Denote this field by K, the integral closure of $\mathbb{Z}$ in K by B, and the discriminant of K by D. Verify that K is cyclic over $\mathbb{Q}$, and that $D = p^2$.

II:(2): Let t be a generator of $(\mathbb{Z}/p\mathbb{Z})^*$ and H be the subgroup $(\mathbb{Z}/p\mathbb{Z})^{*3}$ of $(\mathbb{Z}/\ \mathbb{Z})^*$. Let us set:

$$\vartheta = \sum_{i \in H} \omega^i, \qquad \vartheta' = \sum_{i \in H} \omega^{ti}, \qquad \vartheta'' = \sum_{i \in H} \omega^{t^2 i}.$$

Verify that $\vartheta, \vartheta', \vartheta''$ are elements of K conjugate over $\mathbb{Q}$ and that $K = \mathbb{Q}(\vartheta) = \mathbb{Q}(\vartheta') = \mathbb{Q}(\vartheta'')$.

II:(3): Calculate $\vartheta + \vartheta' + \vartheta''$ and $\vartheta\vartheta' + \vartheta\vartheta'' + \vartheta'\vartheta''$.

II:(4): To calculate the product $\vartheta\vartheta'\vartheta''$, set $r = \frac{p-1}{3}$, and assume that in the expression:

$$\omega^{t^3} \sum_{i,j \in H} \omega^{ti+t^2 j}$$

there are k terms equal to one (therefore there are $k \times r$ terms equal to one in $\vartheta\vartheta'\vartheta''$).

Verify that the minimal polynomial of ϑ is then:

$$P(x) = x^3 + x^2 - rx + \frac{r^2 - kp}{3} .$$

Let Δ be its discriminant. Show that we have $\Delta = p^2 a^2$ with $a \in \mathbb{N}$.

In calculating Δ one finds that k must satisfy an equation of the second degree the discriminant of which is the square of an integral number b'. Then verify that we have $p = \frac{27a^2 + b^2}{4}$ and that:

$$P(x) = x^3 + x^2 - \frac{p-1}{3} x - \frac{p(3+b) - 1}{27} ,$$

a,b being the integers defined in Question I.

N.B. The discriminant of the polynomial $x^3 + a_2 x^2 + a_1 x + a_0$ is:

$$\Delta = a_2^2 a_1^2 + 18a_0 a_1 a_2 - 4a_2^3 a_0 - 4a_1^3 - 27a_0^2 .$$

III:(1): Let q be a prime number distinct from p.

Verify that the following conditions are equivalent:

(i): q lies in K and is a product of three distinct prime ideals;

(ii): The number of prime factors of q in $\mathbb{Q}'$ is divisible by three;

(iii): $q^{(p-1)/3} \equiv 1 \bmod p$;

(iv): There exists $x \in \mathbb{Z}$ such that $q \equiv x^3 \bmod p$.

III:(2): Verify that for $q \not\equiv p$ the equation $P(x) \equiv 0 \bmod q$ cannot have a triple root in $\mathbb{Z}/q\mathbb{Z}$. From this deduce that q decomposes in K if and only if the congruence $P(x) \equiv 0 \bmod q$ has a solution.

III:(3): Show directly that 2 decomposes in K if and only if a,b are even. Deduce from this that in order to be able to write a prime number p in the form $p = 27x^2 + y^2$, with $x,y \in \mathbb{Z}$, it is necessary and sufficient that $p \equiv 1 \bmod 3$ and $2^{(p-1)/3} \equiv 1 \bmod p$.

III:(4): Show that $3\vartheta + 1$ is a root of the polynomial $x^3 - 3px - pb$. From this deduce that if $q \geqslant 5$ is a prime number dividing b, then $q^{(p-1)/3} \equiv 1 \bmod p$.

III:(5): Show that if 3 divides a, $3^{(p-1)/3} \equiv 1 \bmod p$.

EXERCISE 4·6: STUDY OF A FIELD $\mathbb{Q}(\sqrt[3]{d})$

The goal of this Problem is the study of certain fields obtained by adjoining to $\mathbb{Q}$ the cube root of an integer which is not the cube of an integer. In order to simplify matters we restrict ourselves to the case where d is square-free and $d \not\equiv \pm 1 \bmod 9$ ($d \equiv \pm 2, \pm 3$, or $\pm 4 \bmod 9$). We write ϑ for a root of the polynomial $f(x) \equiv x^3 - d$ and K for the field $\mathbb{Q}(\vartheta)$.

I: INTEGERS

(1): Calculate the discriminant $\Delta(1,\vartheta,\vartheta^2)$.

(2): Show that the prime factors of d are totally ramified in K. Show that the same is true for 3 (consider the polynomials $f(x), f(x+1), f(x-1)$).

I:(3): From the preceding deduce that $1, \vartheta, \vartheta^2$ is a basis of the ring O_K of the integers of K, and calculate the discriminant D of K.

II: DECOMPOSITION IN K OF PRIME NUMBERS

Notations: If $p/3d$ we shall write $p = p_p^3$.

II:(1): Let p be a prime number which does not divide $3d$. Show that in order for p to be inert in K it is necessary and sufficient that the congruence $d \equiv x^3 \bmod p$ be impossible:

II:(2): Show that if $p \equiv -1 \bmod 3$, p decomposes in K. In addition verify that -3 is not a square in $\mathbb{Q}_p$. From this deduce that a prime number $p \equiv -1 \bmod 3$ decomposes in K into the form $p_p p'_p$, where p_p is a prime ideal of degree one and p_p is a prime ideal of degree two.

II:(3): Let $q \equiv 1 \bmod 3$ be a prime number not dividing $3d$. Show that q remains inert or becomes in K the product of three prime ideals of degree one (we can show that -3 is a square in $\mathbb{Q}_p$).

III: IDEAL CLASSES

Recall, with the usual notations, that the Minkowski constant is:

$$M = \left(\frac{4}{\pi}\right)^{r_2} \frac{n!}{n^n} \sqrt{|D|}.$$

III:(1): Show that:

$$N_{K/\mathbb{Q}}(a + b\vartheta + c\vartheta^2) = a^3 + db^3 + d^2c^3 - 3dabc.$$

III:(2): Show that if a prime number does not remain inert in K it has a principal prime factor if and only if the equation:

$$a^3 + db^3 + d^2c^3 - 3dabc = p$$

has a solution with $a,b,c \in \mathbb{Z}$.

III:(3): APPLICATION: Verify that for $d = 2$, O_k is a principal ideal domain

IV: STUDY OF THE CASE $d = 7$

IV:(1): Let $m \equiv \pm 2$ or $\pm 3 \bmod 7$ be an integer.
Show that the equation:

$$N(a + b\vartheta + c\vartheta^2) = \pm m$$

has no solutions with $a,b,c \in \mathbb{Z}$. Verify that p_2p_3 is principal. From this deduce that the class number of K is equal to three.

Denote the class of p_3 by h (thus $p_3 \in h$, $p_2 \in h^2$).

IV:(2): Let $p \equiv -1 \bmod 3$ be a prime number.
By considering the equations:

$$N(a + b\vartheta + c\vartheta^2) = p, 3p, 9p,$$

show that p_p and p'_p are principal ideals if and only if $p \equiv \pm 1 \bmod 7$. In general determine the class of p_p by distinguishing the two cases $p \equiv \pm 2$ and $\pm 3 \bmod 7$. (Inspiration will be found in the method used for $p \equiv \pm 1 \bmod 7$).

IV:(3): Let $7 \nmid q \equiv +1 \bmod 3$ be a prime number which decomposes in K.
By using the method of the preceding Question show that the prime factors of q are in the same ideal class and determine

this class by distinguishing the three cases $q \equiv \pm 1, \pm 2, \pm 3 \bmod 7$.

EXERCISE 4·7: CLASS FIELD OF $\mathbb{Q}(\sqrt{65})$

Let $\mathbb{Q}$ be the rational number field, $k = \mathbb{Q}(\sqrt{65})$ and $\omega = \sqrt{65}$.

(1): Show that the ideal (2) splits in k into the form $(2) = p_1 p_2$ where p_1, p_2 are prime ideals. Show that p_1 is generated by 2 and $\frac{\omega + 1}{2}$ and p_2 by 2 and $\frac{\omega - 1}{2}$.

Show that p_1^2 is a principal ideal generated by $4 + \frac{\omega + 1}{2}$.

(2): Show that the ideal p_1 is not principal (assume that p_2 is principal and generated by $2a + b\left(\frac{\omega + 1}{2}\right)$, where a, b are integers, write $N_{k/\mathbb{Q}}\left(2a + b\left(\frac{\omega + 1}{2}\right)\right) = \pm 2$, and reduce to an equation of the form $8 = x^2 - 65y^2$, which does not have a solution in integers).

(3): Show that the group of ideal classes of k is of order two.

(4): Show that in k there exists a unit u with norm over $\mathbb{Q}$ equal to -1.

(5): Let ε be a fundamental unit of k such that the completions of $k(\sqrt{\varepsilon})$ for the Archimedean absolute value are isomorphic to the real number field $\mathbb{R}$.

Show that there exists an ideal of k which is ramified in $k(\sqrt{\varepsilon})$ (it suffices to notice that $k(\sqrt{\varepsilon}) = k(\sqrt{u})$ or $k(\sqrt{\varepsilon}) = k(\sqrt{-u})$).

(6): Let K be a quadratic extension of k whose completions at the Archimedean absolute values are isomorphic to $\mathbb{R}$ and whose discriminant over k is (1) (that is to say, there are no ideals of k ramified in K).

Letting ε always be a fundamental unit of k, show that ε and $-\varepsilon$ are not squares in K.

(7): Let us denote the Galois group of K over K by $(1, \sigma)$.

Show that for every unit a of K, $a + \sigma(a) \neq 0$.

(8): Show that there exists a non-zero integer $b \in K$ such that $b = -\sigma(b)$.

(9): Show that the principal ideal (b) is the extension of a non-principal ideal A of k, that is to say, that if O_k and O_K denote the integral closures of $\mathbb{Z}$ in k and K then:

$$A.O_K = b.O_k.$$

(10): Let K_1, K_2 be two fields having the properties in Question (6) above, show that $K_1 = K_2$.

(11): Show that $\mathbb{Q}(\sqrt{5}, \sqrt{13})$ is the only field K that has the properties of Question (6) above.

EXERCISE 4·8: STUDY OF THE NORM IN AN UNRAMIFIED EXTENSION OF A LOCAL FIELD

Throughout the Problem k denotes a *local field* (that is to say, a complete non-Archimedean valued field with discrete valuation and finite residue field), and K denotes a finite extension of k, furnished with the absolute value extending that of k, that is *unramified* (that is to say, $e(K|k) = 1$).

Notations: (*cf.* The introduction of Chapter 10 on p-Adic Analysis): $[K:k] = n$, π the uniformiser of k; $|\cdot|$ the absolute value in k and K; $O_k = \{x \in k : |x| \leqslant 1\}$; $O_K = \{x \in K : |x| \leqslant 1\}$; $\bar{k}$ (*resp.* $\bar{K}$) the residue field of k (*resp.* K) ($\bar{k}$ will be identified with a subgroup of $\bar{K}$); $x \mapsto \bar{x}$ the canonical mapping of O_K into $\bar{K}$; $\mathrm{card}\bar{k} = q = p^r$.

(1) Verify that the extensions $K|k$ following satisfy the conditions above:

(1) (a) $k = \mathbb{Q}_p$ (the p-adic number field); $K = \mathbb{Q}_p(\alpha)$, α be-

ing a root of the polynomial $X^2 - d$ (d a rational integer not divisible by p and not congruent to a square modulo p).

(1) (b) $k = \mathbb{Q}_5$; $K = \mathbb{Q}_5(\beta)$, β being a root of the polynomial $X^4 - 2$.

(2) Prove the following results:

(2) (a) $[\bar{K}:\bar{k}] = n$.

(2) (b) Every element of $\bar{K}$ is a root of the polynomial $X^{q^n} - X$; K is a splitting field over k of the polynomial $X^{q^n} - X$.

(2) (c) $\bar{K}|\bar{k}$ is Galois; its Galois group $G(\bar{K}|\bar{k})$ is cyclic and generated by the element τ defined by:

$\tau(\omega) = \omega^q$ for all $\omega \in \bar{K}$.

(Recall that the multiplicative group of non-zero elements of a finite field is cyclic).

(3) Show that K is a splitting field over k of the polynomial $X^{q^n} - X$, and that there exists a root ξ of this polynomial such that $K = k(\xi)$. From this deduce that the extension $K|k$ is Galois.

(4) (a) Let $\sigma \in G(K|k)$ be a k-automorphism of K. Show that the mapping $\bar{\sigma}$, which to $\bar{x} \in \bar{K}$ associates $\bar{\sigma}(\bar{x}) = \overline{\sigma x}$ (for an element $x \in O_K$), is well defined and that it is a $\bar{k}$-automorphism of $\bar{K}$.

(4) (b) Show that the mapping $\Phi: G(K|k) \to G(\bar{K}|\bar{k})$ defined by $\Phi(\sigma) = \bar{\sigma}$ is an isomorphism.

(4) (c) From this deduce that the group $G(K|k)$ is cyclic and generated by the element σ^* determined uniquely by $|\sigma^*(x) - x^q| < 1$ for all $x \in O_K$.

(4) (d) Describe σ^* for the examples (1)(a) and (1)(b).

(5) (a) Let $m \geqslant 1$ be a rational integer and let $\mu_m \in 1 + \pi^m O_K$, that is to say $\mu_m = 1 + a\pi^m$ $(a \in O_K)$.
Show that:

$$Nm_{K|k}(\mu_m) \equiv 1 + \mathrm{Tr}_{K|k}(a).\pi^m \pmod{\pi^{m+1}}.$$

(5) (b) Show that the mapping $\psi: O_K \to \bar{k}$, which to an element $a \in O_K$ associates $\psi(a) = \overline{\mathrm{Tr}_{K|k}(a)}$, is surjective. From this deduce the following result: For any $\lambda_m \in 1 + \pi^m O_k$ there exists $\mu_m \in 1 + \pi^m O_K$ such that:

$$Nm_{K|k}(\mu_m) \equiv \lambda_m \pmod{\pi^{m+1}},$$

which can be written:

$$Nm_{K|k}(\mu_m) = \frac{\lambda_m}{\lambda_{m+1}}, \quad \text{where } \lambda_{m+1} \in 1 + \pi^{m+1} O_K.$$

(5) (c) Show that:

$$Nm_{K|k}(1 + \pi O_K) = 1 + \pi O_k \quad \text{and} \quad Nm_{K|k}(U_K) = U_k.$$

(U_k (*resp.* U_K) denotes the multiplicative group of invertible elements of O_k (*resp.* O_K)).

(5) (d) Let d be as in Question (1)(a). Show that every p-adic number x with absolute value 1 can be written:

$$x = a^2 - db^2 \quad \text{with } a, b \in \mathbb{Q}_p.$$

(6) Denote by k^* (*resp.* K^*) the multiplicative group of non-zero elements of k (*resp.* K).
Show that the group $k^*/Nm_{K|k}(K^*)$ is a finite group isomorphic to $\mathrm{Gal}(K|k)$.

EXERCISE 4·9: PURELY TRANSCENDENTAL EXTENSIONS

I Let K be the field of rational functions in an indeterminate X over the field $\mathbb{Q}$ of rational numbers.

I (1) In the ring $K[Y]$ show that the polynomial $Y^2 + X^2 + 1$ is irreducible.

I (2) Let E be the extension of K generated by a root of this polynomial.
Show that E is not a purely transcendental extension of K.

I (3) Let $i \in \mathbb{C}$ be a root of $X^2 + 1$. Show that $E(i)$ is a purely transcendental extension of $\mathbb{Q}(i)$.

I (4) In the ring $K[Y]$ show that the polynomial $Y^2 - X^2 + 1$ is irreducible. Let F be the extension generated by a root of this polynomial. Is F a purely transcendental extension of $\mathbb{Q}$?

II Let $L = \mathbb{C}(X)$ be the field of rational functions in an indeterminate X over the complex numbers.

II (1) Show that the polynomial $Y^3 + X^3 + 1$ is irreducible in the ring $L[Y]$.

II (2) Let M be a splitting field of this polynomial over L.
Show that M is not a purely transcendental extension of $\mathbb{C}$.

SOLUTIONS

SOLUTION 4·1: I:(1): Every extension of a finite field is separable; it is known that the trace is then surjective ([Bou] Chapter V §10, or [Joly] Chapter 1).

This result can be proved directly; obviously it suffices to show the existence of an element of K with non-zero trace. If the degree n of the extension is prime to p it suffices to notice that $T(1) = n$, and therefore $T(1) \neq 0$.

In the general case let us set $n = n_0 p^m$, n_0 being prime to p; let L be the fixed field of σ^{n_0}; then $[K:L] = p^m$ and $[L:K] = n_0$ ([Lang 1] Chapter 8). We have just observed that the trace in the extension L/K is surjective; as $\mathrm{Tr}_{K/k} = \mathrm{Tr}_{L/k}\,\mathrm{Tr}_{K/L}$ ([Bou] Chapter V) it suffices to show that the trace of the extension K/L is surjective. For the same reason, by considering the extensions intermediate between K and L, we can reduce to an extension of degree p.

Therefore let us assume that K is an extension of degree p of a finite field L, and let us show that there exists an element of K with non-zero trace over L. Let α be a generator of the cyclic group K^* (cf., [Sam] Chapter 1), that is to say an element of K of order $q'^p - 1$ (where q' denotes the cardinality of L).

Let:

$$P(X) = X^p + a_1 X^{p_1} + \cdots + a_i X^{p-i} + \cdots + a_p$$

be the irreducible polynomial of α over L. We cannot have $a_1 = a_2 = \cdots = a_{p-1} = 0$, otherwise (by writing τ for the L-automorphism of K defined by $\tau(x) = x^{q'}$) we would have $\left(\frac{\tau\alpha}{\alpha}\right)^p = 1$ which is absurd, because $\left(\frac{\tau\alpha}{\alpha}\right)^p = \alpha^{p(q'-1)} \neq 1$, since $p(q'-1) < q'^p - 1$. Consequently, if i denotes the smallest index such that $a_i \neq 0$, we have:

$$\mathrm{Tr}_{K/L}(\alpha^i) = (-1)^{i+1} i a_i \neq 0.$$

I:(2): If $\vartheta \in K$ is of the form $\vartheta = x - \sigma(x)$ then obviously $T(\vartheta) = T(x) - T(\sigma(x)) = 0$.

Conversely, let ϑ be an element with vanishing trace, and u an element with non-zero trace. Set:

$$x = \frac{1}{T(u)} (\vartheta\sigma(u) + \cdots + (\sum_{0 \leqslant i \leqslant j-1} \sigma^i(\vartheta))\sigma^j(u) + \cdots$$

$$+ (\sum_{0 \leqslant i \leqslant n-2} \sigma^i(\vartheta))\sigma^{n-1}(u)).$$

We have:

$$\sigma(x) = \frac{1}{T(u)} (\sigma(\vartheta)\sigma^2(u) + \cdots + (\sum_{0 \leqslant i \leqslant j-1} \sigma^{i+1}(\vartheta))\sigma^{j+1}(u) + \cdots$$

$$+ (\sum_{0 \leqslant i \leqslant n-2} \sigma^{i+1}(\vartheta))u).$$

Whence:

$$x - \sigma(x) = \frac{1}{T(u)} (\vartheta(\sum_{1 \leqslant j \leqslant n-1} \sigma^j(u)) - (\sum_{1 \leqslant j \leqslant n-1} \sigma^j(\vartheta))u) =$$

$$= \frac{1}{T(u)} (\vartheta(T(u) - u) - (T(\vartheta - \vartheta)u) = \vartheta.$$

If $x_1 \in K$ is such that $x_1 - \sigma x_1 = x - \sigma x$, then the element $\lambda = x_1 - x$ is such that $\lambda = \sigma(\lambda)$, and therefore $\lambda \in k$.

REMARK: The above proof applies in fact to every cyclic extension of an arbitrary field k ([Bou] Chapter V §11) (additive form of Hilbert's "Theorem 90").

II:(1): If ϑ is an element of the form $\vartheta = \frac{x}{\sigma(x)}$, where $x \in K$ is non-zero, then:

$$N(\vartheta) = \frac{N(x)}{N(\sigma(x))} = 1.$$

Conversely, let $\vartheta \in K$ have unit norm. We have:

$$N(\vartheta) = \vartheta^{1+q+\cdots+q^{n-1}} = \vartheta^{(q^n-1)/(q-1)} = 1.$$

Let α be a generator of the cyclic group K^* and let us set $\vartheta = \alpha^i$. As α has order $q^n - 1$ in K^*, the equality:

$$\alpha^{i(q^n-1)/(q-1)} = 1$$

implies that $q - 1$ divides i. Let us set $i = m(q - 1)$. Then we have:

$$\vartheta = \alpha^{m(q-1)} = \left(\frac{\sigma(\alpha)}{\alpha}\right)^m$$

and therefore if $x = \alpha^{-m}$:

$$\vartheta = \frac{x}{\sigma(x)}.$$

If $x_1 \in K^*$ is such that:

$$\frac{x_1}{\sigma(x_1)} = \frac{x}{\sigma(x)},$$

then $\lambda = \frac{x_1}{x}$ satisfies $\sigma(\lambda) = \lambda$, and therefore $\lambda \in k$.

REMARK: For every cyclic extension K of a field k the elements of K with unit norm are the elements of the form $\frac{x}{\sigma(x)}$ (where $x \in K^*$ and σ is the generator of $\mathrm{Gal}(K|k)$) (Hilbert's "Theorem 90", cf., [Bou] Chapter V §11, or [Mac] Chapter 2 §5).

II (2) The image under the norm of the multiplicative group K^* is a multiplicative subgroup $N(K^*)$ of k^*. Let us denote by $\ker N$ the set of elements of K^* of unit norm; it is a subgroup of K^* with cardinality $\frac{\mathrm{card} K^*}{\mathrm{card} k^*}$ by the preceding problem. Since there is an isomorphism $N(K^*) \simeq K^*/\ker N$ we have:

$$\mathrm{card} N(K^*) = \frac{\mathrm{card} K^*}{\mathrm{card}\ker N},$$

and therefore:

$$\mathrm{card} N(K^*) = \mathrm{card} k^*.$$

From this it follows that $N(K^*) = k^*$ (cf., [Joly] Chapter 1).

III (1) A is evidently a multiplicative subgroup of K^* containing k^*. It is clear that φ is a homomorphism of groups; if $\alpha' = \lambda\alpha$, with $\lambda \in k^*$, then $\varphi(\alpha') = \varphi(\alpha)$, and therefore by passage to the quotient φ defines a homomorphism $\bar{\varphi}: A/k^* \to \mu_n$. The homomorphism $\bar{\varphi}$ is surjective, for $\sigma(\alpha) = \alpha$ implies $\alpha \in h^*$; $\bar{\varphi}$ is surjective, because an element $\zeta \in \mu_n$ has norm one, hence there exists (by Question II(1)) an $\alpha \in K$ such that $\zeta = \frac{\alpha}{\sigma(\alpha)}$ and then α is in A.

III (2) Let $\alpha \in K^*$ be such that $\alpha^n \in k^*$, and let d be the order of $\varphi(\alpha)$ in μ_n. By Question III(1) the integer d is also the smallest integer greater than zero such that $\alpha^d \in k^*$. The conjugates $\sigma^i(\alpha) = \zeta^i\alpha$ $(0 \leq i \leq n-1)$ (where $\zeta = \varphi(\alpha)$) satisfy $\sigma^i(\alpha) = \sigma^j(\alpha)$ if and only if $i - j$ is a multiple of d. Since α has d distinct conjugates, the extension $k(\alpha)|\alpha$ has degree d.

If ξ is an n-th primitive root of unity, we have:

$$X^n - \alpha^n = \prod_{0 \leq i \leq \frac{n}{d}-1} (X^d - \xi^i \alpha^d),$$

and the polynomials $X^d - \xi^i \alpha^d$ are irreducible polynomials of $k[X]$.

III:(3): The mapping ψ is evidently a group homomorphism such that $\psi(\lambda\alpha) = \lambda^n \psi(\alpha)$ (where $\alpha \in A$, $\lambda \in k^*$). By passage to the quotient one therefore obtains a homomorphism:

$$\bar{\psi}: A/k^* \to k^*/k^{*n}.$$

The homomorphism $\bar{\psi}$ is injective, for $\varphi(\alpha) = \alpha^n$ is in k^{*n} if and only if $\alpha \in k^*$; on the other hand k^*/k^{*n} is a cyclic group isomorphic to μ_n, and therefore the finite groups $\bar{\psi}(A/k^*)$ and k^*/k^* , having the same cardinality (Question III(1)), coincide. Consequently every element of k is the n-th power of an element of k.

REMARK: The study above (Question III) is a particular case of Kümmer Theory, that is to say, K/k is abelian of degree n, and k is a field, whose characteristic does not divide n, which contains the n-th roots of unity (cf., [Mac] Chapter 2 §7).

SOLUTION 4·2: (1): Let $t \in \pi$; we have $t^p = t$. Therefore:

$$P(\alpha + t) = (\alpha + t)^p - (\alpha + t) - a = \alpha^p - \alpha - a = P(\alpha) = 0.$$

If α is a root of $P(X)$, $(\alpha + t)_{t \in \pi}$ is the set of roots of $P(X)$.

(2): Let us assume that $P(X)$ possesses an irreducible divisor $Q(X)$ of degree $r \geq 2$. $Q(X)$ is equal to a product of r polynomials $X - (\alpha + t)$, where $t \in \pi$. The terms of degree $r - 1$ of $Q(X)$ therefore has a coefficient of the form $-r\alpha + u$, where $u \in \pi$. But then, if $r < p$, $\alpha \in k$ and $Q(X)$ is not irreducible.

Hence $r = p$. Consequently, $P(X)$ is irreducible in $k[X]$ if and only if $P(X)$ has no root in k; in fact by the preceding either $P(X)$ is irreducible or $P(X)$ has all its roots in k.

(3) If $P(X)$ is irreducible on k, with α denoting a root of $P(X)$, one sees (Question (1)) that all the roots of $P(X)$ belong to $k(\alpha)$; $k(\alpha)$ is therefore a normal extension of k ([Lang 1] VII §3). As α is separable over k the extension $k(\alpha)/k$ is separable ([Lang 1] VII §4). Consequently $k(\alpha)/k$ is a Galois extension of prime degree p, therefore cyclic.

(4) Let $K = k(\alpha)$ and let σ be the generator of $\mathrm{Gal}(K|k)$ such that $\sigma(\alpha) = \alpha + 1$. Let us look for the elements $\beta \in K$ for which the irreducible polynomial on k is of the form $\mathrm{Irr}(\beta,X) = X^2 - X - b$. We have $\sigma(\beta) = \beta + t$, where $t \in \pi$, $t \neq 0$, whence $\sigma\left(\frac{\beta}{t}\right) = \frac{\beta}{t} + 1$, and therefore $\sigma\left(\frac{\beta}{t} - \alpha\right) = \frac{\beta}{t} - \alpha$. Consequently $\frac{\beta}{t} - \alpha = \frac{c}{t}$, where $c \in k$. β is therefore a root of the polynomial:

$$\left(\frac{X - c}{t}\right)^p - \left(\frac{X - c}{t}\right) - a.$$

Whence:

$$\mathrm{Irr}(\beta,X) = (X - c)^p - (X - c) - at = X^p - X - (at + c^p - c),$$

and hence:

$$b = at + c^p - c.$$

Conversely, if β is a root of the polynomial:

$$(X - c)^p - (X - c) - at$$

we have $\beta = t\alpha + c$, and therefore $\beta \in K$ and generates K on k.

(5) Let K be a cyclic extension of k of degree p. We have $\mathrm{Tr}_{K|k}(-1) = 0$. Consequently (by the additive form of Hilbert's

Theorem 90, cf., Problem One) there exists $\alpha \in K$ such that $\sigma\alpha - \alpha = 1$ (σ being a generator of $\mathrm{Gal}(K|k)$). Whence:

$$\sigma^i\alpha = \alpha + i \qquad (0 \leqslant i \leqslant p-1).$$

As α has p distinct conjugates we have $K = k(\alpha)$. Furthermore, α is a root of a polynomial $X^p - X - a$, where $a \in k$; in fact:

$$\sigma(\alpha^p - \alpha) = (\alpha+1)^p - (\alpha+1) = \alpha^p - \alpha,$$

and therefore:

$$(\alpha^p - \alpha) \in k.$$

SOLUTION 4·3:

PRELIMINARY QUESTION: The splitting field over $\mathbb{Q}$ of the polynomial $X^3 - 2$ is the field $L = \mathbb{Q}(\sqrt[3]{2}, \sqrt{-3})$. Let us denote by φ the $\mathbb{Q}(\sqrt[3]{2})$-automorphism of L such that $\varphi(\sqrt{-3}) = -\sqrt{3}$, by ψ the $\mathbb{Q}(\sqrt{-3})$-automorphism of L such that $\psi(\sqrt[3]{2}) = j\sqrt[3]{2}$ ([Mac] Chapter 1). One verifies that $\psi\varphi(\sqrt[3]{2}) = j\sqrt[3]{2}$ and $\varphi\psi(\sqrt[3]{2}) = j^2\sqrt[3]{2}$. Consequently the group $\mathrm{Gal}(L|\mathbb{Q})$ is non-abelian. This is a group of order six whose distinct elements are:

$$e = \mathrm{id}_L, \psi, \psi^2, \varphi, \varphi\psi, \varphi\psi^2.$$

In terms of generators and relations $\mathrm{Gal}(L|\mathbb{Q})$ is defined by generators φ, ψ and relations:

$$\varphi^2 = \psi^3 = e, \qquad \varphi\psi = \psi^2\varphi.$$

Up to isomorphism such a group is the only non-abelian group of order six (the dihedral group of order $2m$ is by definition the group defined by the generators φ, ψ and relations:

$$\varphi^2 = \psi^m = e, \qquad \varphi\psi = \psi^{-1}\varphi).$$

(1) The extension $\mathbb{Q}(\sqrt{-3m})|\mathbb{Q}$ is evidently Galois of degree two since m is different from -3 and square-free as well.

Let A denote the ring of integers of k and let p be a prime number split in k ([Sam] Chapter 5), that is to say that $pA = \mathfrak{p}\mathfrak{p}'$, where $\mathfrak{p},\mathfrak{p}'$ are prime ideals distinct from A.

The image under the $\mathbb{Q}$-automorphism f of k of the prime ideal $\mathfrak{p}$ is evidently a prime ideal of A dividing p, hence we have $f(\mathfrak{p}) = \mathfrak{p}$ or $f(\mathfrak{p}) = \mathfrak{p}'$. If we had $f(\mathfrak{p}) = \mathfrak{p}$, then $f(\mathfrak{p}') = \mathfrak{p}'$. Let, then, x be an element of $\mathfrak{p}'$ not in $\mathfrak{p}$. $N_{K|\mathbb{Q}}(x) = x\bar{x} \in pA$, and therefore $x\bar{x} \in \mathfrak{p}$ without either x or $\bar{x}$ belonging to $\mathfrak{p}$, which is absurd (cf., [Mac] Chapter 5 §4). Consequently $f(\mathfrak{p}) = \mathfrak{p}'$, and thus $f(\mathfrak{p}') = \mathfrak{p}$.

Let μ be an element of $\mathfrak{p}$ that is neither in $\mathfrak{p}^2$ nor in $\mathfrak{p}\mathfrak{p}'$, ($\mathfrak{p}^2 \cup \mathfrak{p}\mathfrak{p}'$ is distinct from $\mathfrak{p}$ since $\mathfrak{p} \cup \mathfrak{p}'$ is distinct from A); this choice of μ implies that $\mu \notin \mathfrak{p}'$. We have $\mu A = \mathfrak{p}\mathfrak{q}$, where the ideal $\mathfrak{q}$ is divisible neither by $\mathfrak{p}$ nor $\mathfrak{p}'$. From this we deduce $\bar{\mu}A = \mathfrak{p}'\mathfrak{q}'$, where $\mathfrak{q}'$ is divisible by neither $\mathfrak{p}$ nor $\mathfrak{p}'$. Therefore $\mu^2\bar{\mu}A = \mathfrak{p}^2\mathfrak{p}'\mathfrak{q}^2\mathfrak{q}'$ is not divisible by $\mathfrak{p}^3$. As a result $\mu^2\bar{\mu}$ is not the cube of an element of k.

(a) Let α be a cube root of $\mu^2\bar{\mu}$ in $\mathbb{C}$. The extension $k(\alpha)/k$ is of degree three and not Galois since k does not contain cube roots of one. The splitting field K of the polynomial $X^3 - \mu^2\bar{\mu}$ on k is the field $K = k(\alpha,j) = k(\alpha,\sqrt{-3})$ of degree six over k.

(b) There exists a $k(\sqrt{-3})$-automorphism σ of K such that $\sigma(\alpha) = j\alpha$ and a $k(\alpha)$-automorphism τ of K such that $\tau(\sqrt{-3}) = -\sqrt{-3}$ ([Mac] Chapter I §2). Since $\sigma(\sqrt{-3}) = \sqrt{-3}$ and $\sigma(\sqrt{-3m}) = \sqrt{-3m}$ we have $\sigma(\sqrt{m}) = \sqrt{m}$. As $\tau(\sqrt{-3}) = -\sqrt{-3}$ and $\tau(\sqrt{-3m}) = \sqrt{-3m}$, we have $\tau(\sqrt{m}) = -\sqrt{m}$.

(c) It is clear that σ is or order three and τ of order two; furthermore, $\sigma\tau = \tau\sigma$ (for $\sigma\tau(\alpha) = j\alpha$ and $\tau\sigma(\alpha) = j^2\alpha$). The group $\langle\sigma,\tau\rangle$ is therefore isomorphic to the dihedral group of order six; this is a subgroup of $\mathrm{Gal}(K|k)$ and so they must be the same since they have the same order. The subgroups of G distinct from G and $\{e\}$ are the subgroups $\langle\sigma\rangle, \langle\tau\rangle, \langle\tau\sigma\rangle, \langle\tau\sigma^2\rangle$. There are thus four intermediate fields between k and K: $k(\sqrt{-3})$, the fixed field of $\langle\sigma\rangle$; $k(\alpha)$, the fixed field of $\langle\tau\rangle$; $k(j\alpha)$, the fixed field of $\langle\tau\sigma\rangle$; $k(j^2\alpha)$, the fixed field of $\langle\tau\sigma^2\rangle$ ([Lang 1] Chapter 8).

(3) The extension $k(\alpha)|\mathbb{Q}$ has degree $[k(\alpha):k][k:\mathbb{Q}] = 6$; therefore α has degree over $\mathbb{Q}$ a divisor of six.

The irreducible polynomial of α over k, $\mathrm{Irr}(\alpha,k) = X^3 - \mu^2\bar{\mu} = P(X)$ does not have its coefficients in $\mathbb{Q}$, for by the choice of μ, $\mu \neq \bar{\mu}$. The polynomial $\mathrm{Irr}(\alpha,\mathbb{Q})$ is therefore a multiple of $P(X)$ and of strictly greater degree; so, necessarily:

$$d^0\,\mathrm{Irr}(\alpha,\mathbb{Q}) = [\mathbb{Q}(\alpha):\mathbb{Q}] = 6.$$

Let $\bar{P}(X) = X^3 - \mu^2\bar{\mu}$ be the image under f of the polynomial $P(X)$. The polynomial $P(X)\bar{P}(X)$ has coefficients in $\mathbb{Q}$, hence it is a multiple of the polynomial $\mathrm{Irr}(\alpha,\mathbb{Q})$. As these two polynomials have the same degree we have:

$$\mathrm{Irr}(\alpha,\mathbb{Q}) = P(X)\bar{P}(X) = (X^3 - \mu^2\bar{\mu})(X^3 - \mu^2\bar{\mu}).$$

We see that the roots of the polynomial $P(X)$ are $\frac{\alpha^2}{\mu}$, $j\,\frac{\alpha^2}{\mu}$, $j^2\,\frac{\alpha^2}{\mu}$; the conjugates of α on $\mathbb{Q}$ are therefore $\alpha, j\alpha, j^2\alpha, \frac{\alpha^2}{\mu}$, $j\,\frac{\alpha^2}{\mu}$, $j^2\,\frac{\alpha^2}{\mu}$. The splitting field on $\mathbb{Q}$ of the polynomial $\mathrm{Irr}(\alpha,\mathbb{Q})$ is thus the field $\mathbb{Q}(\alpha,j,\mu) = k(\alpha,j) = K$ (cf., Question (2)(a)). Consequently ([Lang 1] Chapter VII) the extension $K|\mathbb{Q}$ is Galois; its degree is $[K:\mathbb{Q}] = [K:k][k:\mathbb{Q}] = 12$.

The field $\mathbb{Q}(\sqrt{m})$ is contained in K, and as $[\mathbb{Q}(\sqrt{m}):\mathbb{Q}] = 2$, $[K:\mathbb{Q}(\sqrt{m})] = 6$. We have $K = \mathbb{Q}(\sqrt{m},\sqrt{-3},\alpha)$. The degree of α over $\mathbb{Q}(\sqrt{m},\sqrt{-3})$ is therefore equal to three and $\mathrm{Irr}(\alpha,\mathbb{Q}(\sqrt{m},\sqrt{-3})) = X^3 - \mu^2\bar{\mu}$. This polynomial is not in $\mathbb{Q}(\sqrt{m})(X)$, consequently $\mathrm{Irr}(\alpha,\mathbb{Q}(\sqrt{m}))$ is a proper multiple of $X^3 - \mu^2\bar{\mu}$. The degree of α

over $\mathbb{Q}(\sqrt{m})$ is therefore greater than or equal to four and divides six, whence:

$$[\mathbb{Q}(\sqrt{m},\alpha):\mathbb{Q}(\sqrt{m})] = 6$$

and:

$$\mathrm{Irr}(\alpha,\mathbb{Q}(\sqrt{m})) = \mathrm{Irr}(\alpha,\mathbb{Q}) = (X^3 - \mu^2\bar{\mu})(X^3 - \bar{\mu}^2\mu).$$

From this we deduce the existence of a $\mathbb{Q}(\sqrt{m})$-automorphism s of K such that $s(\alpha) = \dfrac{\alpha^2}{\mu}$, whence $s(\mu^2\bar{\mu}) = s(\bar{\mu}^2\mu)$. Therefore since $\mu\bar{\mu} \in \mathbb{Q}$, $s(\mu) = \bar{\mu}$ and $s(\sqrt{-3m}) = -\sqrt{-3m}$. We have $s(\sqrt{-3}) = -\sqrt{-3}$. Moreover, $s^2(\alpha) = s\left(\dfrac{\alpha^2}{\mu}\right) = \dfrac{\alpha^4}{\mu^2\bar{\mu}} = \alpha$. As $K = \mathbb{Q}(\alpha,\sqrt{-3})$ we can conclude $s^2 = \mathrm{id}_K$.

(4): The group $\mathrm{G} = \mathrm{Gal}(K|\mathbb{Q})$ has order twelve; it has subgroups $G = \langle\sigma,\tau\rangle$, of order six, and $\langle s\rangle$ which has order two. The automorphism s is not one of the three elements of order two of G, $\tau,\tau\sigma,\tau\sigma^2$, because these leave $\sqrt{-3m}$ invariant. Therefore $s \notin G$, and so $\mathrm{G} = G \cup sG$.

To show that G is the direct product of $\langle s\rangle$ and G it suffices to show that $\sigma s = s\sigma$ and $\tau s = s\tau$, which is easily established by looking at the action of these automorphisms on $\sqrt{m}$, $\sqrt{-3}$ and α. G is therefore the direct product of G and $\langle s\rangle$.

Let N be the fixed field of s. The extension $N|\mathbb{Q}$ has degree six, and it is Galois since $\langle s\rangle$ is a normal subgroup of G; N contain $\mathbb{Q}(\sqrt{m})$. Furthermore, $\mathrm{Gal}(N|\mathbb{Q}) \simeq \mathrm{G}/\langle s\rangle$ ([Lang 1] Chapter VIII) and therefore $\mathrm{Gal}(N|\mathbb{Q}) \simeq G$. Thus we have obtained an extension N of $\mathbb{Q}(\sqrt{m})$ that is Galois over $\mathbb{Q}$, and such that $\mathrm{Gal}(N|\mathbb{Q})$ is isomorphic to the dihedral group of order six.

(Cf., J. Martinet and J.J. Payan: 'Sur les extensions cubiques non-galoisiennes des rationels et leur clôture galoisienne', *J. reine angew. Math.*, **228**, (1967), 15-37).

SOLUTION 4·4 (1) As the composition of two Galois extensions is Galois ([Lang 1] Chapter VIII), and as k_1, k_2 are quadratic and thus Galois, $K = k_1k_2$ is therefore Galois.

I (2) Since $k_1 \cap k_2 = \mathbb{Q}$ we have $[K:\mathbb{Q}] = [k_1:\mathbb{Q}][k_2:\mathbb{Q}] = 4$

I (3) From $K = k_2(\vartheta_1)$ it follows that every element of K can be written uniquely as $\lambda + \mu\vartheta_1$, $\lambda,\mu \in k_2$, and that the mapping $\sigma : \lambda + \mu\vartheta_1 \mapsto \lambda - \mu\vartheta_1$ is a k_2-automorphism of K (and therefore a $\mathbb{Q}$-automorphism of K). Similarly one sees that τ is the non-trivial k_1-automorphism of K. It is clear that $\mathrm{Gal}(K|\mathbb{Q})$ is made up of the $\mathbb{Q}$-automorphisms $\mathrm{id}_K, \sigma, \tau, \sigma\tau$. $\mathrm{Gal}(K/\mathbb{Q})$ is therefore isomorphic to the Klein four group.

I (4) Let k_3 be the subfield of K formed by the elements of K invariant under $\sigma\tau = \tau\sigma$; k_3 has degree two on $\mathbb{Q}$. This is the only quadratic subfield of K outside of k_1 and k_2, because $\mathrm{Gal}(K/\mathbb{Q})$ only possesses three subgroups of index two.

$\mathbb{Q}(\sqrt{m_1m_2})$ is contained in K, is quadratic over $\mathbb{Q}$, and is distinct from k_1 and k_2, hence we have $k_3 = \mathbb{Q}(\sqrt{m_1m_2})$. Let us denote by m_3 the square-free part of m_1m_2. It is clear that $k_3 = \mathbb{Q}(\sqrt{m_3})$ and that $\Delta_{k_3/\mathbb{Q}} = m_3$ (*resp.* $4m_3$) if $m_3 \equiv 1 \bmod 4$ (*resp.* $m_3 \not\equiv 1 \bmod 4$).

I (5) The calculation of $\Delta_{k_3/\mathbb{Q}}$ shows that $\Delta_{k_1/\mathbb{Q}}, \Delta_{k_2/\mathbb{Q}}, \Delta_{k_3/\mathbb{Q}}$ can have no other common prime divisor than two. A prime ideal (p) of $\mathbb{Q}$, with $p \neq 2$ cannot, therefore, be ramified in k_1, k_2, k_3 simultaneously.

The example $m_1 = 2$, $m_2 = -1$, and thus $m_3 = -2$, shows that the same does not hold for 2.

I (6) Let $(1,\alpha_i)$ be a $\mathbb{Z}$-basis of O_{k_i}. We have:

$$\Delta_{k_i|Q} = \det\begin{bmatrix} 1 & \alpha_i^{(1)} \\ 1 & \alpha_i^{(2)} \end{bmatrix}^2 .$$

It is known that Δ_{K/k_j} (if $j \neq i$) divides $\Delta_{k_i/\mathbb{Q}}$ (by the review at the beginning of the section). From this it follows that if $j \neq i$, Δ_{K/k_j} divides $\Delta_{k_i/\mathbb{Q}}$. Let (p) be a prime ideal of $\mathbb{Q}$ ramified in $K/\mathbb{Q}$. Let us assume it is not ramified in $k_j/\mathbb{Q}$. At least one of its prime divisors in k_j is therefore ramified in K/k_j and consequently divides Δ_{K/k_j}. As a result p divides $\Delta_{k_i/\mathbb{Q}}$ for $i \neq j$, and therefore p is ramified in the two extensions $k_i/\mathbb{Q}$, with $i \neq j$.

II (1) The class number of O_{k_1}, the ring of Gaussian integers, is equal to one. The O_{k_1}-module O_K, which is of finite type and torsionfree over a principal ideal domain, is therefore free ([Sam] Chapter 1).

II (2) We have $k_3 = \mathbb{Q}(\sqrt{-m_2})$ and $\Delta_{k_3|\mathbb{Q}} = -4m_2$.

The ideals of O_{k_3} which are possibly ramified in the extension K/k_3 are the prime divisors of (2) in O_{k_3}; indeed, as Δ_{K/k_3} divides $\Delta_{k_1/\mathbb{Q}} = -4$, only a divisor of two can be ramified in the extension K/k_3. As Δ_{K/k_3} divides $\Delta_{k_2/\mathbb{Q}} = m_2$, which is prime to two, the prime divisors of two in O_{k_3} are not ramified in the extension K/k_3. No prime ideal of O_{k_3} is ramified in the extension K/k_3.

III As the extension $K/\mathbb{Q}$ is Galois a prime ideal (p) of $\mathbb{Q}$ splits in O_K into the form:

$$pO_K = \left(\prod_{i=1}^{g} R_i\right)^e,$$

with the degree of R_i equal to f and $efg = 4$ ([Sam] Chapter 6). As the group $\mathrm{Gal}(K/\mathbb{Q})$ is not cyclic p cannot be inert in the extension of $K/\mathbb{Q}$ (that is to say, $e = g = 1$) ([Sam] Chapter 6).

On the other hand, a prime number $p \neq 2$ is not ramified in one of the extensions $k_i/\mathbb{Q}$ ($i = 1,2,3$) (Question I(5) above). Consequently the extension K/k_j ($j \neq i$) is not ramified over p ([Wei] 3.4.6). Therefore if $p \neq 2$ one has $e = 1$ or $e = 2$.

It is then easily shown that, depending on the values of p, one of the following must hold:

$$pO_K = PP^{\sigma}P^{\tau}P^{\sigma\tau} \qquad (e = 1, f = 1, g = 4),$$

$$pO_K = PP' \qquad (e = 1, f = 2, g = 2),$$

$$pO_K = P^2P'^2 \qquad (e = 2, f = 1, g = 2),$$

$$pO_K = P^2 \qquad (e = 2, f = 2, g = 1).$$

In addition, 2 may have the decomposition:

$$2O_K = P^4 \qquad (e = 4, f = 1, g = 1).$$

SOLUTION 4·5 I (1) The discriminant of the field is -3, and the Minkowski constant is $M = \frac{4}{\pi}\frac{1}{2}\sqrt{3}$. The ring A is therefore a principal ideal domain (moreover, it is known to be Euclidean, cf., [Har] Chapter XII).

Since $p \equiv 1 \bmod 3$ we have $\left(\frac{p}{3}\right) = 1$. From this (by the Quadratic Reciprocity Law, [Sam] Chapter 5) we have:

$$\left(\frac{-1}{p}\right) = \left(\frac{-1}{p}\right)\left(\frac{3}{p}\right) = \left(\frac{-1}{p}\right)\left(\frac{p}{3}\right)(-1)^{(p-1)/2} = 1.$$

As a result p is split in the extension $k/\mathbb{Q}$. Let us write $pA = \mathfrak{p}\bar{\mathfrak{p}}$. As A is a principal ideal domain the ideal $\mathfrak{p}$ is generated by an element of A that can be written $\alpha_0 + j\beta_0$, $\alpha_0, \beta_0 \in \mathbb{Z}$, since $(1,j)$ is a basis of the $\mathbb{Z}$-module A. We have:

$$p\mathbb{Z} = N_{k/\mathbb{Q}}(\mathfrak{p}) ,$$

and from this it follows that:

$$p = N_{k/\mathbb{Q}}(\alpha_0 + j\beta_0).$$

I (2) Let (α,β) be a pair of rational integers such that $N_{k/\mathbb{Q}}(\alpha + j\beta) = p$; the ideal $(\alpha + j\beta)A$ is equal to $\mathfrak{p}$ or to $\bar{\mathfrak{p}}$; as the only units ε of the ring A are $1, \pm j, \pm j^2$ there are only a finite number of pairs (α,β), given by:

$$\alpha + j\beta = \varepsilon(\alpha_0 + j\beta_0) \quad \text{and} \quad \alpha + j\beta = \varepsilon(\alpha_0 + j^2\beta_0).$$

A pair of rational integers (x,y) such that $p = \frac{3x^2 + y^2}{4}$ is a a pair such that $N_{k/\mathbb{Q}} \frac{(y + x\sqrt{-3})}{2} = p$ (where $\sqrt{-3} = 2j + 1$). Therefore $p = \frac{3x^2 + y^2}{4}$ is equivalent to $N_{k/\mathbb{Q}} \frac{((y - x) + 2jx}{2} = p$. Hence there are a finite number of pairs (x,y). Starting from the above pairs (α,β) they are given by:

$$y - x = 2\alpha, \qquad x = \beta.$$

One looks for pairs (x,y) such that $x \equiv 0 \bmod 3$ and $y \equiv 1 \bmod 3$, and this reduces to looking for pairs (α,β) such that $\beta \equiv 0 \bmod 3$ and $\alpha \equiv -1 \bmod 3$. Twelve pairs (α,β) arise, by multiplication by -1 and by symmetry $(\alpha,\beta) \mapsto (\beta,\alpha)$, from the three pairs $(\alpha_0,\beta_0), (-\beta_0,\alpha_0 - \beta_0), (\beta_0 - \alpha_0, -\alpha_0)$. One knows that that α_0 and β_0 satisfy the congruence $\alpha_0^2 + \beta_0^2 - \alpha_0\beta_0 \equiv 1 \bmod 3$. From this one deduces, by enumerating all possible cases, that there are exactly two pairs (α,β) such that $\beta \equiv 0 \bmod 3$ and $\alpha \equiv -1 \bmod 3$. Hence, by imposing the additional condition $\beta > 0$, we have the existence and uniqueness of a pair (α,β).

II (1) The extension $\mathbb{Q}'/\mathbb{Q}$ is cyclic of degree $p - 1$ ([Sam] Chapters 2 and 6, or [Wei] Chapter 7, or [Has] 27); as $p \equiv$

1 mod 3 its Galois group has an unique subgroup of index three, hence there is an unique subfield K of $\mathbb{Q}'$ of degree three over $\mathbb{Q}$ that is cyclic. On the other hand p is totally and tamely ramified in $\mathbb{Q}'/\mathbb{Q}$, and similarly in K, whence $|D| = p^{3-1} = p^2$ (cf., [Wei] Chapitre 3 and the references above: p is the only ramified prime in $K/\mathbb{Q}$). Furthermore, K is totally real, for it is contained in the maximal real subfield $\mathbb{Q}'_0$ of $\mathbb{Q}'$, which has degree $\frac{p-1}{2}$ over $\mathbb{Q}$. Thus the discriminant D is positive, and therefore $D = p^2$.

II (2) Let us denote by $i \to s_i$ the isomorphism of $(\mathbb{Z}/p\mathbb{Z})^*$ onto the Galois group of $\mathbb{Q}'/\mathbb{Q}$ defined by $s_i(\omega) = \omega^i$. The group $\mathrm{Gal}(\mathbb{Q}'/K)$ is the subgroup H' of $\mathrm{Gal}(\mathbb{Q}'/\mathbb{Q})$, the image under this isomorphism of the group $H = (\mathbb{Z}/p\mathbb{Z})^{*3}$.

Since $\vartheta = \sum_{s_i \in H'} s_i(\omega)$, ϑ is invariant under every element of H'. The same holds for ϑ' and ϑ'', therefore $\vartheta, \vartheta', \vartheta'' \in K$. They are conjugate, for $s_t(\vartheta) = \vartheta'$ and $s_{t^2}(\vartheta) = \vartheta''$. From this it follows that $\mathbb{Q}(\vartheta) = \mathbb{Q}(\vartheta') = \mathbb{Q}(\vartheta'') \subset K$. It remains to see that $\vartheta \notin \mathbb{Q}$. If this were the case we would have $\vartheta = \vartheta' = \vartheta''$, whence $3\vartheta = \sum_{1 \leq i \leq p-1} \omega^i = -1$ (for ω is a root of $\sum_{0 \leq i \leq p-1} X^i = 0$), and ϑ would not be integral.

II (3) We have just seen that $\vartheta + \vartheta' + \vartheta'' = -1$. On the other hand we have:

$$\vartheta\vartheta' + \vartheta'\vartheta'' + \vartheta''\vartheta = \sum_{i,j \in H} \omega^{i+tj} + \omega^{ti+t^2 j}.$$

We have a sum of powers of ω containing each term ω^h the same number of times, in fact this sum can be written $\sum_{1 \leq i \leq p-1} n_i \omega^i$, with n_i an integer, and it is invariant under every element s of $\mathrm{Gal}(\mathbb{Q}'/\mathbb{Q})$. From this one deduces (for example by considering the linear system in the corresponding n_i's $(1 \leq i \leq p-1)$, that $n_1 = n_2 = \cdots = n_{p-1}$. This sum consists of $3((p-1)/3)^2$

terms; thus we have:

$$\vartheta\vartheta' + \vartheta'\vartheta'' + \vartheta''\vartheta = \frac{1}{p-1}\, 3 \left(\frac{p-1}{3}\right)^2 \sum_{1\leqslant i\leqslant p-1} \omega^i$$

$$= \frac{p-1}{3} \sum_{1\leqslant i\leqslant p-1} \omega^i = -\frac{p-1}{3}\,.$$

II (4) We have:

$$\vartheta\vartheta'\vartheta'' = \sum_{h,i,j\in H} \omega^{h+ti+t^2j},$$

and we consider the terms in this sum equal to $1, \omega, \ldots, \omega^{p-1}$. The number of terms equal to one in the expression:

$$\omega^h \sum_{i,j\in H} \omega^{ti+t^2j} \qquad (h\in H)$$

is independent of the choice of $h \in H$, as is seen by applying the automorphism s_u, where $u \in H$. With the notations of the problem there are therefore $k - r$ terms equal to one in $\vartheta\vartheta'\vartheta''$. This product comprises $\left(\frac{p-1}{3}\right)^3$ terms; hence there remain $\left(\frac{p-1}{3}\right)^3 - kr = r^2 - kr$ other terms, and we see that each of these occur the same number of times. Thus we have:

$$\vartheta\vartheta'\vartheta'' = kr - \frac{1}{p-1}(r^3 - kr) = kr - \frac{r^2-k}{3},$$

$$\vartheta\vartheta'\vartheta'' = \frac{kp - r^2}{3}\,.$$

The minimal polynomial of ϑ is thus the polynomial:

$$P(x) = x^3 + x^2 - rx + \frac{r^2 - kp}{3}\,.$$

It is known that the discriminant of P is equal to the product of p^2, the discriminant of the field, with the square of an integer a; let us calculate it directly. We have:

$$\Delta = r^2 - 18r\,\frac{r^2 - kp}{3} - 4\,\frac{r^2 - kp}{3} + 4r^3 - 27\,\frac{r^2 - kp}{3}$$

$$= -\,3k^2p^2 + 6r^2 + 6r + \frac{4}{3}\,kp - (3r^4 + 2r^3 + \frac{1}{3}\,r^2).$$

One finds the second degree equation:

$$3p^2k^2 - 6r^2 + 6r + \frac{4}{3} + (3r^4 + 2r^3 + \frac{1}{3}\,r^2) + p^2a^2 = 0.$$

As kp is a rational number the discriminant of this equation is the square of a number b' defined up to sign, which after expanding leads to the relation:

$$48r^3 + 48r^2 + 16r + \frac{16}{9} - 12p^2a^2 = b'^2.$$

Replacing r by $\frac{p-1}{3}$ then yields:

$$\frac{16}{9}\,p^3 = b'^2 + 12p^2a^2,$$

or again:

$$4p = \left(\frac{3b'}{2p}\right)^2 + 27a^2$$

The number $\frac{3b'}{2p}$ is an integer b, and we have the relation $4p = b^2 + 27a^2$. On the other hand, the calculation of kp gives:

$$kp = \frac{1}{6}\,(6r^2 + 6r + \frac{4}{3} + b') = r^2 + r + \frac{2}{9} + \frac{b'}{6}$$

$$= \frac{p^2 - 2p + 1}{9} + \frac{p-1}{3} + \frac{2}{9} + \frac{pb}{9} = \frac{p^2 + p + pb}{9}.$$

Whence:

$$r^2 - kp = \frac{(p - 1)^2 - p^2 - p - pb}{9} = \frac{-3p - pb + 1}{9},$$

which gives the result.

III (1) (i) <=> (ii): q is a product of three factors in K if and only if the decomposition field of α in Q' contains K ([Mac]) Chapter 5, or [Che] Chapter 2), that is to say, if the index of the decomposition group of q in G is divisible by three (the index is the number of prime factors of q in Q').

(ii) <=> (iii): Let f be the smallest positive integer satisfying the congruence $q^f \equiv 1 \bmod p$, and let $g = \frac{p - 1}{f}$. The theory of cyclotomic fields shows that q in $\mathbb{Q}'$ is the product of g prime ideals of degree f ([Wei] Chapter 7). We have:

$$q^{(p-1)/3} \equiv 1 \bmod p,$$

if and only if f divides $\frac{p - 1}{3}$, which is equivalent to $3|g$.

(iii) <=> (iv): If $q \equiv x^3 \bmod p$ we have:

$$q^{(p-1)/3} \equiv x^{p-1} \equiv 1 \bmod p,$$

by Fermat's Little Theorem. Conversely, let r be an integer whose class modulo p generates $(\mathbb{Z}/p\mathbb{Z})^*$. For a suitable integer m we have has $q \equiv r^m \bmod p$, whence:

$$r^{\frac{p-1}{3}m} \equiv 1 \bmod p;$$

$p - 1$ then divides $\frac{p - 1}{3}m$, therefore 3 divides m, and we can take $x = r^{m/3}$.

III (2) Let us assume that P has a triple root α modulo q.

There we have:

$$X^3 + X^2 - \frac{p-1}{3}X - \frac{p(3+b)-1}{27} \equiv (X-\alpha)^3 \bmod q,$$

which implies the congruences $3\alpha \equiv -1 \bmod q$ and $3\alpha^2 \equiv -\frac{p-1}{3}$ mod q. From this we deduce $p = q$.

As the extension $K/\mathbb{Q}$ is Galois a prime number q decomposes in K into a product of prime ideals in either of the following ways:

$$(q) = q_1^3, \qquad (q) = q_1 q_1' q_1'', \qquad \text{or} \qquad (q) = q_3,$$

the q_i's being ideals of inertial degree i. The number p decomposes following the first form, and a prime number $q \neq p$, not being ramified, decomposes in one of the remaining ways.

Let $q \neq p$. If the congruence $P(x) \equiv 0 \bmod q$ has a solution at least one of the roots is simple, and is recovered as a simple root in the q-adic number field. q then has a prime factor of degree one in K, and q splits completely. If, on the contrary, the congruence does not have a solution modulo q, P is irreducible in the q-adic number field, and therefore q is inert in K ([CaF] Chapter 2, or [Wei] Chapter 2).

REMARK: Modulo p, P has a triple root equal to $-\frac{1}{3}$.

III (3) The integer $\frac{p-1}{3}$ is even, and $\frac{p(3+b)-1}{27} \equiv b$ mod 2. Therefore $P(X) \equiv X^3 + X^2 + b \bmod 2$, and hence $P(0) \equiv P(1) \equiv b \bmod 2$. P therefore has a root modulo 2 if and only if b (and thus a) is even. The rest of the Question immediately follows from Question III(1).

REMARK: If P has a root mod 2, 0 is a double root and 1 is a simple root.

III (4) Let us set $P_1(X) = 27P\,\frac{X-1}{3}$. The roots of P_1 are $3\vartheta + 1, 3\vartheta' + 1, 3\vartheta'' + 1$. It is immediate that:

$$P_1(X) = X^3 - 3pX - pb.$$

For $q \neq 3$, P and P_1 have the same number of roots modulo q. From this it immediately follows that for $q \neq 3$ the condition "q divides b" is sufficient for q to split, and therefore for:

$$q^{(p-1)/3} \equiv 1 \bmod p.$$

III (5) Let us set $b = 1 + 3b'$. Without difficulty one shows the congruences:

$$\frac{p-1}{3} \equiv -(b' + 1) \bmod \varepsilon$$

and:

$$\frac{p(3+b)-1}{27} \equiv b'^3 - b'^2 + b' + a^2 \bmod 3.$$

Hence we have:

$$P(X) \equiv X^3 + X^2 + (b' + 1)X - (a^2 + b'^3 - b'^2 + b') \bmod 3.$$

If 3 divides a we see that $P(0) \equiv 0 \bmod 3$ if $b' \equiv 0,-1 \bmod 3$, and then that $P(1)$ is zero for $b' \equiv 1 \bmod 3$. Consequently P has a root modulo 3. From this one deduces that 3 splits in K, and therefore that $3^{(p-1)/3} \equiv 1 \bmod p$. In addition we see that 3 does not divide a, and P has no root modulo 3. Consequently for $q = 2$ and $q = 3$ the condition "q divides ab" is necessary and sufficient for the congruence $q^{(p-1)/3} \equiv 1 \bmod p$ to be satisfied.

REMARK 1: Question III(1) can be solved using the Cubic Reciprocity Law (cf., [Mor] Chapter 15).

REMARK 2: For arbitrary q, the condition "q divides ab" is sufficient for the congruence $q^{(p-1)/3} \equiv 1 \bmod p$ to hold. To see this consider the polynomial $Q(X) = X^3 - pX - pa$. Its discriminant is p^2b^2; the splitting field K' of the polynomial Q on $\mathbb{Q}$ is therefore an abelian field of dgree 3 over $\mathbb{Q}$. The prime number p is totally ramified in K' (because q is an Eisenstein polynomial for p); 2 is not ramified since $Q(X) \equiv X(X-1)^2 \bmod 2$ when 2 divides b. If q is a prime factor of b other than 2 we have $q \neq 3$.

$$\frac{8}{a^3} Q\left(\frac{3a}{2} X\right) \equiv X^3 - 3X - 2 \bmod q,$$

a congruence which has the simple root $X = 2$, hence q is split. Consequently the field K' has p^2 as discriminant; as it is an abelian field over $\mathbb{Q}$ it is contained in $\mathbb{Q}'$ (Kronecker-Weber Theorem, cf., [Wei] Chapter 7); from this it follows that $K' = K$.

One concludes by noticing that modulo the prime factors q of a, $Q(X) \equiv X(X^2 - p)$, and thus q is decomposed in K.

REMARK 3: The equivalence "q divides ab" if and only if $q^{(p-1)/3} \equiv 1 \bmod p$" is false in general, but is valid for small values of q (notably $q = 5$).

SOLUTION 4·6 I (1) The discriminant of $X^3 + pX + q$ is $-4p^3 - 27q^2$, whence $\Delta(1,\vartheta,\vartheta^2) = -27d^2$.

I (2) If p divides d the polynomial $X^3 - d$ is an Eisenstein polynomial for p ([Wei] Chapter 3). From this one deduces that p is totally ramified and that $1,\vartheta,\vartheta^2$ is a local basis at p.

If 3 does not divide d then:

$$f(X + 1) = X^3 + 3X^2 + 3X + 1 - d,$$

and:

$$f(X - 1) = X^3 - 3X^2 + 3X - (1 + d).$$

Therefore one of these polynomials for which the constant term is divisible by 3 is an Eisenstein polynomial at 3, which completes the proof.

I:(3): The lattice with basis $1,\vartheta,\vartheta^2$ coincides locally with the ring of integers O_K for every prime number (for 3 notice that $\mathbb{Z}[\vartheta+1] = \mathbb{Z}[\vartheta-1] = \mathbb{Z}[\vartheta]$). Therefore $O_K = \mathbb{Z}[\vartheta]$ and $D = \Delta(1,\vartheta,\vartheta^2) = -27d^2$.

REMARK 1: In the general case d can be assumed to have no cubic factors, which allows us to write $d = fg^2$, with f,g relatively prime and square-free. $\frac{\vartheta^2}{g}$ is then a root of the polynomial, which is an Eisenstein polynomial at p for every prime divisor of g. Thus when $d \equiv \pm 1 \bmod 9$ $1,\vartheta,\frac{\vartheta^2}{g}$ is an integral basis of K, and hence:

$$D = \Delta(1,\vartheta,\frac{\vartheta^2}{g}) = g^{-2}\Delta(1,\vartheta,\vartheta^2) = -27(fg)^2.$$

Moreover, by replacing ϑ with $\frac{\vartheta^2}{g}$ we can assume that 3 does not divide g. Moreover we can further assume $g \equiv 1 \bmod 3$.

REMARK 2: Let us assume $d \equiv \pm 1 \bmod 9$. It is known that $1,\vartheta,\frac{\vartheta^2}{g}$ is a basis locally for every prime number $p \neq 3$. As the exponents of 3 in the discriminant D of K and the discriminant $\Delta(1,\vartheta,\frac{\vartheta^2}{g})$ is not even, we have:

$$D = -27(fg)^2 \quad \text{or} \quad D = -3(fg)^2.$$

Thus in all cases 3 is ramified in K. Hence:

$$(3) = \mathfrak{p}^3 \quad \text{or} \quad (3) = \mathfrak{p}'\mathfrak{p}''^2$$

($\mathfrak{p},\mathfrak{p}',\mathfrak{p}''$ being prime ideals of degree one). In the first case the extension is not tamely ramified, it then follows that $D = -27(fg)^2$ ([Wei] Chapter 7). In the second case the extension is tamely ramified, we then have $D = -3(fg)^2$.

Let us show that the first case is impossible. Begin by changing d to $-d$, we may also assume $d \equiv 1 \bmod 9$. We have $\vartheta^3 = d$, therefore $\vartheta^3 \equiv 1 \bmod p^6$, which may be written:

$$(\vartheta - 1)^3 + 3\vartheta(\vartheta - 1) \equiv 0 \bmod p^6,$$

whence:

$$(\vartheta - 1)((\vartheta - 1)^2 + 3\vartheta) \equiv 0 \bmod p^6.$$

That implies $\vartheta - 1 \equiv 0 \bmod p$, and therefore $\vartheta - 1 \equiv 0 \bmod p^2$ (otherwise $(\vartheta - 1)^3 + \vartheta(\vartheta - 1) \not\equiv 0 \bmod p^4$). Consequently $(\vartheta - 1)^3 \equiv 3\vartheta$ is divisible by p^3 but not p^4. Hence we have $(\vartheta - 1) \equiv 0 \bmod p^3$, therefore $\frac{\vartheta - 1}{3}$ is integral in K, which is impossible, because:

$$\Delta\left(1, \frac{\vartheta - 1}{3}, \left(\frac{\vartheta - 1}{3}\right)^2\right) = \frac{1}{27^2}\,\Delta(1,\vartheta,\vartheta^2)$$

is not an integer.

We are therefore in the second case. The identity:

$$\vartheta^3 - d \equiv \left(\frac{(\vartheta - 1)}{(\vartheta - 1)^2} + 3\vartheta\right) \bmod 3^2$$

proves the congruence $\vartheta \equiv 1 \bmod p'p''$. But there exists an integer of the form $\frac{1}{3}(a + b\vartheta + c\frac{\vartheta^2}{g})$ with a,b,c not all divisible by 3. As we assume that $g \equiv 1 \bmod 3$,

$$a + b\vartheta + c\frac{\vartheta^2}{g} \equiv a + b + c \bmod p'p'',$$

and necessarily therefore $a + b + c \equiv 0 \bmod 3$.

Since $\vartheta \not\equiv \pm 1 \bmod 3$ by the resoning used in the first case. none of the a,b,c is divisible by three. From this it follows that $\frac{1}{3}(1 + \vartheta + \frac{\vartheta^2}{g})$ is integral, and consequently that $1,\vartheta,$

$\frac{1}{3}(1 + \vartheta + \frac{\vartheta^2}{g})$ is an integral basis.

II (1) If the congruence $x^3 - d \equiv 0 \bmod p$ has a root in $\mathbb{Z}$ this root is simple (since $3x^2 \not\equiv 0 \bmod p$), and thus lifts into $\mathbb{Q}_p$ as a simple root of $x^3 - d$. This implies the existence of a prime factor of degree one of p ([Wei] Chapter 2, or [CaF] Chapter 2). Conversely, if p splits in K it has a prime factor of degree one; the polynomial $X^3 - d$ then has a root in $\mathbb{Q}_p$, whence a root in $\mathbb{Z}$ modulo p. Therefore p is inert in K if and only if the congruence $x^3 \equiv d \bmod d$ does not have a solution in $\mathbb{Z}$.

II (2) Let $p \equiv -1 \bmod 3$. The mapping $x \mapsto x^3$ of $(\mathbb{Z}/p\mathbb{Z})^*$ into itself is an injective homomorphism since $\mathrm{card}(\mathbb{Z}/p\mathbb{Z})^* = p - 1$ is not divisible by three. It is therefore surjective, that is to say, there exists $x_0^3 \equiv d \bmod p$. Hence p splits in K by Questtion II(1).

Since 3 does not divide $p - 1$, -3 is not a square in $\mathbb{Q}_p$; for if $p \not\equiv 2$, $-3 \equiv \lambda^2 \bmod p$ implies:

$$\left(\frac{-1 + \lambda}{2}\right)^3 \equiv 1 \bmod p,$$

and so an element of order three in $(\mathbb{Z}/p\mathbb{Z})^*$; -3 is not a square in $\mathbb{Q}_2$, for $-3 \not\equiv 1 \bmod 8$.

The discriminant $D = -27d^2$ is therefore not a square in $\mathbb{Q}_2$, consequently $X^3 - d$ has only one root in $\mathbb{Q}_p$, which gives the result.

II (3) If $q \equiv 1 \bmod 3$, $(\mathbb{Z}/q\mathbb{Z})^*$ has an element of order three; let ρ be this element. Then:

$$(\rho - 1)(\rho^2 + \rho + 1) = 0 \quad \text{and} \quad (2\rho + 1)^2 = 4(\rho^2 + \rho + 1) = -3.$$

The polynomial $X^2 + 3$ is therefore the product of two polynomials of degree one in $\mathbb{Z}/q\mathbb{Z}[X]$ and so (by Hensel's Lemma, cf., [Wei] Chapter 3) in $\mathbb{Q}_q[X]$. From this it follows that -3 is a square in $\mathbb{Q}_p$, so D is a square in $\mathbb{Q}_q$.

There are, therefore, two possible cases: either the polynomial $X^3 - d$ does not have a root in $\mathbb{Q}_q$ and q is inert, or the polynomial $X^3 - d$ is the product of three polynomials of degree one of $\mathbb{Q}_q[X]$, and q in K is the product of three prime ideals of degree one.

III (1) Let $f = a + b\vartheta + c\vartheta^2$. The conjugates of f in an algebraic closure of K are:

$$f' = a + bj\vartheta + cj^2\vartheta^2 \quad \text{and} \quad f'' = a + bj^2\vartheta + cj\vartheta^2$$

$$(j^3 = 1,\ j \neq 1).$$

We immediately have:

$$f'f'' = a^2 - bcd - (ab - dc^2)\vartheta - (ac - db^2)\vartheta^2,$$

whence:

$$N_{K/\mathbb{Q}}(f) = ff'f'' = a^3 + db^3 + d^2c^3 - 3dabc.$$

REMARK: One also has $\mathrm{Tr}_{K/\mathbb{Q}}(f'f'') = 3(a^2 - bcd)$, which permits the verfication of the result of Remark 2.

III (2) If p has a principal prime factor $\mathfrak{p}$ in K it has a principal prime factor of degree one, for if $\mathfrak{p}$ is of degree two $p\mathfrak{p}^{-1}$ is principal and of degree one. Therefore let $\mathfrak{p} = (\pi)$ be a principal prime factor of degree one. The norm of $\mathfrak{p}$ is the prime ideal generated by $N_{K/\mathbb{Q}}(\pi)$. Hence we have $N(\pi) = \pm p$, and so $N(\pm\pi) = p$. The equation:

$$a^3 + db^3 + d^2c^3 - 3dabc = p$$

has therefore a solution. Conversely, if:

$$a^3 + db^3 + d^2c^3 - 3dabc = p$$

has a solution in $\mathbb{Z}$, let us set:

$$\pi = a + b\vartheta + c\vartheta^2;$$

we have $N_{K/\mathbb{Q}}(\pi) = p$. As π is an integer, the principal ideal generated by π is a prime ideal of degree one dividing p.

III (3) We have:

$$M = \frac{4}{\pi}\frac{6}{27}3d\sqrt{3} < 1.48d.$$

For $d = 2$, $M < 3$. But $(2) = (\vartheta)^3$. As every class contains an ideal with norm at most M, we deduce that O_K is principal. (One shows similarly that O_K is principal for $d = 3$ or $d = 5$; if $d = 5$ one uses the equalities $N(2 - \vartheta) = 3$, $N(1 + \vartheta) = 6$).
$N(1 + \vartheta) = 6$).

IV (1) For $a \in \mathbb{Z}$, $a \not\equiv 0 \bmod 7$, we have $a^3 \equiv \pm 1 \bmod 7$, hence the impossibility of the congruence:

$$a^3 + 7b^3 + 7^2c^3 - 21abc \equiv m \bmod 7$$

whenever $m \equiv \pm 2$ or $\pm 3 \bmod 7$. In particular, by applying this result to $m = 3$ we see (Question III(2) above) that the ideal p_3 is not principal (we have $(3) = p_3^3$).

By Question II(2) above we have $(2) = p_2 p_2'$. By taking $a = -1$, $b = 1$, $c = 0$ in Question III(2) above it is seen that $N(\vartheta - 1) = 6$. Therefore $(\vartheta - 1) = p_2 p_3$ holds. On the other hand, for $d = 7$ $M < 11$. Let us denote the class of p_3 by h; we have $p_2 \in h^2$, $p_2' \in h$. As $N_{K/\mathbb{Q}}(2 + \vartheta) = 15$ we have $(2 + \vartheta) = p_3 p_5$, whence $p_5 \in h^2$, $p_5' \in h$. Moreover, p_2^2 and p_2' are the only ideals of norm 4; $p_2 p_3$ is the only ideal of norm 6; p_7 is the only ideal of norm 7, since $(7) = (\vartheta)^3$; p_2^3 and (2) are the only ideals of norm 8; p_3^2 is the only ideal of norm 9; lastly p_2p_5 is the only ideal of norm 10.

One can therefore conclude that the class number is three, and the class group is cyclic, generated by the class of p_3.

IV:(2): By the preceding, one and only one of the three ideals $p_p, p_3p_p, p_3^2p_p$ is principal. If $p \equiv \pm$ mod 7 the equations $N_{K/\mathbb{Q}}(f) = 3p$ and $N_{K/\mathbb{Q}}(f) = 9p$ (where $f = a + b\vartheta + c\vartheta^2$) are impossible. Therefore p_3p_p and $p_3^2p_p$ are not principal. From this it follows that p_p is principal, and therefore p'_p as well, since $p_pp'_p = (p)$. Conversely, if p_p is principal the equation $N_{K/\mathbb{Q}}(f) = p$ has a solution, which implies $p \not\equiv \pm 1$ mod 7. If $p \equiv \pm 2$ mod 7 we see that p_p and $p_3^2p_p$ are not principal, and hence p_3p_p is, from which we conclude $p_p \in h^2$ and $p'_p \in h$. If $p \equiv \pm 3$ mod 7 we see that p_p and p_3p_p are not principal, and hence $p_3^2p_p$ is, and so $p_p \in h$, $p'_p \in h^2$.

IV:(3): Let us write $qO_K = qq'q''$ and let $h^r, h^{r'}, h^{r''}$ be the respective classes of q, q', q''. The products $p_3^{2r}q, p_3^{2r'}q', p_3^{2r''}q''$ are principal. From this it follows that the equations:

$$a^3 + 7b^3 + 7^2c^3 - 27abc \equiv 3^{2r}q \qquad (resp.\ 3^{2r'}q,\ 3^{2r''}q)$$

have solutions. Thus we have:

$$3^{2r}q \equiv 2^rq \equiv \pm 1 \bmod 7 \text{ (analogous congruences in } r' \text{ and } r''),$$

whence:

$$4^r \equiv \pm q \bmod 7, \qquad 4^{r'} \equiv \pm q \bmod 7, \qquad 4^{r''} \equiv \pm q \bmod 7.$$

Since 4 has order three in $(\mathbb{Z}/7\mathbb{Z})^*$ it is seen that $r \equiv r' \equiv r''$ mod 3, hence the fact that q, q', q'' are in the same class. We can then apply the method of the preceding question. In particular we find that q, q', q'' are principal if and only if $q \equiv \pm 1$ mod 7.

The results of these two questions can be expressed in the following way: Let p ($p \neq 3$ and $p \neq 7$ be a prime number split in K, and let $\mathfrak{p}$ be a prime factor of p in K of degree one. Let $a_{\mathfrak{p}}$ be a representative modulo 7 of $\frac{p^2 - 1}{2}$, $\mathfrak{p}_{\mathfrak{p}}$ then belongs to the class $h^{-a_{\mathfrak{p}}}$.

EXAMPLES: Looking modulo 13 at the polynomial $X^3 - 7$ we see that thirteen is inert. On the other hand, modulo 19 we have $4^3 - 7 = 57 \equiv 0$, consequently nineteen is the product of three factors of degree one, the class of which is h^2.

COMMENTARY: (Using Class-Field Theory): First of all let us recall the following result: Let L be an extension of $\mathbb{Q}$. In order that there be a positive integer f such that all prime numbers prime to f and congruent to a modulo f have the same decomposition in L, $L/\mathbb{Q}$ must be abelian. Furthermore, only the prime divisors of f can be ramified in L (cf., [Che] Introduction, [CaF] Chapter XI).

As a result, the extension $K/\mathbb{Q}$ of the Problem not being abelian, the fact that a prime number p congruent to one modulo three is inert or split in K cannot be decided with the aid of congruences modulo an integer f.

Furthermore, Class-Field Theory (cf., references above) enables us, for all number fields, to construct a bijection between the subgroups of the class group of K and the unramified abelian extensions of K (at every prime ideal and at infinity), the prime ideals of the subgroups considered being precisely those which are completely split in the corresponding unramified extension, the relative Galois group being isomorphic to the subgroup.

The results of Part IV of this Exercise show that the principal prime ideals of K (other than $\mathfrak{p}_7$) are the prime ideals for which the norm to $\mathbb{Q}$ is congruent to ±1 modulo 7. As the group of the ideal classes of K is of order three, class field theory says that the field K has an abelian extension C of degree three (the "class field" of K) in which the prime ideals of K that split completely are exactly the principal prime ideals of K.

Let us see whether there is a field L, abelian and of degree three over $\mathbb{Q}$, such that $\mathbb{C}$ coincides with $K.L$, the field composed of K and of L.

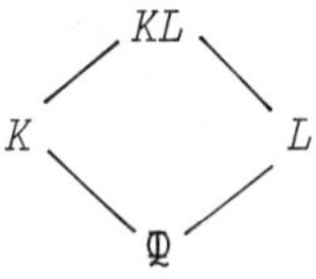

Let L be an abelian field and let it be of degree three over $\mathbb{Q}$. Let $\mathfrak{p}$ be a prime ideal of K over p of degree one. The prime ideal $\mathfrak{p}$ is completely split in $K.L$ if and only if p is completely split or ramified in L (in fact if $\mathfrak{P}$ is a prime divisor of $\mathfrak{p}$ in $K.L$ $\mathfrak{p}$ is completely split if and only if $\mathfrak{P}$ is of degree one in the extension $K.L/\mathbb{Q}$). The field L therefore satisfies the following conditions: 7 must be the only prime number ramified in L, and the prime numbers p split in L must be the prime numbers p congruent to ± 1 modulo 7. Exercise 4·5 shows that there is only one field L possible, the cubic subfield of the field of seventh roots of unity.

Let us show that $K.L$ coincides with C; it comes down to showing that the extension $K.L/K$ is unramified. Let $p \neq 7$ and let $\mathfrak{p}$ be a prime ideal of K over 8. p is unramified in the extension $L/\mathbb{Q}$, and therefore ([Wei] Chapter 3) $\mathfrak{p}$ is unramified in the extension $K.L/K$. With the fields K and $K.L$ being real it remains to verify that the prime ideal $\mathfrak{p}_7$ is unramified in the extension $K.L/K$.

Let $k = \mathbb{Q}(\sqrt{-3})$, $N = K(\sqrt{-3})$. The Galois group G of NL/k is the product of two groups of order three. Let $\mathfrak{p}$ be an ideal of k over 7. Since the quotient of the inertia group G_0 of 7 by its higher ramification subgroup G_1 is cyclic, and as $G_1 = \{1\}$ because 7 is prime to 9 (cf., [Ser 2] Chapter IV), NL/k is not totally ramified at $\mathfrak{p}$. Therefore there exists an unramified extension L_1 of degree three of k at $\mathfrak{p}$; as $N.L = N.L_1$, NL/N is not ramified at the ideals over $\mathfrak{p}$. From this one easily shows that $K.L/K$ is unramified at $\mathfrak{p}_7$, which shows $K.L = C$.

The results of Part IV concerning the determination of the class of an arbitrary prime ideal of K are connected with a reciprocity law.

SOLUTION 4·7 (1) It is known that the integral closure of $\mathbb{Z}$ in k is $\mathbb{Z}[\vartheta]$, where $\vartheta = \frac{1}{2}(\omega - 1)$. The irreducible polynomial of ϑ over $\mathbb{Q}$ is $X^2 + X - 16$. The image of this polynomial in $\mathbb{Z}/2\mathbb{Z}[X]$ has two roots in $\mathbb{Z}/2\mathbb{Z}$, 0 and 1. It follows then ([CaF] Chapter III, Appendix) that the ideal (2) splits into two prime ideals, P_1 generated by 2 and $\vartheta - 1$, and P_2 generated by 2 and ϑ. So letting:

$$P_1 = (2, \tfrac{1}{2}(\omega + 1)), \qquad P_2 = (2, \tfrac{1}{2}(\omega - 1)).$$

Hence we have:

$$P_1^2 = (4, \omega + 1, \tfrac{1}{4}(\omega + 1)^2);$$

now,

$$\tfrac{1}{4}(\omega + 1)^2 = 16 + \tfrac{1}{2}(\omega + 1) = 3.4 + 4 + \tfrac{1}{2}(\omega + 1),$$

therefore:

$$4 + \tfrac{1}{2}(\omega + 1) \in P_1^2.$$

Moreover:

$$4 = (4 + \tfrac{1}{2}(\omega + 1))(4 - \tfrac{1}{2}(\omega - 1)),$$

and therefore:

$$4 \in (4 + \tfrac{1}{2}(\omega + 1)),$$

whence:

$$\tfrac{1}{2}(\omega + 1) \in (4 + \tfrac{1}{2}(\omega + 1)).$$

Consequently:

$$P_1^2 \subset (4 + \tfrac{1}{2}(\omega + 1)),$$

and therefore we have the equality:

$$P_1^2 = (4 + \tfrac{1}{2}(\omega + 1)).$$

(2): If $P_1 = (2a + b\frac{1}{2}(\omega + 1))$, by applying the non-trivial automorphism of Gal($k/\mathbb{Q}$) we would have:

$$P_2 = (2a + b\tfrac{1}{2}(- \omega + 1)),$$

whence:

$$(2) = P_1P_2 = ((2a + \frac{1}{2}\, b)^2 - \frac{65}{4}\, b^2).$$

So there exists a unit $\varepsilon \in \mathbb{Z}$ such that:

$$2\varepsilon = (2a + \frac{1}{2}\, b)^2 - \frac{65}{4}\, b^2,$$

and thus:

$$(2a + \frac{1}{2}\, b)^2 - \frac{65}{4}\, b^2 = \pm\, 2.$$

Now, the equation $x^2 - 65y^2 = \pm 8$ has no integral solutions because $x^2 - 65y^2$ is congruent to 0,1 or 4 modulo 5. Consequently P is not principal.

(3): Given a number field k it is known that in each ideal class there exists an integral ideal B such that:

$$N_{k/\mathbb{Q}}(B) \leqslant \left(\frac{4}{\pi}\right)^{r_2} \frac{n!}{n^n}\, |d|^{\frac{1}{2}},$$

(where n is the degree of the field k, d is the discriminant of k, and $2r_2$ is the number of complex conjugate fields of k, cf., [Sam] Chapter IV §4). Here $r_2 = 0$, $n = 2$, $d = 65$, thus:

$$N_{k/\mathbb{Q}}(B) \leqslant 4.$$

It therefore remains to determine the ideals with norm 2,3,4.

The ideals with norm 2 are the ideals P_1 and P_2. There is no ideal of norm 3 because 3 is inert in the extension $k/\mathbb{Q}$. The ideals with norm 4 are the ideals $2, P_1^2, P_2^2$.

We know (Questions (1) and (2) above) that $Cl(P_1) \neq 1$ and $Cl(P_1^2) = 1$. Furthermore, since $P_1P_2 = (2)$, we have:

$$Cl(P_2) = Cl(P_1) \quad \text{and} \quad Cl(P_1^2) = Cl(P_2^2) = Cl(P_1P_2) = 1.$$

The ideal class group of k is therefore of order two; $Cl(P_1)$ is a generator of this group.

(4): Let u be a unit of k. Let us set:

$$u = a + b\tfrac{1}{2}(\omega + 1) \quad (a,b \in \mathbb{Z}); \qquad N_{k/\mathbb{Q}}(u) = (a + \tfrac{1}{2}b)^2 - \frac{65}{4}b^2.$$

We have $N_{k/\mathbb{Q}}(u) = -1$ if and only if:

$$(a + \frac{1}{2}b)^2 - \frac{65}{4}b^2 = -1.$$

Since the equation $x^2 - 65y^2 = -1$ has $x = 8$, $y = 1$ as a solution, tion, we can take:

$$u = 8 + \omega.$$

(5): If ε is a fundamental unit, then $u = \pm\varepsilon^n$. Since $N_{k/\mathbb{Q}}(u) = -1$, $N_{k/\mathbb{Q}}(\varepsilon) = -1$ and n is odd, $n = 2k + 1$. Hence we have:

$$\pm u = \varepsilon^{2k+1} \quad \text{and} \quad \sqrt{\pm u} = \varepsilon^k\sqrt{\varepsilon}.$$

Consequently, $k(\sqrt{\varepsilon}) = k(\sqrt{u})$, where $k(\sqrt{\varepsilon}) = k(\sqrt{-u})$. Since $k(\sqrt{\varepsilon})$ is real, $k(\sqrt{\varepsilon}) = k(\sqrt{8 + \omega})$ necessarily. The discriminant of $\overline{8 + \omega}$ in the extension $k(\sqrt{8 + \omega})/k$ is equal to $4(8 + \omega)$. A ramified prime ideal in the extension $k(\sqrt{8 + \omega})/k$ must be a divisor of 2.

Let us show that P_1 is ramified in this extension. Let α =

$\sqrt{8+\omega} - 1$. The irreducible polynomial of α over k is the polynomial:

$$X^2 + 2X - 7 - \omega = X^2 + 2X - 6 - 2\tfrac{1}{2}(\omega + 1).$$

It is clear that P_1-adic valuation of $-6 - 2\cdot\frac{1}{2}\cdot(\omega+1)$ is equal to unity (in fact P_1^2 divides $2\cdot\frac{1}{2}\cdot(\omega+1)$ and 6 is divisible by P_1 but not by P_1^2). The irreducible polynomial of α is therefore an Eisenstein polynomial for P_1, consequently ([Wei] Chapter 3) P_1 is ramified in $k(\sqrt{8+\omega})$.

The P_2-adic valuation of $-7-\omega = -8 - 2\frac{1}{2}(\omega - 1)$ is equal to 2 (in fact $2\frac{1}{2}(\omega-1)$ is divisible by P_2^2 and not by P_2^3, and P_2^3 divides 8). The irreducible polynomial over k of $\frac{1}{2}\alpha = \frac{1}{2}(\sqrt{8+\omega} - 1)$ is the polynomial $X^2 + X - 2 - \frac{1}{4}(\omega - 1)$, where $-2 - \frac{1}{4}(\omega - 1)$ therefore has P_2-adic valuation zero. This polynomial remains irreducible in $k_{P_2}[X]$ (where k_{P_2} is the P_2-adic completion of k) for k_{P_2} is isomorphic to the 2-adic number field $\mathbb{Q}_2$, and thus if $x \in k_{P_2}$ $x^2 + x \equiv 0 \bmod P_2$. Consequently P_2 is either inert or ramified in $k(\sqrt{8+\omega})$. Now, the discriminant of $\frac{\alpha}{2}$ in this extension is equal to $8+\omega$, which is an unit. Consequently P_2 is not ramified.

There is therefore a single ramified ideal in the extension $k(\sqrt{8+\omega})$; it is the ideal P_1.

(6): If ε or $-\varepsilon$ were squares in K we would have $K = k(\sqrt{3})$ or $K = k(\sqrt{-3})$. Thus the ideal P_1 of k would be ramified in K, which is contrary to the definition of K.

(7): If a is an unit of K such that $a + \sigma(a) = 0$, a is therefore a root of $X^2 + a\sigma(a)$. $a\sigma(a)$ is an unit of k, and we have:

$$a\sigma(a) = \pm\,\varepsilon^{2k} \quad \text{or} \quad a\sigma(a) = \pm\,\varepsilon^{2k+1}.$$

One cannot have $a\sigma(a) = \varepsilon^{2k}$ because $k(a)$ would not be real. If we had $a\sigma(a) = -\varepsilon^{2k}$, then $a = \pm\varepsilon^{k}$, and consequently $a\sigma(a) = \varepsilon^{2k}$, and so a contradiction. One cannot have $a\sigma(a) = \pm\varepsilon^{2k+1}$, for this implies $(a\varepsilon^{-k})^2 = \pm\varepsilon$, which is impossible by Question (6) above.

(8) Let $c \in K$ be an integer that does not belong to k. Set $b = \sigma c - c$. Then $b \neq 0$ and $\sigma b = -b$.

(9) Let us consider the decomposition of the ideal bO_K into a product of prime ideals of O_K. A prime ideal of O_K is either an ideal P above an inert ideal of O_k or an ideal q over a split ideal of O_K. We have $\sigma P = P$ and $\sigma q \neq q$. Since the ideal bO_K is invariant under σ, and ideal q and its conjugate σq appear with the same exponent in the decomposition of bO_K.

Now P (*resp.* $q.\sigma(q)$) is the extension of an inert (*resp.* split) prime ideal of O_k. Consequently bO_K is the extension of an ideal A of O_k, $bO_K = AO_K$. If A were a principal ideal, that is to say $A = xO_k$ (with $x \in O_k$) one would have $bO_K = xO_K$. There would thus exist an unit a of K such that $x = ab$. But since $\sigma x = x$, $\sigma b = -b$, and $b \neq 0$, this implies $a + \sigma(a) = 0$, which is impossible, by Question (7) above.

(10) Let us assume that there exist two distinct fields K_1 and K_2 possessing the properties of Question (6) aobve. Let us write K_3 for the third field of degree two over k contained in the composition $K_1.K_2$; let σ_1 (*resp.* σ_2) be the K_1- (*resp.* K_2-) automorphism of $K_1.K_2$. Let b_1 (*resp.* b_2) be a non-zero element of O_{K_1} (*resp.* O_{K_2}) such that $\sigma_2(b_1) = -b_1$ (*resp.* $\sigma_1(b_2) = -b_2$). There exists a non-principal ideal A_1 (*resp.* A_2) of O_K such that:

$$b_1O_{K_1} = A_1O_{K_1} \quad (\textit{resp. } b_2O_{K_2} = A_2O_{K_2}).$$

Since O has only two ideal classes there exists an element $y \in k$ such that $A_1 = yA_2$. Consequently $A_1O_{K_1K_2} = yA_2O_{K_1K_2}$, whence $b_1O_{K_1K_2} = yb_2O_{K_1K_2}$. Therefore there exists an unit u of K_1K_2

such that $b_1u = b_2y$. We have $\sigma_1\sigma_2(b_1) = -b_1$, $\sigma_1\sigma_2(b_2) = -b_2$, $\sigma_1\sigma_2(y) = y$. Consequently $\sigma_1\sigma_2(u) = u$, and therefore $u \in K_3$. Now, $\sigma_1(b_1) = b_1$, $\sigma_1(b_2) = -b_2$, $\sigma_1(y) = y$, which implies that $\sigma_1(u) = u$.

It is clear that the field K_3 possesses the properties of Question (6) above (in fact, the extension K_1K_2/k is unramified for every prime ideal of k). By Question (7) above no unit of K_3 has a vanishing trace over k, which contradicts the preceding result.

(11) The field $\mathbb{Q}(\sqrt{5},\sqrt{13})$ contains the field k; $\mathbb{Q}(\sqrt{5},\sqrt{13})$ is real. Let us verify that no prime ideal of k is ramified in $\mathbb{Q}(\sqrt{5},\sqrt{13})$.

Let q be a prime ideal of k and let q be the prime number of $\mathbb{Z}$ such that $q \cap \mathbb{Z} = q\mathbb{Z}$. If q is ramified in $\mathbb{Q}(\sqrt{5},\sqrt{13})$, then q is ramified in the extension $\mathbb{Q}(\sqrt{5},\sqrt{13})/\mathbb{Q}$. Consequently q is ramified either in the extension $\mathbb{Q}(\sqrt{5})/\mathbb{Q}$ or in the extension $\mathbb{Q}(\sqrt{13})/\mathbb{Q}$. Now, 5 is the only ramified prime in $\mathbb{Q}(\sqrt{5})$; 13 is the only ramified prime in $\mathbb{Q}(\sqrt{13})$; 5 and 13 are the only ramified primes in $\mathbb{Q}(\sqrt{65}) = k$. Consequently the ramification index of 5 in the extension $\mathbb{Q}(\sqrt{5},\sqrt{13})$ coincides with its ramification index in the extension $k/\mathbb{Q}$. The prime ideal over 5 in k is therefore unramified in the extension $\mathbb{Q}(\sqrt{5},\sqrt{13})/k$. The same argument shows that the prime ideal over 13 in k is unramified in $\mathbb{Q}(\sqrt{5},\sqrt{13})$. Thus the field $\mathbb{Q}(\sqrt{5},\sqrt{13})$ satisfies the conditions of Question (6) above, and by Question (10) above it is unique. (One calls $\mathbb{Q}(\sqrt{5},\sqrt{13})$ the "class field" of k, cf., [Che] Introduction).

SOLUTION 4·8 (1) (a) The polynomial $X^2 - d$ is irreducible in $k[X]$ because its image in $\bar{k}[X] = \mathbb{F}_p[X]$ is irreducible. The extension of k by the polynomial $K^2 - d$ is therefore of degree two, and it is an unramified extension because the degree of the residual extension is two (cf., [Wei] Chapter 3).

(1) (b) Since 2 is not a square modulo 5, $X^4 - 2$ is irreducible in $\mathbb{F}_5[X]$. Consequently, the extension of $k = \mathbb{Q}$ by $X^4 - 2$ is of degree four, and is unramified because the degree of the residual extension is four.

(2) (a) This follows immediately from the definition.

(2) (b) This is because K is isomorphic to the finite field $\mathbb{F}_{q^n}$.

(2) (c) This is a property of extensions of finite fields. (Cf., [Joly] Chapitre 1).

(3) In $K[X]$ the polynomial $X^{q^n} - X$ factorises into a product of factors of the first degree, all relatively prime. Hensel's Lemma ([Wei] Chapitre 2) shows that $X^{q^n} - X$ also factorises in $K[X]$ into a product of factors of the first degree. Let ξ be a root in K of the polynomial $X^{q^n-1} - 1$ that is primitive (that is to say, of order $q^n - 1$). The extension $k(\xi)$ coincides with K, for these two extensions have the same residue field, and the extension K/k is unramified. K is therefore a splitting field of $X^{q^n} - X$ on k; the extension K/k is Galois.

(4) (a) If $x,y \in O_K$ are such that $x - y$, we have $|x - y| < 1$, whence $|\sigma(x - y)| = |x - y| = 1$, and $\overline{\sigma x} = \overline{\sigma y}$. The mapping $\bar\sigma : \bar x \mapsto \overline{\sigma x}$ is therefore well defined and is a $\bar k$-automorphism of $\bar k$.

(4) (b) We have:

$$\overline{\sigma\tau}(\bar x) = \overline{\sigma\tau(x)} = \bar\sigma(\tau(x)) = \bar\sigma(\bar\tau(\bar x)),$$

and therefore Φ is a homomorphism of $G(K/k)$ into $G(\bar K/\bar k)$. If $\bar\sigma = \mathrm{id}_{\bar K}$, for all $x \in O_K$ we have $|\sigma(x) - x| < 1$, and in particular $|\sigma(\xi) - \xi| < 1$, which implies $\sigma(\xi) = \xi$, and hence $\sigma = \mathrm{id}_K$. The homomorphism Φ is therefore injective, and consequently surjective, because $G(K/k)$ and $G(\bar{K}/\bar{k})$ have the same order.

(4) (c) $G(K|k)$ being isomorphic to $G(\bar{K}|\bar{k})$ is cyclic and generated by the element σ^*, whose image under Φ is the generator τ of $G(\bar{K}|\bar{k})$ (Question (2)(c) above). σ^* is therefore the k-automorphism of K such that $\bar{\sigma}^*(\bar{x}) = \bar{x}^q$ (for all $\bar{x} \in \bar{K}$), that is to say, $|\sigma^*(x) - x^q| < 1$ (for all $x \in O_K$).

(4) (d) In example (1)(a) we clearly have $\sigma^*(\sqrt{d}) = -\sqrt{d}$, which suffices to define σ^*. In example (1)(b) we have $k = \mathbb{Q}_5$, $K = \mathbb{Q}_5(\sqrt[4]{2})$. σ^* is defined if $\sigma^*(\sqrt[4]{2})$ is known. It is clear that $\sigma^*(\sqrt[4]{2}) = i\sqrt[4]{2}$, where i is a square root of -1. Of the two square roots of -1 in k one chooses the one determined by:

$$|i\sqrt[4]{2} - \sqrt[4]{2^5}| < 1$$

that is to say:

$$|i - 1| < 1,$$

(one has $i^2 - 2^2 = (i - 2)(i + 2) = -5$, which shows that the two square roots of -1 in $\mathbb{Q}_5$ are congruent, modulo the maximal ideal, one to 2, the other to -2).

(5) (a) If $\mu_m = 1 + a\pi^m$ (with $a \in O_K$) we have:

$$N_{K/k}(\mu_m) = \prod_{\sigma \in G(K|k)} (1 + \sigma(a)\pi^m) \equiv 1 + \mathrm{Tr}_{K/k}(a)\pi^m \bmod \pi^{m+1}.$$

(5) (c) We have:

$$\overline{\mathrm{Tr}_{K/k}(a)} = \overline{\sum_{\sigma \in G(K/k)} \sigma(a)} = \sum_{\sigma \in G(K/k)} \overline{\sigma(a)} = \sum_{\bar{\sigma} \in G(\bar{K},\bar{k})} \bar{\sigma}(\bar{a})$$

$$= \mathrm{Tr}_{\bar{K}/\bar{k}}(\bar{a}).$$

Since the mapping $\mathrm{Tr}_{\bar{K}/\bar{k}}:\bar{K} \to \bar{k}$ is surjective (cf., Exercise 4·1) the mapping ψ is surjective. Consequently, for any $\lambda_m \in 1 + \pi^m O_K$, that is to say $\lambda_m = 1 + b\pi^m$ with $b \in O_k$, there exists $a \in O_K$ such that $\overline{\mathrm{Tr}_{K/k}(a)} = b$, and therefore if $\mu_m = 1 + a\pi^m$ we have:

$$Nm_{K|k}(\mu_m) \equiv 1 + b\pi^m \bmod \pi^{m+1},$$

and therefore:

$$Nm_{K|k}(\mu_m) \equiv \lambda_m \bmod \pi^m.$$

(5) (c) We evidently have $Nm_{K|k}(1 + \pi O_K) \subset 1 + \pi O_k$. Let $\lambda = \lambda_1$ belong to $1 + \pi O_k$; by Question (5)(b) preceding there exists $\mu_1 \in 1 + \pi O_K$ such that $Nm_{K|k}(\mu_1) = \frac{\lambda_1}{\lambda_2}$ where $\lambda_2 \in 1 + \pi^2 O_K$. By iterating this process one shows the existence of a sequence $(\lambda_m)_{m\geqslant 1}$ of O_k and of a sequence $(\mu_m)_{m\geqslant 1}$ of O_K such that:

$$\lambda_1 = \lambda \quad \text{and} \quad \lambda_m \in 1 + \pi^m O_K, \qquad \mu_m \in 1 + \pi^m O_K,$$

and:

$$Nm_{K|k}(\mu_m) = \frac{\lambda_1}{\lambda_{m+1}}.$$

From this one deduces:

$$Nm_{K|k}(\mu_1\mu_2\cdots\mu_m) = \frac{\lambda_1}{\lambda_{m+1}}.$$

The sequence with general term $u_m = \mu_1\mu_2\cdots\mu_m$ converges towards an element $\mu \in 1 + \pi O_K$, because $u_m \equiv u_{m+1} \bmod \pi^m$. We have $Nm_{K/k}(\mu) = \lambda$, since the sequence with general term $\frac{\lambda_1}{\lambda_{m+1}}$ converges towards $\lambda_1 = \lambda$, and because the norm is a continuous mapping, whence the equality:

$$Nm_{K/k}(1 + \pi O_K) = 1 + \pi O_k(1).$$

Let V_K (*resp.* V_k) be the multiplicative group of $(q^n - 1)$-th roots of unity in K (*resp.* of $(q - 1)$-th roots of unity in k). We have:

$$\left.\begin{aligned} U_K &= V_K \times (1 + \pi O_K) \quad \text{(direct product)}, \\ U_k &= V_k \times (1 + \pi O_k) \quad \text{(direct product)} \end{aligned}\right\} \tag{1}$$

([Has] II §15). V_K is isomorphic to the multiplicative group $\bar{K}^*$ and V_k is isomorphic to $\bar{k}^*$; since the norm in an extension of finite fields is surjective (cf., Exercise 4·1), from this it follows that:

$$Nm_{K|k}(V_K) = V_k. \tag{2}$$

Relations (1) and (2) imply:

$$Nm_{K|k}(U_K) = U_k.$$

(5):(d): This is obvious because $a^2 - db^2$ is the norm of $a + b\sqrt{d}$ and because every unit $x \in \mathbb{Q}_p$ is the norm of a unit of $\mathbb{Q}_p(\sqrt{d})$.

(6): We have the equalities:

$$K^* = \pi^{\mathbb{Z}} \times U_K \quad \text{(direct product)},$$

$$k^* = \pi^{\mathbb{Z}} \times U_k \quad \text{(direct product)}.$$

From this we have:

$$Nm_{K|k}(K^*) = \pi^{n\mathbb{Z}} \times Nm_{K|k}(U_K) = \pi^{n\mathbb{Z}} \times U_k,$$

whence:

$$k^*/Nm_{K|k}(K^*) \simeq \mathbb{Z}/n\mathbb{Z}.$$

Hence we have just shown that if $K|k$ is an unramified extension of local fields, the group of norms of K is a multiplicative subgroup, of finite index, of the group k^*, the quotient being isomorphic to the Galois group of the extension. This is a particular case of local class field theory (cf., [Ser2] Chapter V, SII, XIV).

SOLUTION 4·9: (cf., [Bou] Chapter V §5):

I:(1): If the polynomial $Y^2 + X^2 + 1$ is not irreducible in $K[Y]$, as it is of degree two it possesses a root in $K = \mathbb{Q}(X)$; furthermore, this root is then a polynomial $P(X)$ of $\mathbb{Q}[X]$, for with the ring $\mathbb{Q}[X]$ being principal it is integrally closed ([Sam] Chapitre 2).

Now the equality $P(X)^2 + X^2 + 1 = 0$ is impossible because the polynomial $X^2 + 1$ is irreducible in $\mathbb{Q}[X]$.

II:(2): The algebraic dimension (or transcendence degree) of E on $\mathbb{Q}$ is one. If E is a purely transcendental extension of $\mathbb{Q}$ ([Bou] Chapitre 5 §5) there then exists $Z \in E$ such that $E = \mathbb{Q}(Z)$. Let $\xi \in E$ be a root of the polynomial $Y^2 + X^2 + 1 \in K[Y]$. Let us write that $X, \xi \in \mathbb{Q}(Z)$. There exist non-zero polynomials $P_1, Q_1, P_2, Q_2 \in \mathbb{Q}[Z]$ such that:

$$X = \frac{P_1(Z)}{Q_1(Z)} \quad \text{and} \quad \xi = \frac{P_2(Z)}{Q_2(Z)},$$

P_1, Q_1 being relatively prime, and P_2, Q_2 as well. The relation $\xi^2 + X^2 + 1 = 0$ can be written:

$$P_2^2Q_1^2 + P_1^2Q_2^2 + Q_1^2Q_2^2 = 0.$$

We see that Q_2 divides Q_1 (since Q_2, P_2 are relatively prime and

Q_2 divides P_2Q_1) and that Q_1 divides Q_2 (since Q_1, P_1 are relatively prime and Q_1 divides P_1Q_2). Hence there exists a polynomial $Q \in \mathbb{Q}[Z]$ such that:

$$P_2^2 + P_1^2 + Q^2 = 0.$$

Examining the terms of higher degree, and the impossibility of solving the equation:

$$a^2 + b^2 + c^2 = 0 \quad \text{with } (a,b,c) \in \mathbb{Q}^3,\ (a,b,c) \neq (0,0,0),$$

gives the desired contradiction.

I (3) The possibility of parametrising the ellipse $X^2 + Y^2 + 1 = 0$ by:

$$X = \frac{i(1 - Z^2)}{1 + Z^2} \quad \text{and} \quad Y = \frac{2iZ}{1 + Z^2}$$

suggests using the identity:

$$\left(\frac{i(1 - Z^2)}{1 + Z^2}\right)^2 + \left(\frac{2iZ}{1 + Z^2}\right)^2 + 1 = 0.$$

Let us look for Z (in $\mathbb{Q}(i, X, \xi)$) such that:

$$X = \frac{i(1 - Z^2)}{1 + Z^2} \quad \text{and} \quad \xi = \frac{2iZ}{1 + Z^2}.$$

We easily obtain $Z = \dfrac{\xi}{i + X}$, which shows that $E(i)$ is a purely transcendental extension of $\mathbb{Q}(i)$:

$$E(i) = \mathbb{Q}(i)\left(\frac{3}{i + X}\right).$$

I (4) If the polynomial $Y^2 + X^2 + 1$ were irreducible in $K[Y]$, since it is of degree two it would have a root in $K = \mathbb{Q}[X]$, and this root would then be a polynomial $P(X) \in \mathbb{Q}[X]$ (cf., Question (1)). Now the equality $P(X)^2 - X^2 + 1 = 0$ is impossible because $X^2 - 1 = (X - 1)(X + 1)$ is not a square in $\mathbb{Q}[X]$.

Finally, F is a purely transcendental extension of Q, because of the identity:

$$\left(\frac{2Z}{1 - Z^2}\right)^2 - \left(\frac{1 + Z^2}{1 - Z^2}\right)^2 + 1 = 0,$$

which shows, if we denote a root of $Y^2 - X^2 + 1$ in F by η, that we have:

$$F = \mathbb{Q}\left(\frac{\eta}{1 + X}\right) .$$

II Let $1, j, j^2$ be the cube roots of unity in $\mathbb{C}$.

II (1) If the polynomial $Y^3 + X^3 + 1$ were not irreducible in $L[Y]$ there would exist $\frac{P}{Q} \in L$ such that:

$$(P(X))^3 + (X^3 + 1)(Q(X))^3 = 0,$$

with P,Q being relatively prime in $\mathbb{C}[X]$. Hence $X + 1, X + j, X + j^2$ divide P, and if:

$$P(X) = (X + 1)(X + j)(X + j^2)P_1(X) = (X^3 + 1)P_1(X),$$

we have:

$$(Q(X))^3 + (X^3 + 1)^2(P_1(X))^3 = 0,$$

and consequently $X^3 + 1$ divides Q, a contradiction.

II (2) If M were a purely transcendental extension of $\mathbb{C}$, there would exist three polynomials $u,v,w \in \mathbb{C}[X]$, not all constant, pairwise relatively prime, and such that $u^3 + v^3 + w^3 = 0$, that is to say, $w^3 = -(u + v)(u + jv)(u + j^2v)$. Let n be the maximum of the degrees of u,v,w, and let us choose these polynomials so that n is a minimum. We may assume that w is of degree n).

If $p \in \mathbb{C}[X]$ is an irreducible polynomial which divides $u + v$, then it does not divide $u + jv$ (because u,v are relatively prime), nor $u + j^2v$, and p divides w^3, therefore p divides w; from this we deduce that p^3 divides $u + v$. Thus $u + v$ is a cube in $\mathbb{C}[X]$: there exists $r \in \mathbb{C}[X]$ such that $u + v = r^3$. Similarly there exist $s,t \in \mathbb{C}[X]$ such that $u + jv = s^3$, $u + j^2v = t^3$, whence:

$$r.s.t = - w.$$

As u,v,w are not all constant, at least two of the polynomials r,s,t are not constants, hence the maximum of the degrees r,s,t is strictly less than n. The relation:

$$r^3 + js^3 + j^2t^3 = 0$$

shows that there exist in $\mathbb{C}[X]$ three polynomials $u_1 = r$, $v_1 = \sqrt[3]{j}s$, $w_1 = \sqrt[3]{j^2}t$, not all constant, and of degree strictly less than n satisfying:

$$u_1^3 + v_1^3 + w_1^3 = 0,$$

contradicting the minimality of n.

CHAPTER 5

Distribution Modulo 1

INTRODUCTION

5.1 EQUIDISTRIBUTION DISCREPANCY: (Exercises 5·1-5·10)

Let $(u_n)_{n\in\mathbb{N}}$ be a sequence of elements in the interval $[0,1]$, and let $[\alpha,\beta]$ be an interval within $[0,1]$. For every integer $N \in \mathbb{N}^*$ we *denote* by $s_N(\alpha,\beta)$ the *number of integers k such that* $0 \leqslant k < N$, $u_k \in [\alpha,\beta]$. The *DISCREPANCY* is the sequence (D_N) defined by:

$$D_N = \sup_{0\leqslant\alpha\leqslant\beta\leqslant 1} \left| \frac{s_N(\alpha,\beta)}{N} - (\beta-\alpha) \right| .$$

A sequence $(u_n)_{n\in\mathbb{N}}$ of elements of the interval $[0,1]$ is said to be *EQUIDISTRIBUTED in* $[0,1]$ if:

$$\text{For all } \alpha,\beta \text{ such that } [\alpha,\beta] \subset [0,1] \quad \lim_{N\to\infty} \frac{s_N(\alpha,\beta)}{N} = \beta - \alpha.$$

A sequence $(u_n)_{n\in\mathbb{N}}$ of real numbers is said to be *EQUIDISTRIBUTED MODULO* 1 if the sequence $(\{u_n\})_{n\in\mathbb{N}}$ of fractional parts is equidistributed in $[0,1]$. (If $x \in \mathbb{R}$ we write $[x] = \sup\{k \in \mathbb{Z}, k \leqslant x\}$ and $\{x\} = x - [x]$).

Let S be a finite set, and let $(u_n)_{n\in\mathbb{N}}$ be a sequence of elements of S. For all $a \in S$ and every integer $N \in \mathbb{N}^*$ let us *denote* by $s_N(a)$ the *number of integers* k *such that* $0 \leqslant k < N$ *and* $u_k = a$. We say that the *sequence* $(u_n)_{n\in\mathbb{N}}$ is *EQUIDISTRIBUTED IN A SET* S if:

$$\text{For all } a \in S \lim_N \frac{s_N(a)}{N} = \frac{1}{\operatorname{card}(S)} .$$

A sequence $(u_n)_{n\in\mathbb{N}}$ of elements of $\mathbb{Z}$ is said to be *EQUIDISTRIBUTED MODULO* q ($q \geqslant 1$ an integer) if the sequence $(\hat{u}_n)_{n\in\mathbb{N}}$ is equidistributed in the set $\mathbb{Z}/q\mathbb{Z}$ (where $\hat{x}$ *denotes* the *class of* x *modulo* q).

5.2 CRITERIA FOR EQUIDISTRIBUTION MODULO 1: (Exercises 5·11-5·20)

The sequence $(u_n)_{n\in\mathbb{N}}$ of elements of $[0,1]$ is equidistributed on $[0,1]$ if and only if for every Riemann integrable function f on $[0,1]$ we have:

$$\frac{1}{N} \sum_{0 \leqslant n < N} f(u_n) \to \int_0^1 f \quad \text{as } N \to \infty.$$

WEYL'S CRITERION: The sequence $(u_n)_{n\in\mathbb{N}}$ of elements of $\mathbb{R}$ is equidistributed modulo 1 if and only if:

$$\text{For all } p \in \mathbb{Z}^* \quad \frac{1}{N} \sum_{0 \leqslant n < N} e_p(u_n) \to 0 \quad \text{as } N \to \infty,$$

where $e_p(x) = e^{2i\pi px}$.

These results lead to:

DEFINITION: Let S be a compact space, μ a positive measure of total weight one on S, $(u_n)_{n\in\mathbb{N}}$ a sequence of elements of S. The sequence $(u_n)_{n\in\mathbb{N}}$ is said to be *DISTRIBUTED IN ACCORDANCE WITH A MEASURE* μ if for every continuous function f of S into $\mathbb{C}$ we have:

$$\frac{1}{N} \sum_{0 \leq n < N} f(u_n) \to \langle \mu, f \rangle \quad \text{as } N \to \infty.$$

5.3 PARTICULAR SEQUENCES: (Exercises 5·21-5·28)

FÉJER'S CRITERION: Let g be a monotonic differentiable continuous function with monotonic derivative. If, as $t \to \infty$, $g(t) \to \infty$, $g'(t) \to \infty$, $tg'(t) \to \infty$, then $(g(n))_{n\in\mathbb{N}}$ is equidistributed modulo 1.

VAN DER CORPUT'S CRITERION: A sufficient condition for the sequence $(u_n)_{n\in\mathbb{N}}$ to be equidistributed is that for any integer $h \geq 1$ the sequence $(u_{n+h} - u_n)_{n\in\mathbb{N}}$ is equidistributed.

5.4 SEQUENCES WITH MORE THAN POLYNOMIAL GROWTH: (Exercises 5·29-5·39)

For almost all $\vartheta > 1$ (in the sense of the Lebesgue measure on $\mathbb{R}$) the sequence (ϑ^n) is equidistributed modulo 1 (a result of Koksma).

If a sequence $(u_n)_{n\in\mathbb{N}}$ satisfies the property: There exists α such that $0 \leq \alpha < 1$ and $\underline{\lim}\, n^\alpha(u_{n+1} - u_n) > 0$, then for almost all λ the sequence $(\lambda u_n)_{n\in\mathbb{N}}$ is equidistributed modulo 1 (a result of H. Weyl).

A real number λ is said to be *NORMAL IN BASE* g ($g \geq 2$ an integer) if the sequence $(\lambda g^n)_{n\in\mathbb{N}}$ is equidistributed modulo 1. Therefore almost all numbers are normal. More generally, the *NORMAL SET* $B((u_n))$ *associated with the sequence* (u_n) is by definition the set of real numbers λ such that $(\lambda u_n)_{n\in\mathbb{N}}$ is equidistributed modulo 1.

It is known that given a set E a necessary and sufficient condition for there to exist a sequence $(u_n)_{n\in\mathbb{N}}$ such that $B((u_n)) = E$ is that: (i) $0 \notin E$, (ii) $\mathbb{Z}^* E \subset E$, (iii) E is a $F_{\sigma\delta}$ (that is to say it is the countable intersection of a countable union of closed sets).

5.5 CONSTRUCTION OF SEQUENCES STARTING FROM GIVEN SEQUENCES: (Exercises 5·40-5·46)

Let $(u_n)_{n\in\mathbb{N}}$ be a given sequence, σ a mapping of $\mathbb{N}$ into $\mathbb{N}$. We propose to deduce distribution properties for the sequence $(u_{\sigma(n)})_{n\in\mathbb{N}}$ from distribution properties of the sequence $(u_n)_{n\in\mathbb{N}}$.

Let σ be an increasing mapping, and let χ be the *associated* *GENERALISED CHARACTERISTIC FUNCTION* of σ, given by $\chi(n) = \operatorname{card}\{\sigma^{-1}(n)\}$. A particularly important situation is where χ is *almost periodic* in the sense of Besicovitch, that is to say it belongs to the closure of the trigonometric polynomials P of the form:

$$P(n) = \sum_{\lambda\in\Lambda} a_\lambda e(\lambda n) \qquad (\Lambda \text{ is the finite part of } \mathbb{R})$$

with the semi-norm:

$$\|f\| = \overline{\lim_{N\to\infty}} \frac{1}{N} \sum_{k<N} |f(k)|.$$

The following result is known: *If for every integer* $\ell \geqslant 1$ *the function which associates* $e(\ell u_n)$ *with* $n \in \mathbb{N}$ *is pseudo-random, and if* χ *is almost periodic with non-zero mean, then the sequence* $(u_{\sigma(n)})_{n\in\mathbb{N}}$ *is equidistributed modulo* 1. (A *PSEUDO-RANDOM FUNCTION* is a function f of $\mathbb{N}$ into $\mathbb{C}$ admitting a correlation function γ defined by:

$$\gamma(p) = \lim_{n\to\infty} \frac{1}{n+1} \sum_{k=0}^{n} \overline{f(k)} f(k+p)$$

satisfying, for example, the condition:

$$\overline{\lim_{n\to\infty}} \frac{1}{n+1} \sum_{k=0}^{n} |\gamma(k)|^2 = 0).$$

A method widely used for the construction of sequences is the method of "blocks", an example of which is the object of Exercise 5·45.

5.6 ERGODIC THEOREM: (Exercises 5·47-5·49)

Let $(S,\mathfrak{A},\mu)$ be a probability space, and let ϑ be a measure-preserving mapping of S into S, that is to say, such that:

$$\text{For all } A \in \mathfrak{A}, \quad \vartheta^{-1}(A) \in \mathfrak{A} \text{ and } \mu(\vartheta^{-1}A) = \mu(A).$$

Then if $f \in L^1(S,\mathfrak{A},\mu)$ for almost all $x \in S$, the quantity:

$$\frac{1}{N} \sum_{k<N} f(\vartheta^k x) \to g(x) \quad \text{as} \quad N \to \infty.$$

The function $g(x) \in L^1(S,\mathfrak{A},\mu)$ (furthermore, the convergence also holds in the norm) and is invariant under ϑ, and:

$$\int g\,d\mu = \int f\,d\mu.$$

In particular, if ϑ is *ergodic* (that is to say, if the only sets invariant under ϑ are of measure zero or measure one) then for almost all $x \in S$:

$$\lim_{N\to\infty} \frac{1}{N} \sum_{k<N} f(\vartheta^k x) = \int f\,d\mu.$$

Let S be a compact topological space, μ a Radon measure of total mass one on S, ϑ a continuous mapping of S into S such that the *unique* Radon measure invariant under ϑ and of total mass one is μ, then if f is a continuous function from S into S we have:

$$\lim_{N\to\infty} \sup_{x\in S} \left| \frac{1}{N} \sum_{k<N} f(\vartheta^n x) - \mu(f) \right| = 0.$$

BIBLIOGRAPHY

[Rau], [Kui], [Cas], [Sal].

PROBLEMS

EXERCISE 5·0: Tony lives very near a bus station from whence two services leave with the same frequency: Route 41, which takes him to his blonde girlfriend Bridget, and Route 63, which takes him to his other girlfriend, Claudia, a brunette. On Saturday mornings Tony gets up at a time that varies with his mood or with the time it takes to go round to one or the other of his girlfriends. But feeling at a loss as to how to decide between the blonde and the brunette, he prefers to trust to luck; so when he arrives at the bus station he takes the first bus that leaves. Doing this he arrives at Claudia's more often.

Can you explain why?

EXERCISE 5·1: Let $(u_n)_{n\in\mathbb{N}}$ be a sequence of elements in $[0,1]$.

Prove that $(u_n)_{n\in\mathbb{N}}$ is equidistributed on $[0,1]$ if and only if $D_N \to 0$ as $N \to \infty$.

EXERCISE 5·2: Let $(u_n)_{n\in\mathbb{N}}$ be a sequence of elements in $[0,1]$, $N \in \mathbb{N}^*$. Set:

$$D_N^* = \sup_{0\leqslant\alpha\leqslant 1} \left| \frac{s_N(0,\alpha)}{N} - \alpha \right| .$$

(1): Show that $D_N^* \leqslant D_N \leqslant 2D_N^*$.

(2): Show that if $u_0 \leqslant u_1 \leqslant \cdots < u_{N-1}$ we have:

$$D_N^* = \sup_{0 \leqslant k < N} \sup\left(\left|u_k - \frac{k}{N}\right| , \left|u_k - \frac{k+1}{N}\right|\right)$$

$$= \frac{1}{2N} + \sup_{0 < k < N} \left|u_k - \frac{2k+1}{2N}\right| .$$

(3): From this deduce that for every sequence $(u_n)_{n \in \mathbb{N}}$ and for every integer $N \geqslant 1$ we have:

$$D_N^* \geqslant \frac{1}{2N} .$$

EXERCISE 5·3: Define a sequence $(u_n)_{n \in \mathbb{N}}$ by setting:

$$u_0 = 0, \quad \text{and if } 2^m \leqslant n < 2^{m+1} \ (m \in \mathbb{N}) \ u_n = u_{n-2^m} + \frac{1}{2^{m+1}} .$$

(1): Show that if $2^m < n \leqslant 2^{m+1}$, on setting $N = 2^m$ we have:

$$s_N(0,\alpha) = 1 + [2^m\alpha] + s_{n-N}\left(0,\alpha - \frac{1}{2^{m+1}}\right) .$$

(2): From this deduce that for any $n \geqslant 1$ we have

$$nD_n^* \leqslant 1 + \frac{3}{2}\left(\frac{\log n}{\log 2}\right) .$$

EXERCISE 5·4: Let $(u_n)_{n \in \mathbb{N}}$ be a sequence of real numbers.
Show the following are equivalent:

(i): $(u_n)_{n \in \mathbb{N}}$ is equidistributed modulo 1;

(ii): For every integer $q \geqslant 1$ the sequence $([qu_n])_{n \in \mathbb{N}}$ is equidistributed modulo q;

(iii): The set of integers $q \geqslant 1$ such that the sequence $([qu_n])_{n\in\mathbb{N}}$ is equidistributed modulo q is infinite.

EXERCISE 5·5: (1): For every pair of integers (N,k) such that $0 \leqslant k \leqslant N$, denote the interval $[k/(N+1),(k+1)/(N+1)]$ by $I(N,k)$. Let $(u_n)_{n\in\mathbb{N}}$ be a sequence of elements in $[0,1]$ such that if:

$$\frac{N(N+1)}{2} \leqslant n < \frac{(N+1)(N+2)}{2},$$

we have:

$$u_n \in I\left(N, n - \frac{N(N+1)}{2}\right).$$

Is the sequence $(u_n)_{n\in\mathbb{N}}$ equidistributed on $[0,1]$?

(2): The same question, but this time calling $I(N,k)$ the interval $[k/2^N,(k+1)/2^N]$ $(0 \leqslant k < 2^N)$ and assuming that if $2^N - 1 \leqslant n < 2^{N+1} - 1$, then $u_n \in I(N, n - 2^N + 1)$.

EXERCISE 5·6: Let p be an integer, $\alpha \in [0,1]$. Evaluate the number of integers n such that $p \leqslant \sqrt{n} \leqslant p + \alpha$, and deduce from this that the sequence $(\sqrt{n})_{n\in\mathbb{N}}$ is equidistributed modulo 1.

EXERCISE 5·7: Is the sequence $(\log(n+1))_{n\in\mathbb{N}}$ equidistributed modulo 1?

EXERCISE 5·8: (1): Let $\alpha \in [0,1]$, and let p,q be relatively prime integers with $q \geqslant 1$.

Evaluate the number of integers n such that:

$$0 \leqslant n \leqslant q \quad \text{and} \quad 0 < n\frac{p}{q} - \left[n\frac{p}{q}\right] \leqslant \alpha.$$

From this deduce bounds for the number of integers such that:

$$0 \leqslant n \leqslant N \quad \text{and} \quad 0 < n\frac{p}{q} - \left[n\frac{p}{q}\right] \leqslant \alpha \qquad (N \geqslant 1).$$

(2): By using a result in Diophantine approximation theory deduce from the preceding (1) that the sequence $(n\vartheta)_{n\in\mathbb{N}}$ is equidistributed modulo 1 when ϑ is irrational.

EXERCISE 5·9: Show that the Fibonacci sequence $(F_n)_{n\in\mathbb{N}}$, where $F_0 = 0$, $F_1 = 1$, and for $n \geqslant 0$ $F_{n+2} = F_{n+1} + F_n$, is equidistributed modulo 5.

EXERCISE 5·10: Let λ be the measure on $\{0,1\}^{\mathbb{N}}$, that is the product of the measure μ on $\{0,1\}$ defined by $\mu(\{0\}) = \mu(\{1\}) = \frac{1}{2}$. On the other hand, let φ_k be the function from $\{0,1\}^{\mathbb{N}}$ into $\{-1,+1\}$ that to $u = (u_n)_{n\in\mathbb{N}}$ associates $\varphi_k(u) = +1$ if $u_k = 0$ and $\varphi_k(u) = -1$ if $u_k = 1$. Set:

$$f_N(u) = \frac{1}{N}\sum_{0\leqslant k<N} \varphi_k(u).$$

(1): Evaluate the quantity $\rho_N = \int |f_N(u)|^2 d\lambda(u)$ and show that the series with the general term (ρ_{N^2}) is convergent.

(2): Show that for almost all u (with respect to the measure λ) $f_{N^2}(u) \to 0$ as $N \to \infty$.

(3): From this deduce that every sequence (with respect to the measure λ) with terms in $\{0,1\}$ is equidistributed in $\{0,1\}$.

EXERCISE 5·11: (1): Let $(u_n)_{n\in\mathbb{N}}$ and $(v_n)_{n\in\mathbb{N}}$ be two sequences of real numbers such that $(u_n - v_n) \to 0$ as $n \to \infty$.

Show that $(u_n)_{n\in\mathbb{N}}$ is equidistributed modulo 1 if and only if $(v_n)_{n\in\mathbb{N}}$ is.

(2): Do the same question, this time assuming $u_n - v_n = \log(n + 1)$. What happens if we have:

$$u_n - v_n = (\log(n + 1))^\delta \quad \text{with } \delta > 1?$$

EXERCISE 5·12: Let $(u_n)_{n\in\mathbb{N}}$ be a sequence of real numbers, and let (N_k) be a sequence of strictly increasing integers such that $N_{k+}/N_k \to 1$ as $k \to \infty$. Assume that:

$$\text{For all } p \in \mathbb{Z}^* \quad \sum_{N_k \leqslant n < N_{k+1}} e_p(u_n) \in o(N_{k+1} - N_k).$$

Show that the sequence $(u_n)_{n\in\mathbb{N}}$ is equidistributed modulo 1.

EXERCISE 5·13: Let a,b,ε be real numbers such that $0 < \varepsilon$, $0 < b - a < b - a + 2\varepsilon < 1$. Denote by F the function with period one, which is also affine on the intervals $[a - \varepsilon,a]$, $[a,b]$, $[b,b + \varepsilon]$, $[b + \varepsilon,a - \varepsilon + 1]$, and which has the value zero on $[b + \varepsilon,a - \varepsilon + 1]$ and the value one on $[a,b]$.

Show that if $F(x) = \sum_{k\in\mathbb{Z}} c_k e_k(x)$ is the expansion of F as a Fourier series, then for $k \neq 0$ we have the upper bound:

$$|c_k| \leqslant \min\left(1, \frac{1}{\pi^2 k^2 \varepsilon}\right).$$

EXERCISE 5·14: Let $(u_n)_{n\in\mathbb{N}}$ be a sequence of real numbers. For every integer q set:

$$\sigma_q(N) = \sum_{0\leqslant n<N} e_q(u_n),$$

and for every $[\alpha,\beta] \subset [0,1]$ denote by $s_N(\alpha,\beta)$ the number of integers n such that:

$$0 \leqslant n < N, \quad \{u_n\} \in [\alpha, \beta].$$

Using the result of the preceding Exercise 5·13, show that for $Q \geqslant 1$ we have the upper bound:

$$|s_N(\alpha,\beta) - N(\beta - \alpha)| < 2 \sum_{1 \leqslant q \leqslant Q} |\sigma_q(N)| + \frac{4N}{\pi}\sqrt{\sum_{k>Q} \frac{|\sigma_q(N)|}{q^2 N}}$$

Thus recover Weyl's Criterion and the result of Exercise 5·1.

EXERCISE 5·15: Using the results of Exercise 5·2 show that if $(u_n)_{n\in\mathbb{N}}$ is a sequence of elements of $[0,1]$ and f is a continuous function from $[0,1]$ into $\mathbb{R}$, we have the inequality:

$$\left|\frac{1}{N} \sum_{0 \leqslant n < N} f(u_n) - \int_0^1 f\right| \leqslant \omega(D_N^*),$$

where ω denotes the modulus of continuity of f, that is to say the function defined for $r > 0$ by:

$$\omega(r) = \sup_{|t-s|<r} |f(t) - f(s)|.$$

EXERCISE 5·16: Let $X = \{0,1\}^{\mathbb{N}}$ be the set of sequences with terms in $\{0,1\}$. Let $u = (u_n)_{n\in\mathbb{N}}$ be an element of X and let σ be the function from $\mathbb{N}$ into $\mathbb{N}$ defined by:

$$\sigma(0) = 0, \qquad \sigma(n + 1) = \sigma(n) + 1 + u_n.$$

A mapping π of X to itself is defined by associating the sequence $\pi(u) = (v_n)_{n\in\mathbb{N}}$ to the sequence $u = (u_n)_{n\in\mathbb{N}}$ such that for all $n \in \mathbb{N}$:

$$v_{\sigma(n)} = 1, \quad \text{and if } \sigma(n + 1) - \sigma(n) = 2 \text{ then } \quad v_{\sigma(n)+1} = 0.$$

(1): Show that the map π has a unique fixed point w (i.e., a point w satisfying $\pi(w) = w$) that is $w = (1,0,1,1,0,1,0,1,\ldots)$.

(2): Show that there exists a measure λ on X such that the sequence $(T^n w)_{n\in\mathbb{N}}$ is distributed in accordance with the measure λ, where T is the mapping of X to itself which to the sequence $u = (u_n)_{n\in\mathbb{N}}$ associates the sequence $Tu = (u_{n+1})_{n\in\mathbb{N}}$.

EXERCISE 5·17: Let R be the set $\{0,1,\ldots,9\}$ and let $u = (u_n)_{n\in\mathbb{N}}$ be a sequence with terms in $R^{\mathbb{N}}$ such that for every integer n, if $a_k\times 10^k + a_{k-1}\times 10^{k-1} + \cdots + a_0$ is the expression of 2^n in base ten, then u_n as a sequence of elements of R begins as $(a_k, a_{k+1}, \ldots, a_0, \ldots)$.

Does the sequence u admit a distribution in $R^{\mathbb{N}}$?

EXERCISE 5·18: Show that:

$$\lim_{N\to\infty} \frac{1}{N}\left|\sum_{0\leqslant n<N} e^{it\log(n+1)}\right| = \left|\int_0^1 e^{it\log x}\,dx\right|.$$

What can be concluded from this for the sequence $(\log(n + 1))_{n\in\mathbb{N}}$ modulo 1?

EXERCISE 5·19: Denoting the n-th prime number by p_n, what can be said about the equidistribution modulo 1 of the sequence $(\log p_n)_{n\in\mathbb{N}}$. (The formula:

$$\lim_{n\to\infty}\left(\frac{p}{n\log n}\right) = 1$$

may be assumed.

EXERCISE 5·20: Let U_3 be the multiplicative group of 3-adic units. When given the 3-adic topology it is a compact group admitting a Haar measure m that can be assumed normalized (i.e., $m(U_3) = 1$).

Show that the sequence $(5^n)_{n\in\mathbb{N}}$ is distributed in accordance with the measure m in U_3. (One could begin by establishing that for $n \geqslant 1$, $5^k \equiv 1 \bmod 3^n$ if and only if $2\times 3^{n-1}$ divides k).

EXERCISE 5·21: Show that if $0 < \gamma < 1$ the sequence $(n^\gamma)_{n\in\mathbb{N}}$ is equidistributed modulo 1 in a form analogous to Féjer's Criterion.

EXERCISE 5·22: Using Exercise 5·18 as inspiration, give a criterion, analogous to Féjer's Criterion, for not being equidistributed modulo 1.

EXERCISE 5·23: Give a necessary and sufficient condition for the sequence $(P(n))_{n\in\mathbb{N}}$ to be equidistributed modulo 1, where P is a generalized polynomial:

$$P(x) = \alpha_1 x^{\gamma_1} + \cdots + \alpha_s x^{\gamma_s} \qquad (\alpha_1,\ldots,\alpha_s,\gamma_1,\ldots,\gamma_s \in \mathbb{R}).$$

EXERCISE 5·24: Let α,β be real numbers. Denote by t_n the cardinality of the set of integers k such that $0 \leqslant k \leqslant n$, $0 \leqslant \{\alpha k\} \leqslant \{\beta k\}$.

What conditions must α,β satisfy in order that:

$$\frac{t_n}{(n+1)} \to \frac{1}{2} \quad \text{as } n \to \infty?$$

EXERCISE 5·25: Show that the function f from D into $\mathbb{C}$, where $D = \{z \in \mathbb{C}, |z| < 1\}$, defined by:

$$f(z) = \sum_{n=0}^{\infty} \{n\vartheta\} z^n$$

admits the circle $|z| = 1$ as a natural boundary if and only if ϑ is irrational.

EXERCISE 5·26.: Using the result of Exercise 5·13 show that the discrepancy of the sequence $(n\vartheta)_{n\in\mathbb{N}}$ satisfies $D_n^* \in O(1/\sqrt{n})$,

(i): for almost all ϑ;

(ii): for all algebraic ϑ.

(One could use the results of Khinchin and Roth on the approximation of real numbers by rational numbers).

EXERCISE 5·27: Let φ be a periodic function from $\mathbb{R}$ into $\mathbb{C}$ with period one that is Riemann integrable. Let α,β be two real numbers such that β does not belong to the vector space generated on $\mathbb{Q}$ by 1 and α.

Show that as $n \to \infty$ the quantity

$$\frac{1}{n+1} \sum_{k=0}^{n} \varphi(k\alpha)e^{2i\pi k\beta}$$

has a limit that can be calculated. From this deduce that if $n_0 < n_1 < \cdots$ denotes the sequence of integers n such that $\{n\alpha\}$ belongs to a given interval I of $[0,1]$, then the sequence $(\beta n_k)_{k\in\mathbb{N}}$ is equidistributed modulo 1 (I is of length greater than zero).

EXERCISE 5·28: Let α,β be two real numbers such that $1,\alpha,\alpha\beta$ are linearly independent on $\mathbb{Q}$.

Show that the sequence $(\beta[\alpha n])_{n\in\mathbb{N}}$ is equidistributed modulo 1.

EXERCISE 5·29: If $(u_n)_{n\in\mathbb{N}}$ satisfies $\lim n(u_{n+1} - u_n) > 0$, does it follow from this that for almost all λ the sequence $(\lambda u_n)_{n\in\mathbb{N}}$ is equidistributed modulo 1?

EXERCISE 5·30: Let ϑ be a normal number in base g.

Show that $\dfrac{\vartheta}{g-1}$ is also normal in base g.

EXERCISE 5·31: Let P be a polynomial with real coefficients and

$g \geqslant 2$ an integer. Compare the following two assertions:

(i): $(xg^n)_{n\in\mathbb{N}}$ is equidistributed modulo 1;

(ii): $(xg^n + P(n))_{n\in\mathbb{N}}$ is equidistributed modulo 1;

($x \neq 0$ is real).

EXERCISE 5·32: Let $g > 1$ be an integer and λ, $0 \leqslant \lambda \leqslant 1$, a real number with expansion in base g:

$$\lambda = \sum_{k=0}^{\infty} \frac{a_k}{g^{k+1}}, \qquad a_k \in \{0,\ldots,g-1\}.$$

Show the following are equivalent:

(i): The sequence $(\lambda g^n)_{n\in\mathbb{N}}$ is equidistributed modulo 1;

(ii): For every integer $s \geqslant 0$ the sequence $((a_n,\ldots,a_{n+s}))_{n\in\mathbb{N}}$ is equidistributed in the finite set G^{s+1}, where $G = \{0,\ldots,g-1\}$.

EXERCISE 5·33: Let $(u_n)_{n\in\mathbb{N}}$ be a sequence of strictly positive real numbers, and let $I_n = [\alpha_n,\beta_n]$ be a sequence of compact intervals of $\mathbb{R}$ such that:

$$\text{For all } n \in \mathbb{N} \quad \alpha_n \frac{u_{n+1}}{u_n} \leqslant \alpha_{n+1} \leqslant \beta_{n+1} \leqslant \beta_n \frac{u_{n+1}}{u_n}.$$

Show that there exists λ such that:

$$\text{For all } n \in \mathbb{N} \quad \lambda u_n \in I_n.$$

EXERCISE 5·34: From Exercise 5·33 preceding deduce that if the sequence (u_n) satisfies:

$$\text{For all } n \in \mathbb{N} \quad \frac{u_{n+1}}{u_n} \geqslant r, \quad \text{where } r > 2,$$

then there exists λ such that:

$$\text{For all } n \in \mathbb{N} \quad \{\lambda u_n\} \in [0, 1/(r-1)],$$

and that by also assuming $r > 3$, the set of λ's such that for any $n \in \mathbb{N}$ we have:

$$\{\lambda u_n\} \in [0, 2/(r-1)]$$

has the power of the continuum.

EXERCISE 5·35: By using the preceding result, Exercise 5·34, for a sequence (v_n), where $v_n = u_{an+b}$ and a,b are suitable integers, prove that if a sequence $(u_n)_{n\in\mathbb{N}}$ of real numbers greater than zero satisfies $\lim \frac{u_{n+1}}{u_n} > 1$, then the complement of $B((u_n))$ is everywhere dense and has the power of the continuum.

EXERCISE 5·36: Let λ, ϑ be real numbers such that $\lambda \neq 0$ and $\vartheta > 1$. Assume that there exists a sequence of integers $(u_n)_{n\in\mathbb{N}}$ and a sequence of real numbers $(\varepsilon_n)_{n\in\mathbb{N}}$ such that:

$$\lambda\vartheta^n = u_n + \varepsilon_n, \qquad \lim_{n\to\infty} \varepsilon_n = 0.$$

Show that starting from a certain index we necessarily have:

$$\left| u_{n+2} - \frac{u_{n+1}^2}{u_n} \right| < \frac{1}{2} .$$

From this deduce that the set of such pairs (λ, ϑ) is countable. Now show that the set of pairs (λ, ϑ) such that the sequence

$(\{\lambda\vartheta^n\})_{n\in\mathbb{N}}$ has a finite number of limit points is countable. Show that the pair $\left(1, \frac{1+\sqrt{5}}{2}\right)$ belongs to this set. (For a generalization to Pisot's numbers, cf., [Sal]).

EXERCISE 5·37: Show that to any sequence $(u_n)_{n\in\mathbb{N}}$ of elements of $[0,1)$ one can associate a real number α such that:

$$\lim_{n\to\infty}(\{n!\alpha\} - u_n) = 0.$$

In particular, show that $\lim_{n\to\infty}\{n!e\} = 0$.

EXERCISE 5·38: Give examples of sequences $(u_n)_{n\in\mathbb{N}}$ of real numbers such that:

$$B((u_n)) = \mathbb{R} - \{0\}, \mathbb{Z} - \{0\}, \mathbb{R} - \mathbb{Q}, \emptyset.$$

EXERCISE 5·39: Give an example of a strictly increasing sequence of integers $(u_n)_{n\in\mathbb{N}}$ such that $B((u_n)) = \mathbb{R} - E$, where E is the vector space over $\mathbb{Q}$ generated by 1 and α (α irrational). You may need to use the results of Exercise 5·27.

EXERCISE 5·40: Show that from every sequence $(u_n)_{n\in\mathbb{N}}$ of elements of $[0,1]$ that is dense in $[0,1]$ one can extract a sequence $(u_{\sigma(n)})_{n\in\mathbb{N}}$ (σ strictly increasing) equidistributed modulo 1.

EXERCISE 5·41: Prove that the function χ from $\mathbb{N}$ to $\mathbb{N}$ is almost periodic in each of the following cases:

(i): $\chi(n) = \text{card}\{k \in \mathbb{N} : [\alpha k + \beta] = n\}$, where $\alpha, \beta \in \mathbb{R}$, $\alpha > 0$;

(ii): χ is the characteristic function of the set E of integers n such that:

There exist $\alpha \in A$ and $\alpha | n$,

A a set of integers such that $\sum_{\alpha \in A} \frac{1}{\alpha} < \infty$. (Example: The set of square-free integers).

EXERCISE 5·42: Let $g \geq 2$ be an integer, λ a normal number in base g. Show that for any integer $\ell \geq 1$ the function which associates $e(\ell\lambda g^n)$ with n is pseudo-random. From this deduce, in particular, that λ is normal in base g^s for any integer $s \geq 1$. Comparing this result with that of Exercise 5·30, deduce from it that if λ is g-normal, then the same is true of every number of the form $r\lambda$, where $r \in \mathbb{Q}^*$.

EXERCISE 5·43: For integral $g \geq 2$ let $B(g)$ be the set of normal numbers with base g.

Under what condition on the integers $g,h \geq 2$ does the equality $B(g) = B(h)$ hold?

EXERCISE 5·44: Prove that if $(u_n)_{n\in\mathbb{N}}$ is a sequence of real numbers such that for every integer $p \geq 1$ $(u_{n+p} - u_n)_{n\in\mathbb{N}}$ is equidistributed modulo 1, then the function which associates $e(\ell u_n)$ to n (where $\ell \in \mathbb{N}^*$) is pseudo-random. In this way recover Van der Corput's Criterion.

EXERCISE 5·45: Let $(N_k)_{k\in\mathbb{N}}$ be a sequence of integers $N_k \geq 1$ such that $N_k \to \infty$ as $k \to \infty$. Let $(u_n)_{n\in\mathbb{N}}$ and $(v_n)_{n\in\mathbb{N}}$ be two sequences of real numbers. Construct a sequence $(w_n)_{n\in\mathbb{N}}$ in the following manner:

$$w_0 = u_0, \ldots, w_{N_0-1} = u_{N_0-1},$$

$$w_{N_0} = v_0, \ldots, w_{N_0+N_1-1} = v_{N_1-1}, \qquad w_{N_0+N_1} = u_0, \ldots .$$

More generally, $(w_0, w_1, \ldots)$ is a sequence of "blocks" $(C_0, C_1, \ldots)$ where:

$$C_{2n} = (u_0,\ldots,u_{N_{2n}}), \qquad C_{2n+1} = (v_0,\ldots,v_{N_{2n+1}}).$$

Prove that if f is a function from $\mathbb{R}$ into $\mathbb{R}$ we have the inequality:

$$\overline{\lim_n} \frac{1}{n+1} \sum_{k=0}^{n} f(w_k)$$

$$\leqslant \sup\left(\overline{\lim_n} \frac{1}{n+1} \sum_{k=0}^{n} f(u_k),\ \overline{\lim_n} \frac{1}{n+1} \sum_{k=0}^{n} f(v_k)\right).$$

EXERCISE 5·46: By using the result of Exercise 5·45 and that of Exercise 5·5(1), show that, given a function f from $[0,1]$ into $\mathbb{R}$, a necessary and sufficient condition for it to be Riemann integrable is that for every sequence $(u_n)_{n\in\mathbb{N}}$ of elements of $[0,1]$ equidistributed on $[0,1]$ the quantity

$$\frac{1}{n+1} \sum_{k=0}^{n} f(u_k)$$

has a finite limit as $n \to \infty$. (Begin by showing that this limit is independent of the sequence chosen).

EXERCISE 5·47: Let $\alpha \in \mathbb{R} - \mathbb{Q}$ and let ϑ be the transformation on $\mathbb{R}$ defined by sending $x \mapsto x + \alpha$.

Show that the only measure on $\mathbb{T}$ invariant under ϑ is Haar measure. Thus recover the result on the equidistribution modulo 1 of the sequence $(n\alpha)_{n\in\mathbb{N}}$.

EXERCISE 5·48: Apply the Ergodic Theorem to the transformation of $\mathbb{T}$ onto $\mathbb{T}$ defined by $x \to gx$, where $g \in \mathbb{Z}^*$.

EXERCISE 5·49: Let m be a positive measure on $\mathbb{T}$ that has total mass one, and let μ be the product measure on $S = \mathbb{T}^{\mathbb{N}}$.

Show that the transformation $\vartheta : S \to S$ defined by $\vartheta((u_n)_{n\in\mathbb{N}}) = (u_{n+1})_{n\in\mathbb{N}}$ preserves the measure μ and is ergodic. From this deduce that with respect to μ almost all the sequences on $\mathbb{T}$ are m-distributed.

SOLUTIONS

SOLUTION 5·0: The buses leave, for example, every thirty minutes, the 63 according to the schedule 9.00, 9.30, 10.00, etc.,..., and the 41 with the schedule 9.05, 9.35, 10.05, etc.,... . On average, five out of six times Tony will visit Claudia. (See also, Landau: *Grundlagen der Mathematik*,...).

SOLUTION 5·1: It is evident that if $D_N \to 0$ then (u_n) is equidistributed on [0,1].

Conversely, let us assume (u_n) is equidistributed on [0,1] and let $\varepsilon > 0$. Let P be an integer. By hypothesis, for every pair of integers (h,k) such that $0 \leqslant h \leqslant k \leqslant P$,

$$\left|\frac{1}{N}\, s_N\left(\frac{h}{P}, \frac{k}{P}\right) - \left(\frac{h}{P} - \frac{k}{P}\right)\right| \to 0 \text{ as } N \to \infty.$$

As the set of such pairs is finite, it follows from this that there exists an integer N_0 such that for all $N \geqslant N_0$ and for every pair (h,k) we have:

$$\frac{1}{N}\left|s_N\left(\frac{h}{P}, \frac{k}{P}\right) - \left(\frac{h}{P} - \frac{k}{P}\right)\right| < \frac{\varepsilon}{2}\,.$$

If $[\alpha,\beta]$ is an arbitrary interval there exists a pair (h,k) with

$0 \leqslant h \leqslant k \leqslant P$ such that:

$$\left[\frac{h}{P}, \frac{k}{P}\right] \supset [\alpha,\beta] \supset \left[\frac{h+1}{P}, \frac{k-1}{P}\right] .$$

From this, for $N \geqslant N_0$, we deduce:

$$\left|\frac{1}{N} s_N(\alpha,\beta) - (\beta - \alpha)\right| < \frac{\varepsilon}{2} + \frac{2}{P} .$$

It suffices to choose P such that $P > \frac{4}{\varepsilon}$ in order to deduce the result.

SOLUTION 5·2:(1): Clearly $D_N^* \leqslant D_N$. Furthermore, for any $\varepsilon > 0$ we have:

$$s_N(\alpha,\beta) \leqslant s_N(0,\beta) - s_N(0,\alpha - \varepsilon).$$

Making $\varepsilon \to 0$ one obtains $D_N \leqslant 2D_N^*$.

(2): The function from $[0,1]$ into $\mathbb{R}$ which with α associates $\varphi(\alpha) = \left|\frac{s_N(0,\alpha)}{N} - \alpha\right|$ is affine on every interval of the form (u_k, u_{k+1}). Its maximum is therefore of the form $\varphi(u_k - 0)$ or $\varphi(u_k + 0)$. Let us assume $u_0 < u_1 < \cdots < u_{N-1}$. We then have:

$$\varphi(u_k - 0) = \left|\frac{k}{N} - u_k\right| \quad \text{and} \quad \varphi(u_k + 0) = \left|\frac{k+1}{N} - u_k\right| .$$

Whence the first result. By passing to the limit this result remains true if $u_0 \leqslant u_1 \leqslant \cdots \leqslant u_{N-1}$.

The second result is obtained by noticing that if u_k is on the same side as $\frac{k+i}{N}$ ($i = 0$ or 1) relative to the centre $\frac{2k+1}{2N}$ of the interval $\left[\frac{k}{N}, \frac{k+1}{N}\right]$, we have:

$$\left|u_k - \frac{k+i}{N}\right| \leqslant \left|u_k - \frac{k+1-i}{N}\right| = \left|u_k - \frac{2k+1}{2N}\right| + \frac{1}{2N} .$$

(3): The inequality $D_N^* \geqslant \frac{1}{2N}$ clearly follows from the preceding inequality. It remains to observe that if $u_k = \frac{2k+1}{2N}$ the equality is obtained.

SOLUTION 5·3: By induction on m and by setting $2^m = N$, we see that:

$$\{u_0, \ldots, u_{N-1}\} = \left\{\frac{k}{N}, \ 0 \leqslant k < N\right\}$$

(and in particular that for all n, $u_k \in [0,1[$).

(1): From this it follows that:

$$s_N(0,\alpha) = 1 + [N\alpha].$$

From the equalities:

$$s_N(0,\alpha) = s_N(0,\alpha) + \text{card}\{k{:}N \leqslant k < n, u_k \in [0,\alpha]\}$$

and:

$$\text{card}\{k{:}N \leqslant k < n, uk \in [0,\alpha]\}$$

$$= \text{card}\{k{:}N \leqslant k < n, u_{k-N} \in [0,\alpha - 1/2N]\}$$

we deduce the desired equality. (By convention one assumes that $s_N(0,\beta) = 0$ if $\beta < 0$).

(2): By induction let us show that if $2^m < n \leqslant 2^{m+1}$, for all α such that $-\frac{1}{2^{m+1}} \leqslant \alpha \leqslant 1$ we then have:

$$|s_n(0,\alpha) - n\alpha| < 1 + \frac{3m}{2}.$$

The formula is evidently true for $m = 0$. Let us assume that it is true for all $n \leqslant 2^m$ and let n be such that $2^m < n \leqslant 2^{m+1}$. The formula is obviously true if $\alpha < 0$. Therefore it suffices to show that it is true for $0 \leqslant \alpha \leqslant 1$ also. Now,

$$|1 + [2^m\alpha] - 2^m\alpha| \leqslant 1$$

and

$$\left|\left(\alpha - \frac{1}{2^{m+1}}\right)(n - 2^m) - \alpha(n - 2^m)\right| \leqslant \frac{1}{2} .$$

By using the formula of (1) above, the result is obtained, whence the result of (2) is obtained immediately.

REMARK: In fact much better upper bounds can be obtained.

SOLUTION 5·4: Clearly (ii) => (iii). Let q,r,s be integers such that $q \geqslant 1$, $0 \leqslant r < s \leqslant q$, and let:

$$t_{q,r,s}(n) = \text{card}\{k:0 \leqslant k < n, r \leqslant [qu_k] < s\}.$$

Clearly,

$$t_{q,r,s}(n) = s\left(\frac{r}{q}, \frac{s}{q} - 0\right),$$

and $[qu_n]$ is equidistributed modulo q if and only if:

$$\frac{1}{n} s_n\left(\frac{r}{q}, \frac{s}{q} - 0\right) \to \frac{s - r}{q} \quad \text{as } n \to \infty.$$

From this it immediately follows that (i) => (ii) and that an argument analogous to the one carried out in Exercise 5·1 enables us to show that (iii) => (i).

SOLUTION 5·5:(1): Let $t_N(\alpha)$ be the number of integers n such that:

$$\frac{N(N+1)}{2} \leqslant n < \frac{(N+1)(N+2)}{2} \quad \text{and} \quad u_n \in [0,\alpha].$$

Evidently if $\alpha \in I(N,k)$, $k \leqslant t_N(\alpha) \leqslant k+2$ and $k \leqslant (N+1) \leqslant k+1$, whence:

$$|t_N(\alpha) - (N+1)\alpha| \leqslant 2.$$

If now:

$$\frac{N(N+1)}{2} \leqslant n < \frac{(N+1)(N+2)}{2} \quad \text{and} \quad M = \frac{N(N+1)}{2},$$

then:

$$|s_n(0,\alpha) - s_M(0,\alpha)| \leqslant n - M \quad \text{and} \quad s_M(0,\alpha) = \sum_{k=0}^{N-1} t_k(\alpha),$$

whence:

$$|s_M(0,\alpha) - M\alpha| \leqslant 2N,$$

whence:

$$|s_n(0,\alpha) - n\alpha| \leqslant 2(n - M) + 2N \leqslant 4N + 2 \leqslant 4\sqrt{2n} + 2,$$

so:

$$D_N^* \leqslant \frac{4\sqrt{2n} + 2}{n} \to 0 \quad \text{as } n \to \infty.$$

The sequence (u_n) is therefore equidistributed on $[0,1]$.

(2): The preceding argument used the fact that the interval $[\frac{1}{2}N(N+1), \frac{1}{2}(N+1)(N+2))$ was of approximate length $\frac{1}{2}N$, and therefore "negligeable" in comparison with the interval $[0,\frac{1}{2}N(N+1))$ (which would allow $s_n(0,\alpha)$ to be approximated by $s_M(0,\alpha)$).

The same is no longer true in the case considered here. For

example, let us assume $n = 2^N - 1 + 2^{N-1}$ and let $M = 2^N - 1$. Then:

$$s_n(0,\tfrac{1}{2}) = s_M(0,\tfrac{1}{2}) + n - M.$$

By the preceding argument one easily shows that when $N \to \infty$:

$$\frac{1}{M} s_M(0,\tfrac{1}{2}) \to \tfrac{1}{2}.$$

From this it follows that:

$$\frac{1}{n} s_n(0,\tfrac{1}{2}) \to \frac{2}{3} \neq \frac{1}{2}.$$

The sequence (u_n) is therefore not equidistributed on $[0,1]$. Neither does it admit a distribution with an arbitrary measure.

SOLUTION 5·6: If $t_p = \mathrm{card}\{n : p \leqslant \sqrt{n} \leqslant p + \alpha\}$, since $p \leqslant \sqrt{n} \leqslant p + \alpha$ is equivalent to $p^2 \leqslant n \leqslant (p + \alpha)^2$ one evidently has:

$$t_p = 1 + [2p\alpha + \alpha^2],$$

so:

$$|t_p - 2p\alpha| \leqslant 2.$$

Now let $n \geqslant 1$ and let $p = [\sqrt{n}]$ and $P = p^2$,

$$|s_n(0,\alpha) - s_P(0,\alpha)| \leqslant n - P \leqslant 2p + 1.$$

Furthermore,

$$s_P(0,\alpha) = \sum_{q=0}^{p-1} t_q,$$

whence:

$$|s_n(0,\alpha) - n\alpha| < 7p + 2 < 7\sqrt{n} + 2.$$

From this one certainly deduces that $D_n^* \to 0$, which completes the proof.

SOLUTION 5·7: If one tries to apply the preceding argument to the sequence $\log(n + 1)$ the same difficulty is met as in passing from the first to the second part of Exercise 5·5. By evaluating $s_n(0,\frac{1}{2})$ for n of the form $[e^{k+\frac{1}{2}}]$ (k an integer) we will show that this quantity does not tend towards $\frac{1}{2}$, and therefore as before the sequence $(\log(n + 1))_{n\in\mathbb{N}}$ is not equidistributed modulo 1, nor does it admit a distribution measure.

SOLUTION 5·8:(1): With p,q being relatively prime, multiplication by p is a bijection of $\mathbb{Z}/q\mathbb{Z}$ onto itself. Hence we have:

$$\operatorname{card}\{n: 0 \leqslant n < q, \left\{\frac{np}{q}\right\} \in [0,\alpha]\},$$

$$= \operatorname{card}\{h: 0 \leqslant h < q, \frac{h}{q} \in [0,\alpha]\} = 1 + [\alpha q].$$

If:

$$T_N = \operatorname{card}\{n: 0 \leqslant n < N, \left\{\frac{np}{q}\right\} \in [0,\alpha]\},$$

then we have:

$$T_q = 1 + [q\alpha].$$

Setting $r = [N/q]$, we deduce:

$$T_{rq} \leqslant T_N \leqslant T_{rq} + q,$$

whence finally we have:

$$|T_N - N\alpha| \leqslant \frac{N}{q} + q.$$

(2): Let $\varepsilon > 0$ be given. By a classic theorem on Diophantine approximation we know that there exist two integers p,q such that:

$$1 \leqslant q \leqslant \frac{\varepsilon}{3} N, \qquad \left|\vartheta - \frac{p}{q}\right| \leqslant \frac{1}{q[N\varepsilon/3]} .$$

In addition, with ε sufficiently small $p \neq 0$ certainly holds, and $(p,q) = 1$ can be assumed.

If $S_N(\alpha)$ and $T_N(\alpha)$ denote respectively the numbers of such integers n such that:

$$0 \leqslant n < N \quad \text{and} \quad \{n\vartheta\} \in [0,\alpha] \quad \text{or} \quad \left\{n \frac{p}{q}\right\} \in [0,\alpha],$$

we clearly have:

$$T\left(\alpha - \frac{N}{q[N\varepsilon/3]}\right) \leqslant S_N(\alpha) \leqslant T_N\left(\alpha + \frac{N}{q[N\varepsilon/3]}\right) ,$$

whence, on using the result of Exercise 5·7:

$$\left|\frac{S_N(\alpha)}{N} - \alpha\right| \leqslant \frac{1}{q} + \frac{q}{N} + \frac{N}{q[N\varepsilon/3]} .$$

ϑ being irrational, when $N \to \infty$ so does q. In particular, whenever N is large enough:

$$q > \frac{3}{\varepsilon} \max\left(1, \frac{N}{[N\varepsilon/3]}\right) ,$$

whence:

$$\left|\frac{S_N(\alpha)}{N} - \alpha\right| < \varepsilon.$$

SOLUTION 5·9: Modulo 5 we have $F_0 = 0$, $F_1 = 1, \ldots, F_{20} = 0$, $F_{21} = 1$. And therefore for $n = 0$ and $n = 1$ $F_{n+20} = F_n$. By induction one deduces from this that the sequence is periodic with period twenty. It only remains to establish by a further direct calculation that whenever $n \in \{0,\ldots,19\}$ F_n takes exactly every value modulo 5 four times.

REMARK: More generally, F_n is periodic modulo 5^k ($k \geq 1$ an integer) with period 4×5^k, and in each period it takes each value modulo 5^k four times, hence is equidistributed modulo 5^k. One can interpret this result in terms of a distribution in $\mathbb{Z}_5$ (5-adic integers). In addition, if F_n is equidistributed modulo q ($q > 1$ an integer) q is necessarily of the form 5^k.

SOLUTION 5·10: (1): Clearly:

$$\int \varphi_k(u)\varphi_h(u)d\lambda(u) = \begin{cases} 1 & \text{if } k = h, \\ 0 & \text{otherwise.} \end{cases}$$

From this one immediately deduces $\rho_N = \frac{1}{N}$, whence the result.

(2): The series:

$$\sum_{N=1}^{\infty} \int |f_{N^2}(u)|^2 d\lambda(u)$$

being convergent, from the Fatou-Lebesgue Theorem it immediately follows that the series $\sum_{N=1}^{\infty} f_{N^2}(u)^2$ is convergent almost everywhere, and in particular that $f_{N^2}(u) \to 0$ as $N \to \infty$.

(3): By noticing that, for $N^2 \leq n < (N+1)^2$:

$$|f_n(u) - f_{N^2}(u)| \leq n - N^2 \leq 1 + 2\sqrt{n},$$

we deduce that for almost all u, with respect to the Lebesgue measure λ:

$$\frac{1}{n} f_n(u) \to 0 \quad \text{as } n \to \infty.$$

Now, the space of functions from $\{0,1\}$ to $\mathbb{R}$ is generated by the constant function which is equal to one and by the function Φ such that $\Phi(0) = 1$, $\Phi(1) = -1$. Thus for almost all u, we have with respect to λ:

$$\frac{1}{N} \sum_{k<n} \Phi(u_k) \to 0 = \int \Phi(x)d\mu(x),$$

whence the result:

SOLUTION 5·11: (1): Let p be an integer, and let $\varepsilon > 0$. There exists $\delta > 0$ such that:

$$|x - y| < \delta \Rightarrow |e_p(x) - e_p(y)| < \frac{\varepsilon}{2}.$$

Hence there exists an integer N_0 such that for $n \geqslant N_0$:

$$|e_p(u_n) - e_p(v_n)| < \frac{\varepsilon}{2},$$

whence, if $N \geqslant 4N_0/\varepsilon$:

$$\left|\frac{1}{N} \sum_{n<N} e_p(u_n) - \frac{1}{N} \sum_{n<N} e_p(v_n)\right| < \frac{\varepsilon}{2} + \frac{2N_0}{N} < \varepsilon.$$

The result follows by applying Weyl's Criterion.

(2): Let us assume $p \geqslant 1$ is an integer and that (v_n) is equidistributed modulo 1. Weyl's Criterion implies that:

$$\frac{1}{N} \sum_{n<N} e_p(v_n) \to 0.$$

In particular, for any $\delta \geqslant 0$ there exists C_0 such that:

$$\text{For all } N > 1 \quad \left|\sum_{n<N} e_p(v_n)\right| < \delta N + C_0,$$

or:

$$\text{For all } M,N,\ N > M \geqslant 0, \quad \left|\sum_{M\leqslant n<M} e_p(v_n)\right| < 2\delta N + 2C_0.$$

Furthermore, there exists $C_1 > 0$ such that:

$$|e_p(x) - e_p(y)| < C_1(x - y).$$

Let then $\varepsilon > 0$ and M be an integer such that $\frac{2C_1}{M} < \frac{\varepsilon}{2}$, let $\delta > 0$ be such that:

$$4\delta \frac{e^{1/M}}{e^{1/M} - 1} < \frac{\varepsilon}{2},$$

and let (N_k) be the sequence defined by:

$$N_k = [e^{k/M}] - 1.$$

Let us assume that the integer N satisfies the inequality $N_k < N \leqslant N_{k+1}$. Then:

$$\left|\sum_{n<N} [e_p(u_n) - e_p(v_n)]\right| \leqslant \sum_{k=0}^{K-1} \left|\sum_{N_k\leqslant n<N_{k+1}} e_p(u_n) - e_p(v_n)\right|$$
$$+ \left|\sum_{N_k\leqslant n<N} [e_p(u_n) - e_p(v_n)]\right| .$$

If n is such that $N_k \leqslant n < N_{k+1}$:

$$\left|\log(n + 1) - \frac{k}{M}\right| < \frac{2}{M},$$

whence:

$$[e_p(u_n) - e_p(v_n)] - e_p(v_n)[e_p\left(\frac{k}{M}\right) - 1] < \frac{2C_1}{M},$$

whence:

$$\left|\sum_{N_k \leqslant n < N_{k+1}} [e_p(u_n) - e_p(v_n)]\right| \leqslant 2\,\frac{C_1}{M}\,(N_{k+1} - N_k) + 2\left|\sum_{N_k \leqslant n < N_{k+1}} e_p(v_n)\right|,$$

so:

$$\left|\sum_{N_k \leqslant n < N_{k+1}} [e_p(u_n) - e_p(v_n)]\right| \leqslant 2\frac{C_1}{M}\; N_{k+1} - N_k) + 4\delta N_{k+1} + 4C_0.$$

Similarly:

$$\left|\sum_{N_k \leqslant n < N} [e_p(u_n) - e_p(v_n)]\right| \leqslant 2\,\frac{C_1}{M}\,(N_{k+1} - N_k) + 4\delta N + 4C_0.$$

But:

$$\sum_{k=0}^{K-1} N_{k+1} \leqslant \sum_{k=0}^{K-1} e^{(k+1)/M} \leqslant \frac{e^{(K+1)/M}}{e^{1/M} - 1} \leqslant \frac{N + 2}{e^{1/M} - 1}\, e^{1/M},$$

and finally:

$$\left|\sum_{n<N} [e_p(u_n) - e_p(v_n)]\right| \leqslant 2\,\frac{C_1 N}{M} + 4\delta(N + 2)\,\frac{e^{1/M}}{e^{1/M} - 1} + 4C_0(K + 1).$$

As $N \to \infty$, $(N + 2)/N \to 1$, $(K + 1)/N \to 0$, therefore as soon as N

is large enough:

$$\left|\frac{1}{N}\sum_{n<N}[e_p(u_n) - e_p(v_n)]\right| < \varepsilon,$$

which implies that $\frac{1}{N}\sum_{n<N} e_p(u_n) \to 0$, and therefore, by Weyl's Criterion, that (u_n) is equidistributed modulo 1. Conversely, it is evident that if (u_n) is equidistributed modulo 1 we show in the same way that the same holds for (v_n).

The essential fact used in this proof is that we have divided the interval [0,1] into intervals $[N_k, N_{k+1})$ such that for $n \in [N_k, N_{k+1})$, $(u_n - v_n)$ is almost constant, the sum $\sum_{N_k \leq N} N_k$ is bounded by CN where $N \to \infty$ (C being a quantity depending only upon the approximation of $u_n - v_n$ sought). When $\delta > 1$ it would therefore be necessary to take a sequence of the type $N_k \sim e^{(k/M)^{1/\delta}}$, which is not increasing fast enough to obtain the result sought. In fact, a criterion close to that of Féjer shows that the sequence $u_n = (\log(n + 1))^\delta$ is equidistributed modulo 1 for all $\delta > 1$; then taking $v_n = 0$ we see that the preceding result does not hold when $\delta > 1$.

SOLUTION 5·12: Let us assume that given $p \in \mathbb{Z}^*$ and $\varepsilon > 0$ there exists K such that:

$$\text{For all } k \geqslant K \quad \left|\sum_{N_k \leqslant n < N_{k+1}} e_p(u_n)\right| < \frac{\varepsilon}{2}(N_{k+1} - N_k).$$

First of all, by induction on k we have:

$$\text{For all } k \geqslant K \quad \left|\sum_{N_k \leqslant n < N_{k+1}} e_p(u_n)\right| < \frac{\varepsilon}{2}(N_{k+1} - N_k).$$

And consequently if $k \geqslant K$, and $N_k < n \leqslant N_{k+1}$,

$$\left|\sum_{N_k<h\leqslant n} e_p(u_n)\right| < \frac{\varepsilon}{2}(N_k - N_K) + N_{k-1} - N_k.$$

So finally, if $k \geqslant K$, $N_k < n \leqslant N_{k+1}$:

$$\left|\sum_{0\leqslant h<n} e_p(u_n)\right| < \frac{\varepsilon}{2} N_k + N_{k+1} - N_k + N_K,$$

so again:

$$\left|\frac{1}{n}\sum_{0\leqslant h<n} e_p(u_n)\right| < \frac{\varepsilon}{2} + \frac{N_K}{n} + \left(\frac{N_{K+1}}{N_k} - 1\right).$$

As soon as n, and therefore k, is large enough:

$$\frac{N_K}{n} < \frac{\varepsilon}{4}, \qquad \frac{N_{k+1}}{N_{kk}} - 1 < \frac{\varepsilon}{4},$$

and thus:

$$\left|\frac{1}{n}\sum_{0\leqslant h<n} e_p(u_n)\right| < \varepsilon.$$

The sequence (u_n) thus satisfies Weyl's Criterion and is therefore equidistributed modulo 1.

SOLUTION 5·13: We have:

$$c_k = \int_{a-\varepsilon}^{1+a-\varepsilon} e(-kx)F(x)\mathrm{d}x$$

$$= \frac{1}{4a^2k^2\varepsilon}\,[e(-ka) - e(-ka + k\varepsilon) + e(-kb) - e(-kb - k\varepsilon)],$$

whence the upper bound:

$$|c_k| \leqslant \frac{1}{\pi^2k^2\varepsilon}.$$

The upper bound $|c_k| \leqslant 1$ evidently results from the fact that:

For all $x \in \mathbb{R}$ $\quad |e(-kx)F(x)| \leqslant 1.$

SOLUTION 5·14: Taking $a = \alpha$, $b = \beta$, $\varepsilon > 0$, we define a function F as in the preceding Exercise 5·14. If $\beta - \alpha + 2\varepsilon < 1$ we then have:

$$s_N(\alpha,\beta) = \sum_{n<N} F(u_n) = \sum_{q\in\mathbb{Z}} c_q\sigma_q(N),$$

whence the upper bound:

$$s_N(\alpha,\beta) \leqslant c_0 N + 2\sum_{q\geqslant 1} |c_q||\sigma_q(N)|,$$

and taking into account $c_0 = \beta - \alpha + \varepsilon$:

$$s_N(\alpha,\beta) - N(\beta - \alpha) \leqslant \varepsilon N + 2\sum_{q\geqslant 1} |c_q||\sigma_q(N)|.$$

If $\beta - \alpha + 2\varepsilon \geqslant 1$ we obviously have:

$$s_N(\alpha,\beta) \leqslant N,$$

and thus:

$$s_N(\alpha,\beta) - N(\beta - \alpha) \leqslant 2\varepsilon N.$$

And therefore in all cases:

$$s_N(\alpha,\beta) - N(\beta - \alpha) \leqslant 2\varepsilon N + 2\sum_{q\geqslant 1} |c_q||\sigma_q(N)|.$$

Replacing the function F by the function G, obtained this time by taking $a = \alpha + \varepsilon$, $b = \beta - \varepsilon$, one would obtain an inequality going in the opposite direction:

$$s_N(\alpha,\beta) - N(\beta - \alpha) \geqslant - 2\varepsilon N - 2 \sum_{q\geqslant 1} |c_q||\sigma_q(N)|.$$

So finally, for the discrepancy we have:

For all $\varepsilon > 0$, for all $N \geqslant 1$,

$$D_N \leqslant 2\varepsilon + 2 \sum_{q\geqslant 1} \min\left(1, \frac{1}{\pi^2 q^2 \varepsilon}\right) \left|\frac{\sigma_q(N)}{N}\right|$$

(by using the upper bound of the preceding Exercise 5·12).

One can further write:

$$D_N \leqslant 2\varepsilon + 2 \sum_{1\leqslant q\leqslant Q} \left|\frac{\sigma_q(N)}{N}\right| + 2 \sum_{q>Q} \frac{1}{\pi^2 q^2 \varepsilon}\left|\frac{\sigma_q(N)}{N}\right| .$$

Taking:

$$\varepsilon = \frac{1}{\pi} \sqrt{\sum_{q<Q} \frac{|\sigma_q(N)/N|}{q^2}} ,$$

one finds:

$$D_N \leqslant 2 \sum_{1\leqslant q\leqslant Q} \left|\frac{\sigma_q(N)}{N}\right| + \frac{4}{\pi} \sqrt{\sum_{q>Q} \frac{|\sigma_q(N)/N|}{q^2}}$$

which is the required result.

Let us notice that $|\sigma_q(N)/N| \leqslant 1$, and that:

$$\sum_{q>Q} \frac{1}{q^2} \leqslant \int_Q^\infty \frac{dx}{x^2} = \frac{1}{Q} ,$$

whence:

$$D_N \leqslant 2 \sum_{1\leqslant q\leqslant Q} \left|\frac{\sigma_q(N)}{N}\right| + \frac{4}{\pi\sqrt{Q}} .$$

In particular, if $\sigma_q(N)/N \to 1$ for a sequence for all $q \geq 1$, for any Q we will have:

$$\overline{\lim_{N\to\infty}} D_N \leq \frac{4}{\pi\sqrt{Q}} .$$

Therefore $D_N \to 0$ as $N \to \infty$, and, in particular, the sequence is equidistributed. Conversely, if a sequence is equidistributed one knows that for all $q \geq 1$ $\sigma_q(N)/N \to \infty$, therefore $D_N \to 0$, which is the result of Exercise 5·1.

SOLUTION 5·15: Let us reorder the first N terms of the sequence (u_n), say: $v_0 \leq v_1 \leq \cdots \leq v_{N-1}$, then we have:

$$\frac{1}{N}\sum_{k<N} f(u_k) = \frac{1}{N}\sum_{k<N} f(v_k) \quad \text{and} \quad \int_0^1 f = \sum_{k=0}^{N-1}\int_{k/N}^{(k+1)/N} f.$$

But as f is continuous there exists $\xi_k \in [k/N,(k+1)/N]$ such that:

$$\int_{k/N}^{(k+1)/N} f = \frac{1}{N} f(\xi_k),$$

whence:

$$\left|\frac{1}{N}\sum_{k<N} f(u_k) - \int_0^1 f\right| \leq \sup_{k<N}|f(v_k) - f(\xi_k)|.$$

But:

$$|v_k - \xi_k| \leq \sup\left(\left|v_k - \frac{k}{N}\right|, \left|v_k - \frac{k+1}{N}\right|\right) \leq D_N^*,$$

by Exercise 5·2. From this it follows that for all k:

$$|f(v_k) - f(\xi_k)| \leq \omega(D_N^*),$$

whence the result.

SOLUTION 5·16: (1): Let us introduce the metric on X defined, if $u \neq v$, by:

$$d(u,v) = 2^{-\nu(u,v)},$$

where $\nu(u,v) = \inf\{n, u_n \neq v_n\}$.

It is easy to see that restricting ρ to the set Y of sequences u such that $u_0 = 1$ gives an operation on Y. More precisely, if u begins with $(1,\dots)$ then $\pi(u)$ begins with $(1,0,\dots)$ and:

$$\text{For all } u,v \in Y \quad d(\pi(u),\pi(v)) \leqslant \tfrac{1}{2}d(u,v).$$

As π restricted to Y is a contraction it admits an unique fixed point. Furthermore, if w is a fixed point of π one has $\pi(w) \in Y$ and as $\pi(w) = w$ from this we deduce that $w \in Y$ — which gives the result we want.

In fact it is easily seen that the mapping π consists of making correspond to the sequence $u = (u_0, u_1, \dots)$ the sequence $w = \pi(u)$ formed by juxtaposing the blocks $v = (B_0, B_1, \dots)$, where B_k is the block $(1,0)$ if $u_k = 1$ and the block (1) if $u_k = 0$. Thus one obtains the beginning of w by these successive substitutions.

$$w = (1,\dots),$$

whence:

$$w = (1,0,\dots),$$

whence:

$$w = (1,0,1,\dots),$$

whence:

$$w = (\underbrace{1,0},\underbrace{1},\underbrace{1,0},\dots),$$

whence:

whence:

$$w = (\underbrace{1,0}_{1},\underbrace{1,1}_{0},\underbrace{0,1}_{1},\underbrace{0,1}_{0},\dots),$$

etc.,... .

(2): To say a distribution measure exists for the sequence $(T^n w)_{n\in\mathbb{N}}$ is equivalent (with the functions depending only upon a finite number of coordinates forming a dense subset of the set of continuous functions on X) to showing that being given a finite sequence $f = (f_0,\dots,f_m)$ the set of $n \in \mathbb{N}$ such that:

$$w_n = f_0, \quad \dots, \quad w_{n+m} = f_m$$

has a density. This is a classic result of permutation theory, the associated matrix here being the matrix $\begin{bmatrix} 0 & 1 \\ 1 & 1 \end{bmatrix}$, the square of which has strictly positive coefficients.

In this particular case this result can be recovered directly. In fact, let $\alpha = \frac{3-\sqrt{5}}{2}$ be the root of the equation $x^2 - 3x + 1 = 0$ lying between 0 and 1. Let X_0, X_1 be the intervals in [0,1] defined by:

$$X_0 = [1-\alpha, 1), \qquad X_1 = [0, 1-\alpha),$$

and let T be the transformation of [0,1) into itself defined by:

$$Tx = \{x + \alpha\}.$$

To every point $x \in [0,1)$ one can associate the sequence $(\nu(T^n x))_{\in\mathbb{N}}$ with terms in {0,1}, where one sets $\nu(x) = 0$ if $x \in X_0$, $\nu(x) = 1$ if $x \in X_1$.

Given a sequence $f = (f_0,\dots,f_m)$ *a priori*, the set of $n \in \mathbb{N}$ such that:

$$\nu(T^n x) = f_0, \quad \ldots, \quad \nu(T^{n+m} x) = f_m,$$

that is to say, furthermore, that:

$$n\alpha \in (X_{f_0} - x) \cap (X_{f_1} - x - \alpha) \cap \ldots \cap (X_{f_m} - x - m) \text{ modulo } 1$$

evidently admits a density, because of the equidistribution of the sequence $(n\alpha)$. Therefore it suffices to show that there exists an element $a \in [0,1)$ such that $w = (\omega(T^n a))_{n\in\mathbb{N}}$.

On noticing that the transformation S induced by T on the interval X_1, that is to say the transformation defined by:

$$S(x) = T^{n(x)}(x) \qquad \text{where} \quad n(x) = \inf\{n > 0, T^n(x) \in X_1\}$$

is again a rotation:

$$Sx = x - \alpha(1 - \alpha) \text{ modulo } (1 - \alpha)$$

such that if Φ denotes the affine mapping defined by:

$$\Phi(0) = 1 - \alpha, \qquad \Phi(1) = 0,$$

we have $\Phi o T = S o \Phi$. It is easily seen that the unique fixed point of Φ (say, $a = \dfrac{1 - \alpha}{2 - \alpha}$) answers the question. (Notice that, quite generally, for all $x \in X_1$ the sequence $(\nu(T^n x)_{n\in\mathbb{N}})$ is obtained from the sequence $(\nu(\Phi^{-1} S^n x))_{n\in\mathbb{N}}$ by the mapping π).

SOLUTION 5·17: In order for u to admit a distribution in $\mathbb{R}^{\mathbb{N}}$ it is necessary and sufficient that for all $s \geqslant 0$ and every finite sequence $(c_0,\ldots,c_s) \in \mathbb{R}^{s+1}$ the set of integers n, where $2^s = a_k 10^k + \cdots + a_0$, such that $a_k = c_0,\ldots, a_{k-s} = c_s$ has a density.

Now, $a_k = c_0,\ldots, a_{k-s} = c_s$ means:

$$c_0 10^k + \cdots + c_s 10^{k-s} \leqslant 2^n < c_0 10^k + \cdots + c_s 10^{k-s} + 10^{k-s},$$

so:

$$\frac{2^n}{10^k} \in [c_0 + \cdots + c_s 10^{-s}, c_0 + \cdots + c_s 10^{-s} + 10^{-s}].$$

so, moreover:

$$n\,\frac{\log 2}{\log 10} - k \in \left[\frac{\log A}{\log 10}, \frac{\log B}{\log 10}\right],$$

on setting $A = c_0 + \cdots + c_s 10^{-s}$ and $B = A + 10^{-s}$. The n's therefore satisfy:

$$\left\{n\,\frac{\log 2}{\log 10}\right\} \in \left[\frac{\log A}{\log 10}, \frac{\log B}{\log 10}\right],$$

and such an n satisfies the question. As $\frac{\log 2}{\log 10}$ is irrational the sequence $\left\{n\,\frac{\log 2}{\log 10}\right\}$ is equidistributed on $[0,1)$ and the set of n's therefore has a density equal to:

$$\frac{\log B - \log A}{\log 10}.$$

SOLUTION 5·18: Let $J = \int_0^1 \exp(it\log x)\,dx$. A classic result in Riemann integration theory shows that:

$$J = \lim_{N\to\infty} \frac{1}{N} \sum_{k=0}^{N-1} \exp\left(it\log\left(\frac{k+1}{N}\right)\right).$$

From this we deduce:

$$\frac{1}{N} \sum_{n<N} e^{it\log(n+1)} = Je^{it\log N} + 0(1).$$

In particular:

$$\lim_{N\to\infty}\left|\frac{1}{N} \sum_{n<N} e^{it\log(n+1)}\right| = |J|.$$

In order to show that for no t is the sequence $(t\log(n+1))$ equidistributed modulo 1 it suffices to show that J is never zero.

The change of variable $x = e^{-u}$ shows that:

$$J = \int_0^{+\infty} e^{-(1+it)u} du = \frac{1}{1+it} \neq 0.$$

SOLUTION 5·19: Let us assume that the sequence $(\log p_n)$ admits a distribution modulo 1. Denote by N_k and M_k the integers defined by:

$$N_k = \inf\{n, p_n > e^k\}, \qquad M_k = \inf\{n, p_n > e^{k-\frac{1}{2}}\}.$$

Now let χ be the periodic function with period one which is equal to one on $[0,\frac{1}{2})$ and to zero on $[\frac{1}{2},1)$. Clearly:

$$\sum_{n<M_k} \chi(\log p_n) = \sum_{n<N_k} \chi(\log p_n),$$

and the hypothesis that $(\log p_n)$ admits a distribution implies that the two sequences:

$$\frac{1}{M_k} \sum_{n<M_k} \chi(\log p_n) \quad \text{and} \quad \frac{1}{N_k} \sum_{n<N_k} \chi(\log p_n)$$

have the same limit ℓ as $k \to \infty$. If this limit is not zero, this implies, in particular:

$$\frac{N_k}{M_k} \to 1 \quad \text{as } k \to \infty.$$

Now, if $\pi(x)$ denotes the number of prime numbers less than or equal to x we know that:

$$\pi(x) \sim \frac{x}{\log x} \quad \text{as } x \to \infty.$$

Therefore, as $k \to \infty$ we have:

$$N_k = \pi(e^k) \sim \frac{e^k}{k} \sim \frac{e^k}{k - \frac{1}{2}} \sim \sqrt{e}M_k,$$

which is a contradiction. It remains to show that if the limit were to exist it would necessarily be different from zero. Now:

$$\sum_{n<M_k} \chi(\log p_n) \geqslant \mathrm{card}\{p : k - 1 \leqslant \log p < k - \tfrac{1}{2}\},$$

and:

$$\mathrm{card}\{p : k - 1 \leqslant \log p < k - \tfrac{1}{2}\} = \pi(e^{k-\frac{1}{2}}) - \pi(e^{k-\frac{1}{2}})$$

$$= \frac{e^{k-\frac{1}{2}}}{k - \frac{1}{2}}\left[1 - \frac{k - \frac{1}{2}}{k - 1} e^{-\frac{1}{2}}\right] + O\left(\frac{e^{k-\frac{1}{2}}}{k - \frac{1}{2}}\right),$$

and so:

$$\varliminf_{k\to\infty} \frac{1}{M_k} \sum_{n<M_k} \chi(\log p_n) \geqslant 1 - e^{-\frac{1}{2}} > 0.$$

SOLUTION 5·20: It suffices to show that for every integer $n \geqslant 1$ the sequence (5^k) is periodic modulo 3^n, and that in the fundamental interval the number of times it has a residue which is invertible modulo 3^n is independent of this residue (it is clear that it is always invertible). As the number of invertible residues modulo 3^n is equal to $2\times3^{n-1}$, it suffices to establish that the fundamental period of the sequence 5^k is equal to $2\times3^{n-1}$.

Therefore let $T(n)$ be the smallest positive integer such that:

$$5^{T(n)} = 1 \bmod 3^n.$$

Clearly, if $5^k = 5^h \bmod 3^n$, $k - h$ is a multiple of $T(n)$. In particular, $T(n)$ divides $T(n+1)$. Now, we have $T(1) = 2$ and $5^{T(1)} = 1 + 8\times 3$. From this one then easily deduces by induction that:

$$T(n) = 2\times 3^{n-1} \quad \text{and} \quad 5^{T(n)} = 1 + A(n)\times 3^n,$$

where 3 does not divide $A(n)$.

SOLUTION 5·21: It suffices to apply Féjer's Criterion to the function $t \to t^{\gamma}$.

SOLUTION 5·22: Let Φ be a strictly increasing function such that there exists $c > 0$ satisfying:

$$\Phi(k+1) - \Phi(k+\tfrac{1}{2}) \geqslant c\Phi(k+\tfrac{1}{2})$$

whenever k is large enough. If φ is the inverse function of Φ, the sequence $(\varphi(n))_{n\in\mathbb{N}}$ cannot be equidistributed modulo 1, for denoting by $S(x)$ the number of integers n such that $n \leqslant x$ and $\varphi(n) - [\varphi(n)] \in [0,\frac{1}{2})$, for any large enough k we would have:

$$S(\Phi(k+1)) = S(\Phi(k+\tfrac{1}{2})),$$

whence:

$$\frac{S(\Phi(k+1))}{\Phi(k+1)} > \frac{1}{1+c}\,\frac{S(\Phi(k+\frac{1}{2}))}{\Phi(k+\frac{1}{2})},$$

therefore $\dfrac{S(\Phi(k+1))}{\Phi(k+1)}$ and $\dfrac{S(\Phi(k+\frac{1}{2}))}{\Phi(k+\frac{1}{2})}$ cannot both converge to $\frac{1}{2}$.

To obtain a criterion analogous to Féjer's, one takes, for example, φ to be differentiable and $t\varphi'(t)$ bounded ($\varphi' \geqslant 0$). Then if $M = \sup t\varphi'(t)$ for any u we have:

$$\int_u^{ue^{1/(2M)}} \varphi'(t)dt \leqslant \frac{1}{2},$$

or again:

$$\varphi(ue^{1/(2M)}) - \varphi(u) \leqslant \frac{1}{3}\,.$$

If $u = \Phi(k + \frac{1}{2})$:

$$\varphi(e^{1/(2M)}\varphi(k + \tfrac{1}{2})) \leqslant k + \tfrac{1}{2} + \tfrac{1}{2} = k + 1,$$

whence:

$$e^{1/(2M)}\varphi(k + \tfrac{1}{2}) \leqslant \Phi(k + 1),$$

which is the relation sought, with $c = e^{1/(2M)} - 1$.

SOLUTION 5·23: By assuming $\gamma_1 > \gamma_2 > \cdots > \gamma_s$, and by applying Van der Corput's Criterion a sufficient number of times we are led to studying sequences of the type:

$$v_n = \alpha_1 p_1 \cdots p_k \gamma_1(\gamma_1 - 1)\cdots(\gamma_1 - k + 1)n^{\gamma_1 - k} + w_n,$$

where w_n tends to a limit, and where $0 < \gamma_1 - k \leqslant 1$, and $p_i \neq 0$ are integers.

If γ_1 is not integral, or if α_1 is irrational, (v_n) is equidistributed modulo 1, hence u_n is also. If γ_1 is integral and α_1 is rational, the sequence $(\alpha_1 n^{\gamma_1})$ is periodic modulo 1. If T is its period it then suffices to show that each sequence of the type:

$$u_{nT+R} - \alpha_1(nT + t)^{\gamma_1} \qquad (t \in \{0,\ldots,T - 1\})$$

is equidistributed modulo 1. By induction one thus shows that (u_n) is equidistributed modulo 1 if one of the γ's is not integral or if one of the α's is irrational. In the contrary case

(u_n) is periodic modulo 1, therefore is not equidistributed. We have thus obtained a necessary and sufficient condition.

SOLUTION 5·24: If $(1,\alpha,\beta)$ are $\mathbb{Q}$-linearly independent the equidistribution of the sequence $(\alpha n,\beta n)_{n\in\mathbb{N}}$ in $\mathbb{R}^2$ mod $\mathbb{Z}^2$ implies that the limit of $t_n/(n+1)$ exists and is equal to the measure of the set of (x,y)'s such that $0 \leqslant x \leqslant y \leqslant 1$, which is equal to $\frac{1}{2}$. If α,β are rational numbers let $\alpha = a/q$, $\beta = b/q$ with $(a,b,q) = 1$, then if we denote by r_n, s_n the integers such that $0 \leqslant r_n < q$, $0 \leqslant s_n < q$, then:

$$r_n = na \pmod q \quad \text{and} \quad s_n = nb \pmod q.$$

We have:

$$\{n\alpha\} = \frac{r_n}{q}, \qquad \{n\beta\} = \frac{s_n}{q},$$

therefore:

$$\{\alpha n\} \leqslant \{\beta n\} \iff r_n \leqslant s_n.$$

Furthermore, $r_{n+q} = r_n$ and $s_{n+q} = s_n$, and if we respectively denote by t'_n, t''_n the number of integers k such that $0 \leqslant k \leqslant n$ and $\{\alpha n\} < \{\beta n\}$ or $\{\alpha n\} = \{\beta n\}$ we see that:

$$t_n = t'_n + t''_n, \qquad \frac{t'_n}{(n+1)} \to \frac{t'}{q}, \quad \frac{t''_n}{(n+1)} \to \frac{t''}{q},$$

$$\frac{t_n}{(n+1)} \to \frac{t}{q} = \frac{(t' + t'')}{q},$$

t', t'' respectively denoting the number of integers n such that $0 \leqslant n < q$ and $r_n < s_n$ or $r_n = s_n$. But $r_{q-n} = q - r_n$ and $s_{q-n} = q - s_n$, therefore if $r_n < s_n$ we have $r_{q-n} > s_{q-n}$, from which we deduce:

$$t'' + 2t' = q,$$

so:

$$t = \frac{1}{2} + \frac{t''}{2q} .$$

If d is the g.c.d. of $b - a$ and q (with the convention $(b - a, q) = q$ if q divides $b - a$) clearly $t'' = d \geqslant 1$ holds, therefore $t_n/(n+1)$ never tends to $\frac{1}{2}$.

If α is rational and β is irrational, then $\{\beta n\}$ is equidistributed modulo 1 and $\{\alpha n\}$ periodically takes the values $\{0, \frac{1}{q}, \ldots, \frac{q-1}{q}\}$, each sequence $(\beta(qn + r))_{n\in\mathbb{N}}$ is equidistributed modulo 1 when r runs through the set of residues modulo q, and for each of these sequences the frequency when $\frac{k}{q} \leqslant \{\beta(nq + r)\}$ is therefore $1 - \frac{h}{q}$. Finally:

$$\frac{t_n}{n+1} \to \frac{1}{q} \sum_{h=0}^{q-1} 1 - \frac{h}{q} = \frac{q-1}{2q} \neq \frac{1}{2}$$

The same is true if α is irrational and β is rational.

Lastly, there remains the case where α, β are both irrational, but where $\alpha, \beta, 1$ are not independent over $\mathbb{Q}$. Arguing as in Exercise 5·8(1) we show that this time $t_n/(n + 1) \to \frac{1}{2}$.

Specifically, $t_n/(n + 1) \to \frac{1}{2}$ if and only if α, β are both irrational.

SOLUTION 5·25: Let $\alpha = p + q\vartheta$, $q \neq 0$, and p, q be integers. Since the set of points $e(\alpha)$ is dense on the circle $|z| = 1$, it suffices to show that $f(\mathrm{re}(\alpha))$ is not bounded whenever $r \to 1$. Now,

$$f(\mathrm{re}(\alpha)) = \sum_{n=0}^{\infty} \{n\vartheta\} r^n e(n\alpha).$$

If we show that:

$$\frac{1}{n+1}\sum_{k=0}^{n}\{n\vartheta\}e\{n\alpha\} \to \lambda \neq 0,$$

we will be done, for by virtue of a classical lemma, as $r \to 1$ we would have:

$$(1-r)f(\mathrm{re}(\alpha)) \to \lambda.$$

Now, with $\{n\vartheta\}$ being equidistributed on $[0,1]$:

$$\frac{1}{n+1}\sum_{k=0}^{n}\{n\vartheta\}e(n\alpha) = \frac{1}{n+1}\sum_{k=0}^{n}\{n\vartheta\}e(nq\vartheta)$$

$$\to \int_0^1 xe(x)\mathrm{d}x = \frac{1}{2\pi iq} \neq 0.$$

SOLUTION 5·26: Let us take $Q = 0$ in the upper bound in Exercise 5·13. We then have:

$$D_n \leqslant \frac{4}{\pi}\sqrt{\sum_{q\geqslant 1}\frac{|\sigma_q(n)|}{q^{2n}}},$$

it then suffices to show that:

$$\sum_{q\geqslant 1}\frac{|\sigma_q(n)|}{q^{2n}} \in O\left(\frac{1}{n}\right).$$

In the case $u_n = n\vartheta$:

$$|\sigma_q(n)| \leqslant \frac{C}{\|q\vartheta\|} \quad \text{if } \|x\| = \min_{k\in\mathbb{Z}}|x-k|.$$

Therefore it is sufficient to show that the series $\sum_{q\geqslant 1}\frac{1}{q^2\|q\vartheta\|}$ is convergent.

Now, by Khinchin's Theorem, for almost all ϑ, $\|q\vartheta\| > q^{-4/3}$ for q sufficiently large, and by a theorum of Roth the same upper bound is valid for all algebraic ϑ.

Let us then consider the sequence $q_0 < q_1 < \cdots$ of integers such that $\|q\vartheta\| < q^{-3/4}$. As the series $\sum_{q\geqslant 1} \frac{1}{q^{5/4}}$ is convergent, it suffices to shown that the series $\sum_{n\geqslant 1} \frac{1}{q_n^2\|q_n\vartheta\|}$ is convergent.

But $\|q_n\vartheta\| < q_n^{-3/4}$ and $\|q_{n+1}\vartheta\| < q_{n+1}^{-3/4}$ implies that:

$$\|(q_{n+1} - q_n)\vartheta\| < 2q_n^{-3/4},$$

therefore starting from a certain index $2q_n^{-3/4} > (q_{n+1} - q_n)^{-4/3}$, so, furthermore:

$$q_{n+1} > q_n + cq_n^{9/16}, \qquad c > 0.$$

Since $9/16 > 1/2$, it follows from this by induction that there exists $\gamma > 0$ such that whenever n is large enough $q_n > \gamma n^2$. But $\|q_n\vartheta\| > q_n^{-4/3}$, therefore $q_n^2\|q_n\vartheta\| > q_n^{2/3} > \gamma^{2/3}n^{4/3}$, and the series is certainly convergent.

SOLUTION 5·27: Let us assume α to be irrational. The given hypothesis then implies that the sequence $(n\alpha,n\beta)$ is equidistributed modulo $\mathbb{Z}^2$ in $\mathbb{R}^2$, from which it follows that:

$$\lim_{n\to\infty} \frac{1}{n+1} \sum_{k=0}^{n} \varphi(k\alpha)e(k\beta) = \int_0^1\!\!\int \varphi(x)e(y)\mathrm{d}x\mathrm{d}y = 0.$$

In particular, let us take φ to be the characteristic function of I and let us replace β by $q\beta$ ($q = 0$). We have:

$$\lim_{n\to\infty} \frac{1}{n+1} \sum_{n_k\leqslant n} e_q(n_k\beta) = 0,$$

and clearly:

$$\lim_{n\to\infty} \frac{1}{n+1} \sum_{n_k \leqslant n} 1 = \ell(I) = \int_0^1 \varphi(x)dx.$$

From the second formula it follows that the sequence (n_k) has a density equal to $\ell(I)$, that is to say (taking $n = n_k$) $\frac{k+1}{n_k+1}$ $\ell(I) \neq 0$. Putting this back into the first formula, for $q \neq 0$ gives:

$$\frac{1}{k+1} \sum_{h=0}^{k} e_q(n_h\beta) \to 0,$$

which implies that the sequence $(n_k\beta)$ is equidistributed modulo 1.

These results clearly hold when α is irrational.

SOLUTION 5·28: Let us assume, to simplify matters, that $\alpha > 1$. We then immediately verify that since we are given two integers n, k,

$$\frac{n}{\alpha} \in (k - \frac{1}{\alpha}, k) \Longleftrightarrow n = [k\alpha].$$

Therefore the sequence (n_k) of $[k\alpha]$'s is the sequence of integers n such that:

$$\left\{\frac{n}{\alpha}\right\} \in (1 - \frac{1}{\alpha}, 1).$$

We can apply the preceding Exercise 5·27 if $1, \frac{1}{\alpha}, \beta$ are Φ-linearly independent, that is to say, if $\alpha, 1, \alpha\beta$ are.

SOLUTION 5·29: The sequence $u_n = \log(n + 1)$ furnishes a counter-example.

SOLUTION 5·30: By Van der Corput's Criterion it suffices to show that:

For all $k > 0$ $\left(\frac{\vartheta}{g-1}(g^{n+k} - g^n)\right)_{n\in\mathbb{N}}$ is equidistributed modulo 1.

Now, this sequence is of the form $(p\vartheta g^n)_{n\in\mathbb{N}}$ with $p = 1 + g + \cdots + g^{k-1} \in \mathbb{Z}^*$, and Weyl's Criterion shows that if ϑ is normal in base g, then so is $p\vartheta$.

SOLUTION 5·31: By induction on the degree of P and by applying Van der Corput's Criterion, it is easily seen that (i) => (ii).

As opposed to this, by taking, for example, $x = 1$ and P as a polynomial with an irrational coefficient (not the constant term) it is seen that (ii) can be true without (i) being true.

SOLUTION 5·32: Let E be the set of functions $f_{s,c}$ of $\mathbb{R}$ into $\mathbb{R}$ that are periodic with period one, which for every pair (s,c), $s \in \mathbb{N}$, $c = (c_0,\ldots,c_s) \in G^{s+1}$ have as restriction to $[0,1)$ the characteristic function of:

$$\left[\frac{c_0}{g} + \cdots + \frac{c_1}{g^{s+1}}, \frac{c_0}{g} + \cdots + \frac{c_1}{g^{s+1}} + \frac{1}{g^{s+1}}\right) .$$

The functions in E are Riemann integrable, and it is easy to see that for every periodic function f with period one that is Riemann integrable, and for every $\varepsilon > 0$, there exist elements g,h of the vector space over $\mathbb{R}$ generated by E such that:

$$g \leqslant f \leqslant h, \qquad \int_0^1 h - \int_0^1 g < \varepsilon.$$

On the other hand:

$$\int_0^1 f_{(s,c)} = \frac{1}{g^{s+1}} ,$$

and if $\lambda \in [0,1)$ has as its expansion:

$$\lambda = \sum_{k=0}^{\infty} \frac{a_k}{g^{k+1}} .$$

The quantity $\sum_{k=0}^{n} f_{s,c}(\lambda g^k)$ is equal to the number of integers k such that:

$$0 \leqslant k \leqslant n \quad \text{and} \quad (a_k,\dots,a_{k+1}) = (c_0,\dots,c_s).$$

If property (i) holds,

$$\lim_{n\to\infty} \frac{1}{n+1} \sum_{k=0}^{n} f_{s,c}(\lambda g^k) = \int_0^1 f_{s,c} = \frac{1}{g^{s+1}} ,$$

and consequently we certainly have (ii). Conversely, if (ii) holds for every function f of E,

$$\lim_{n\to\infty} \frac{1}{n+1} \sum_{k=0}^{n} f(\lambda g \;) = \int_0^1 f.$$

This is still true for every f of the generated vector space, therefore by a classical argument (see Section 5.2 of the Introduction) it is true for every Riemann integrable function f, which show that (i) holds.

SOLUTION 5·33: Let $J_n = [\alpha_n/u_n, \beta_n/u_n]$. The hypothesis implies $J_{n+1} \subset J_n$. The decreasing closed sets $(J_n)_{n\in\mathbb{N}}$ have a non-empty intersection. Every $\lambda \in \bigcap_{n\in\mathbb{N}} J_n$ works.

SOLUTION 5·34: Let p_0 be an integer, let us set $\alpha_0 = p_0$, $\beta_0 = p_0 + \frac{1}{r-1}$. We will define $[\alpha_n,\beta_n]$ by induction so that $\alpha_n = p_n \in \mathbb{Z}$ and $\beta_n = \alpha_n + \frac{1}{r-1}$. Let us assume that $[\alpha_n,\beta_n]$ has been defined. The interval $\left[\alpha_n \frac{u_{n+1}}{u_n}, \beta_n \frac{u_{n+1}}{u_n}\right]$ has length:

$$(\beta_n - \alpha_n)\frac{u_{n+1}}{u_n} \geqslant r(\beta_n - \alpha_n) = 1 + \frac{1}{r-1},$$

therefore there exists an integer $p_{n+1} \in \left[\alpha_n \frac{u_{n+1}}{u_n}, \beta_n \frac{u_{n+1}}{u_n}\right)$, and:

$$p_{n+1} + \frac{1}{r-1} < \beta_n \frac{u_{n+1}}{u_n}.$$

If we set $\alpha_{n+1} = p_{n+1}$, $\beta_{n+1} = p_{n+1} + \frac{1}{r-1}$ the relations from the preceding Exercise 5·33 certainly hold. But then there exists λ such that $\lambda u_n \in \left(p_n, p_n + \frac{1}{r-1}\right)$ for any $n \in \mathbb{N}$. (In addition we see that there exists such a λ in every interval $[p/u_0, p + \left(\frac{1}{r-1}\right)/u_0]$ for any integer p. Since $p_n \in \mathbb{Z}$ and $\frac{1}{r-1} < 1$, we certainly have $\{\lambda u_n\} \leqslant \frac{1}{r-1}$.

If $r > 3$ let us give a sequence $\varepsilon = (\varepsilon_n)_{n\in\mathbb{N}}$, where $\varepsilon_n \in \{0,1\}$ and let us define the intervals $I_n(\varepsilon)$ by induction in the following way:

$$\text{Let } p_0 \in \mathbb{N},\ \alpha_0 = p_0,\ \beta_0 = p_0 + \frac{2}{r-1},$$

$$I_0(\varepsilon) = \left[p_0, p_0 + \frac{2}{r-1}\right].$$

Assume:

$$I_n(\varepsilon) = \left[p_n(\varepsilon), p_n(\varepsilon) + \frac{2}{r-1}\right]$$

is defined. Then let q be the integer

$$q_n \in \left[p_n(\varepsilon)\frac{u_{n+1}}{u_n}, p_n(\varepsilon)\frac{u_{n+1}}{u_n} + 1\right).$$

Take

$$I_{n+1}(\varepsilon) = \left[q_n + \varepsilon_n, q_n + \varepsilon_n + \frac{2}{r-1}\right],$$

and then we certainly have $I_{n+1}(\varepsilon) \subset I_n(\varepsilon)$, for:

$$\left[\frac{u_{n+1}}{u_n} p_n(\varepsilon), \frac{u_{n+1}}{u_n} p_n(\varepsilon) + \frac{u_{n+1}}{u_n} \frac{2}{r-1}\right] \text{ has length equal to}$$

$$\frac{2}{r-1} = 2r + \frac{2}{r-1}.$$

On the other hand, if we set $J_n(\varepsilon) = [p_n(\varepsilon)/u_n, \left(p_n(\varepsilon) + \frac{2}{r-1}\right)/u_n]$ it is easily seen that the condition $r > 3$ implies that if $\varepsilon \neq \eta$ there exists n such that $J_n(\varepsilon) \cap J_n(\eta) = \emptyset$. The points λ associated in this way with two distinct sequences ε and η will be distinct. The set in question therefore has the cardinality of the continuum.

SOLUTION 5·35: By hypothesis there exists $\rho > 1$ and N such that for all $n \geq N$ $u_{n+1}/u_n \geq \rho$. Then let $b \geq N$ and let a be such that $\rho^a = r > 3$.

By the preceding Exercise 5·34 the set of λ's such that $\{\lambda v_n\} \in [0, 2/(r-1)]$ has the cardinality of the continuum. In addition there exists such a λ on every interval of the form $[p/v_0, \left(p + \frac{2}{r-1}\right)/v_0]$. The sequence $(an + b)$ has a density $1/a$ in the set $\mathbb{N}$, from which it follows that the sequence (λu_n) cannot be equidistributed modulo 1 if $\frac{2}{r-1} < \frac{1}{a}$ (because for this sequence:

$$\underline{\lim} \frac{1}{n+1} s_n\left(0, \frac{2}{r-1}\right) \geq \frac{1}{a}).$$

The condition $\frac{2}{r-1} < \frac{1}{a}$ holds as soon as a is chosen large enough for $\rho^a > 2a + 1$, which is always possible.

Therefore the set of λ's such that (λu_n) is not equidistributed modulo 1 has the cardinality of the continuum. In addition there exists such a λ in every interval $[p/u_b, \left(p + \frac{2}{r-1}\right)/u_p]$, where $p \in \mathbb{Z}$. Since b, and therefore u_b, can be chosen arbitrarily large, the set of such λ's is dense in $\mathbb{R}$.

SOLUTION 5·36: We have:

$$u_{n+2}u_n - u_{n+1}^2 = -\lambda\vartheta^n(\varepsilon_{n+2} - 2\vartheta\varepsilon_{n+1} + \vartheta^2\varepsilon_n) + \varepsilon_{n+2}\varepsilon_n - \varepsilon_n^2.$$

As $n \to \infty$,

$$\frac{\lambda\vartheta^n}{u_n} \to 1, \qquad \varepsilon_{n+2} - 2\vartheta\varepsilon_{n+1} + \vartheta^2\varepsilon_n \to 0,$$

$$\varepsilon_{n+2}\varepsilon_n - \varepsilon_n^2 \to 0.$$

Hence:

$$\left|u_{n+2} - \frac{u_{n+1}^2}{u_n}\right| \to 0;$$

in particular, whenever n is large enough:

$$\left|u_{n+2} - \frac{u_{n+1}^2}{u_n}\right| < \frac{1}{2}.$$

With every pair (λ,ϑ) let us associate one of the triples (n,u_n,u_{n+1}) such that:

$$\text{For all } k \geqslant n \qquad \left|u_{k+2} - \frac{u_{k+1}^2}{u_k}\right| < \frac{1}{2}.$$

The set of triples $(n,a,b) \in \mathbb{N}^2$ being countable, it suffices to prove that the correspondence so defined is injective. Now, as (n,a,b) is given, if there exists (λ,α) and (μ,β) such that:

$$\begin{cases} \lambda\alpha^n = u_n + \varepsilon_n, \\ \mu\beta^n = v_n + \eta_n, \end{cases} \qquad u_n, v_n \in \mathbb{N}, \qquad \varepsilon_n, \eta_n \to 0,$$

if:

$$\text{For all } k \geq n \quad \left| u_{k+2} \quad \frac{u_{k+1}^2}{u_k} \right| < \frac{1}{2}, \qquad \left| v_{k+2} - \frac{v_{k+1}^2}{v_k} \right| < \frac{1}{2},$$

and if:

$$u_n = v_n = a, \qquad u_{n+1} = v_{n+1} = b.$$

As there exists an unique integer c satisfying $\left| c - \frac{b^2}{a} \right| < \frac{1}{2}$, we see that we must have $u_{n+2} = v_{n+2} = c$. Proceeding by induc- by induction we show that:

$$\text{For all } k \geq n \qquad u_k = v_k.$$

But then:

$$\alpha = \lim_{k\to\infty} \frac{u_{k+1}}{u_k} = \lim_{k\to\infty} \frac{v_{k+1}}{v_k} = \beta,$$

and:

$$= \lim_{k\to\infty} \frac{u_k}{\alpha^k} = \lim_{k\to\infty} \frac{v_k}{\beta^k} = \mu.$$

The correspondence is certainly injective and the set of pairs (λ,ϑ) is countable.

REMARK: Starting from a pair (a,b), where $b > a$, we show that the sequence (u_n) defined by $u_0 = a$, $u_1 = b$, and for $n \geq 0$ by $-\frac{1}{2} \leq u_{n+2} - \frac{u_{n+1}^2}{u_n} < \frac{1}{2}$ $(u_n \in \mathbb{N})$, is such that $\frac{u_{n+1}}{u_n} \to \alpha$.

The set of such α's is countable. Nevertheless, it is not known how to characterize it other than by its definition.

If $\vartheta = \frac{1 + \sqrt{5}}{2}$ the other root of the equation $x^2 - x - 1 = 0$ for ϑ is $\vartheta' = \frac{1 - \sqrt{5}}{2}$. If we set $u_n = \vartheta^n + \vartheta'^n$ we have $u_0 = 2$, $u_1 = 1$, and (u_n) satisfies the recurrence relation (of the Fibonacci sequence) $u_{n+2} = u_{n+1} + u_n$. In particular u_n is integral. As $\vartheta^n = u_n - \vartheta'^n$ and as $|\vartheta'| < 1$, it follows from this that $\vartheta^n = u_n + \varepsilon_n$, with $\varepsilon_n \to 0$.

SOLUTION 5·37: Let us specify a sequence (A_n, a_n) in the following way: set $A_0 = 0$, and, with A_n having been defined, set:

$$a_n = u_n - \{A_n\}, \qquad A_{n+1} = (n + 1)(A_n + a_n).$$

We then have:

$$\text{For all } n \in \mathbb{N} \quad |a_n| \leqslant 1, \qquad A_n + a_n = [A_n] + u_n.$$

If we set $\alpha = \sum_{n=0}^{\infty} \frac{a_n}{n!}$ a classical majorisation shows:

$$\left|\alpha - \sum_{k=0}^{n} \frac{a_k}{k!}\right| \leqslant \frac{1}{n(n!)},$$

whence:

$$\left|n!\alpha - \sum_{k=0}^{n} a_k \frac{n!}{k!}\right| \to 0.$$

It is easily seen that:

$$\sum_{k=0}^{n} a_k \frac{n!}{k!} = A_n + a_n = [A_n] + u_n,$$

whence the result. In particular we see that:

$$0 \leqslant n!e - [n!e] < \frac{1}{n}\ .$$

SOLUTION 5·38:

By Exercise 5·6, for example, if $u_n = n$, $B((u_n)) = \mathbb{R} - \{0\}$.

By Exercise 5·8(2), if $u_n = n$, $B((u_n)) = \mathbb{R} - \mathbb{Z}$.

On the other hand, if $u_n = 0$, $B((u_n)) = \emptyset$.

Finally, if (u_n) is a sequence of elements of [0,1] equidistributed on [0,1], let $\lambda \in \mathbb{R}^*$, $q \in \mathbb{Z}^*$, and we have:

$$\lim_{n\to\infty} \frac{1}{n+1} \sum_{k=0}^{n} e_q(\lambda u_k) = \int_0^1 e(\lambda q x)\mathrm{d}x = \frac{e(\lambda q) - 1}{\lambda q}\ .$$

This quantity vanishes for all $q \in \mathbb{Z}^*$ if and only if $\lambda \in \mathbb{Z}^*$, therefore $B((u_n)) = \mathbb{Z}^*$.

SOLUTION 5·39: If (n_k) is the sequence of integers defined in Exercise 5·27 we already know that if $\lambda \in \mathbb{R} - E$, (λn_k) is equidistributed modulo 1. Then let $\lambda \in E$, there exist integers $a,b,q \geqslant 1$ such that $\lambda q = a\alpha + b$. Keeping the same notation we then have:

$$\frac{1}{k+1} \sum_{h=1}^{k} (e\lambda q n_h) = \frac{n_k + 1}{k+1} \frac{1}{n_k + 1} \sum_{n=1}^{n_k} e(a\alpha n)\varphi(n\alpha)$$

$$\to \frac{1}{\ell(I)} \int_0^1 e_a(x)\varphi(x)\mathrm{d}x$$

(for $\alpha \notin \mathbb{Q} \Rightarrow (n\alpha)$ is equidistributed modulo 1).

If $I = [0,\xi]$, where $\xi \notin \mathbb{Q}$, the integral has the value ξ if $a = 0$ and $\dfrac{e(a\xi) - 1}{2\pi i a}$ if $a \neq 0$, in any case it is a number different from zero, and the sequence (λn_k) cannot be equidistributed. Therefore we certainly have $B((n_k)) = \mathbb{R} - E$.

SOLUTION 5·40: Using the result of Exercise 5·5(1) this is obvious. One may also be given a sequence $(v_n)_{n\in\mathbb{N}}$ of elements of $[0,1]$ equidistributed on $[0,1]$; one then takes:

$$\sigma(n) = \inf\left\{k : k > \sigma(n-1), |u_k - v_k| < \frac{1}{n+1}\right\}.$$

SOLUTION 5·41: (i): Since $\chi(n) = \sum_{r=0}^{q-1} \chi_r(n)$, where

$$\chi_r(n) = \text{card}\{k : [\alpha qk + \beta + r\alpha] = n\},$$

We can always assume that $\alpha > 1$ (taking q large enough). Then let φ be the periodic function of $\mathbb{R}$ into $\mathbb{R}$ with period one, the restriction of which to the interval $(\beta/(\alpha - 1), \beta/\alpha]$ is the characteristic function of the interval $(\beta/(\alpha - 1), \beta/\alpha]$, φ is clearly almost periodic and $\chi(n) = \varphi(n/\alpha)$, therefore χ is as well.

(ii): Since χ being almost periodic implies that $1 - \chi$ is almost periodic, we can limit ourselves to proving it for the characteristic function χ of E^C.

Let $1 \leqslant a_0 < a_1 < \cdots < a_k < \cdots$ be the set A, and let us set:

$$\varphi_k(x) = 1 - \frac{1}{a_k}\sum_{r=0}^{a_k-1} e\left(\frac{rx}{a}\right), \qquad \chi_k(x) = \prod_{h=0}^{k} \varphi_h(x);$$

by construction χ_k is a trigonometric polynomial. Clearly we have $\chi_k \geqslant \chi$, and the set J of $n \in \mathbb{N}$ such that $|\chi_k(n) - \chi(n)| = 1$ is therefore contained in the union $\bigcup_{h>k} J_h$ where J_h is the set of multiplies of a_h. J_h has density $1/a_h$, and therefore:

$$\|\chi - \chi_k\| \leqslant \sum_{h>k} \frac{1}{a_h} \to 0 \quad \text{as } k \to \infty.$$

In particular, as the series $\sum_{a\geqslant 1} \frac{1}{a^2}$ is convergent, we can take the set of square-free integers for E.

SOLUTION 5·42: If $\varphi(n) = e(\ell\lambda g^n)$, $\varphi(n+k)\overline{\varphi(n)} = e(p\lambda g^n)$ with $p = \ell(g^k - 1)$, its mean is therefore zero by an earlier remark (Exercise 5·30) if $k \neq 0$. From this it follows (Exercise 5·30) that if $h \neq 0$:

$$\lim_{n\to\infty} \frac{1}{n+1} \sum_{k=0}^{n} \varphi(n+k)\overline{\varphi(n)} = 0,$$

which shows that φ is pseudo-random.

If $\sigma(n) = s(n)$ the associated characteristic function is periodic and therefore almost periodic, and by virtue of the result given in the Introduction the sequence $(\lambda g^{\sigma(n)})_{n\in\mathbb{N}}$ is equidistributed modulo 1, therefore λ is g^s-normal. Conversely, it is easy to see that every g^s-normal number is g-normal.

Using the preceding result and Exercise 5·30, it follows that if λ is g-normal so is every number of the form $\dfrac{\lambda p}{g^t(g^s - 1)}$, where s,t,p are integers satisfying $s \geqslant 1$, $t \geqslant 0$, $p \neq 0$. As every rational number is of this form, the result is proved.

SOLUTION 5·43: First of all let us assume that the ratio $\log g/\log h$ is rational, say, for example:

$$\frac{\log g}{\log h} = \frac{b}{a} \quad \text{with } a \geqslant 1,\ b \geqslant 1,\ (a,b) = 1.$$

One then has $g^a = h^b$. If p is a prime, let $\alpha(p),\beta(p)$ be the exponents defined by:

$$\alpha(p) = \sup\{\alpha : p^\alpha \text{ divides } g\}, \qquad \beta(p) = \sup\{\beta : p^\beta \text{ divides } h\}.$$

Then we have $a\alpha(p) = b\beta(p)$, and consequently a divides $\beta(p)$, b divides $\alpha(p)$, and if in addition $\gamma(p) = \frac{B(p)}{a}$, then $\gamma(p) = \frac{\alpha(p)}{b}$.

If r denotes the number $\prod_p p^{\gamma(p)}$ we then have $g = r^b$, $h = r^a$, and by the result of Exercise 5·42 above:

$$B(g) = B(r) = B(h).$$

Conversely, it can be shown that if the ratio $\log g/\log h$ is irrational there exist g-normal numbers which are not h-normal. The proofs known to date are quite delicate.

SOLUTION 5·44: In the same way as above, if $\varphi(n) = e(\ell u_n)$, $\ell \in \mathbb{N}^*$, for $p \neq 0$ we have:

$$\lim_n \frac{1}{n+1} \sum_{k=0}^{n} \varphi(k+p)\overline{\varphi(k)} = 0,$$

therefore φ is pseudo-random for $\ell \in \mathbb{N}^*$, and in particular the sequence $(u_n)_{n\in\mathbb{N}}$ is equidistributed modulo 1.

SOLUTION 5·45: Let:

$$A = \sup\left(\overline{\lim} \frac{1}{n+1} \sum_{k=0}^{n} f(u_k),\ \overline{\lim} \frac{1}{n+1} \sum_{k=0}^{n} f(v_k)\right).$$

Then for any $B > A$ there exists C such that:

$$\text{For all } n \in \mathbb{N} \quad \sum_{k=0}^{n} f(u_k) \leqslant B(n+1) + C$$

$$\text{and} \quad \sum_{k=0}^{n} f(v_k) \leqslant B(n+1) + C.$$

If we set $M_0 = 0$, $M_1 = N_0$, $M_2 = N_0 + N_1$, $\ldots$, $M_k = N_0 + \cdots + N_{k-1}$, then if $M_k < n \leqslant M_{k+1}$ we have:

$$\sum_{h<n} f(w_h) = \sum_{M_k \leqslant h<n} f(w_h) + \sum_{r=0}^{k-1} \sum_{M_r \leqslant h<M_{r+1}} f(w_h),$$

whence:

$$\sum_{h<n} f(w_h) < Bn + (k + 1)C$$

But the hypothesis $N_k \to \infty$ implies that $\frac{k+1}{M} \to 0$, whence:

$$\overline{\lim} \frac{1}{n} \sum_{h<n} f(w_h) \leqslant B.$$

Since this can be done for all $B > A$ we certainly have:

$$\overline{\lim} \frac{1}{n} \sum_{h<n} f(w_h) \leqslant A.$$

SOLUTION 5·46: Taking $\overline{\lim} \frac{1}{n} \sum_{h<n} f(w_h)$ and $-\overline{\lim} \frac{1}{n} \sum_{h<n} (-f)(w_h)$ we see that if (w_h) is the sequence constructed by starting from (u_n) and (v_n), as in the preceding Exercise 5·45, we have for every real function (and therefore for every function having values in $\mathbb{C}$) f:

If: $\lim_{n\to\infty} \frac{1}{n} \sum_{h<n} f(u_h)$ and $\lim_{n\to\infty} \frac{1}{n} \sum_{h<n} f(v_h)$ exist and are the same,

Then: $\lim_{n\to\infty} \frac{1}{n} \sum_{h<n} f(w_h)$ also exists and is equal to their common values.

In particular, if (u_n) and (v_n) are equidistributed modulo 1, it follows from Weyl's Theorem that (w_n) is also equidistributed modulo 1.

On the other hand, if:

$$\frac{1}{n} \sum_{h<n} f(u_h) \to \lambda \quad \text{and} \quad \frac{1}{n} \sum_{h<n} f(v_h) \to \mu,$$

for all $\varepsilon > 0$ there exists C such that:

$$\text{For all } n \in \mathbb{N} \quad \left|\sum_{h<n} f(u_h) - \lambda n\right| < \varepsilon n + C,$$

$$\left|\sum_{h<n} f(v_h) - \mu n\right| < \varepsilon n + C.$$

We then have:

$$\left|\sum_{n<M_{2k}} f(w_h) - \lambda(N_1 + N_3 + \cdots + N_{2k-1}) - \mu(N_2 + N_4 + \cdots + N_{2k})\right|$$

$$< \varepsilon M_{2k} + 2KC,$$

and:

$$\left|\sum_{n<M_{2k}} f(w_h) - \lambda(N_1 + \cdots + N_{2k+1}) - \mu(N_2 + \cdots + N_{2k})\right|$$

$$< \varepsilon M_{2k+1} + (2k + 1)C.$$

If the sequence (N_k) is chosen so that:

$$\frac{N_1 + N_2 + \cdots + N_{k-1}}{N_k}$$

(for example $N_k = k!$), it follows from this that:

$$\lim \frac{1}{M_{2k}} \sum_{h<M_{2k}} f(w_h) = \lambda \quad \text{and} \quad \lim \frac{1}{M_{2k+1}} \sum_{h<M_{2k+1}} f(w_h) = \mu;$$

in particular, if $\lambda \neq \mu$, $\frac{1}{n} \sum_{h<n} f(w_h)$ does not have a limit as $n \to \infty$.

In this Exercise, if $\frac{1}{n} \sum_{\alpha<n} f(w_h)$ has a limit for every equi-

distributed sequence, it follows that this limit is independent of the sequence (w_h) chosen. Let ℓ be this limit. Let us now construct a sequence (u_n) as in Exercise 5·5(1) by taking u_n in $I(N, n-N)$ such that:

$$f(u_n) > \sup_{x \in I(N,n-N)} f(x) - \varepsilon.$$

If for all N we denote by S_N the quantity $\frac{1}{N+1} \sum_{k=0}^{N} \sup_{I(N,k)} f(x)$, by considering the limit of $\frac{2}{N(N+1)} \sum_{n<\frac{1}{2}N(N+1)} f(u_n)$ we deduce that we have the inequality:

$$\underline{\lim} \frac{1 \times S_0 + \cdots + NS_{N-1}}{\frac{1}{2}N(N+1)} - \varepsilon \leqslant \ell \leqslant \overline{\lim} \frac{1 \times S_0 + \cdots + NS_{N-1}}{\frac{1}{2}N(N+1)},$$

and so:

$$\frac{1 \times S_0 + \cdots + NS_{N-1}}{\frac{1}{2}N(N+1)} \to \ell;$$

but it is known that $S_N \geqslant \overline{\int} f$. Hence we have $\overline{\int} f \leqslant \ell$; by an analogous argument we deduce $\underline{\int} f \geqslant \ell$; consequently f is Riemann integrable, and its integral is equal to ℓ.

SOLUTION 5·47: As Haar measure is (up to a proportionality factor) the only measure invariant under translation, it suffices to show if μ is a measure invariant under ϑ, and β a real number, then:

$$\int_0^1 f(x)\,d\mu(x) = \int_0^1 f(x+\beta)\,d\mu(x).$$

Now, as α is irrational it follows that for all $\varepsilon > 0$ there exists n such that $\|n\alpha - \beta\| < \varepsilon$ (whence $\|x\| = \min_{k\in\mathbb{Z}}|x - k|$). f is uniformly continuous, therefore for all $\varepsilon > 0$ there exists $\eta > 0$ such that $|x - y| < \eta \Rightarrow |f(x) - f(y)| < \varepsilon$. From this it follows that for all $\eta > 0$ there exists n such that:

$$\sup_{x\in\mathbb{R}}|f(x - \beta) - f(\vartheta\, x)| < \eta.$$

Since:

$$\int_0^1 f(\vartheta\, x)\mathrm{d}\mu(x) = \int_0^1 f(x)\mathrm{d}\mu(x),$$

we certainly have the inequality sought.

SOLUTION 5·48: We must show that the transformation ϑ of $\mathbb{T}$ into $\mathbb{T}$ defined by $\vartheta(x) = gx$ preserves the Haar measure on $\mathbb{T}$.

If $[\alpha,\beta] \subset [0,1)$, the inverse image under ϑ of $[\alpha,\beta]$ is made of the g intervals $[\alpha/g + h/g, \beta/g + h/g]$, $h = 0,\ldots,g - 1$, and the measure of the union is the same as that of $[\alpha,\beta]$. This is also true for every finite union of intervals, therefore also, by passing to the limit, for every Borel set.

On the other hand, we must show that the transformation ϑ is ergodic. Let then A be a set invariant under ϑ. For any $\varepsilon > 0$ there exists a set B made up of a finite union of intervals of the type:

$$\left[\sum_{k=0}^{n} \frac{c_k}{g^{k+1}}, \sum_{k=0}^{n} \frac{c_k}{g^{k+1}} + \frac{1}{g^{n+1}}\right)$$

such that $m(A\Delta B) < \varepsilon$ (denoting the Haar measure by m and the symmetric difference by Δ). We then have for all integral $N \geqslant 1$:

$$\vartheta^{-N}(A\Delta B) = (\vartheta^{-N}A)\Delta(\vartheta^{-N}B) = A\Delta(\vartheta^{-N}B),$$

therefore:

$$m(A\Delta\vartheta^{-N}B) = m(A\Delta B) < \varepsilon.$$

If N is chosen larger than the largest of the n's entering into the definition of the intervals that form B, we have:

$$m(B\cap\vartheta^{-N}B) = m(B)m(\vartheta^{-N}B) = m(B)^2.$$

Since:

$$(A\cap B)\Delta(B\cap\vartheta^{-N}B) = B\cap(A\Delta\vartheta^{-N}B) \subset A\Delta\vartheta^{-N}B,$$

we have:

$$|m(A\cap B) - m(B\cap\vartheta^{-N}B)| < \varepsilon,$$

and similarly:

$$(A\cap B)\Delta(A\cap A) = A\cap(A\Delta B) \subset A\Delta B,$$

hence:

$$|m(A\cap B) - m(A)| < \varepsilon.$$

So finally:

$$|m(A) - m(B\cap\vartheta^{-N}B)| < 2\varepsilon.$$

On the other hand, $|m(A) - m(B)| < \varepsilon$, therefore $|m(A)^2 - m(B)^2| < 2\varepsilon$, or again:

$$|m(A) - m(A)^2| < 4\varepsilon.$$

As this holds for any $\varepsilon > 0$ we have $m(A) = m(A)^2$, so $m(A) = 0$ or 1.

SOLUTION 5·49: It suffices to verify that ϑ preserves the measure on finite unions of the "cylinders" $A_1 \times \cdots \times A_n \times \mathbb{T} \times \mathbb{T} \times \cdots$, this is clearly true. The ergodicity is proved exactly as in the preceding Exercise 5·48, by approximating every invariant set by a finite union of cylinders. From this it follows that for any continuous f of $\mathbb{T}$ into $\mathbb{C}$ there exists a set $A \subset S$ such that $\mu(A) = 1$ and for any $(u_n)_{n\in\mathbb{N}} \in A$:

$$\frac{1}{n+1}\sum_{k=0}^{n} f(u_k) \to \langle m,f\rangle .$$

Let us take a countable set of functions, say (f_n), dense with respect to uniform convergence (or even such that the vector space they generate is dense) in the set of continuous functions. To each function f_n we associate a set A_n, and if $A = \cap A_n$ we have $\mu(A) = 1$. But if $(u_n)_{n\in\mathbb{N}} \in A$, then for any continuous f and any $\varepsilon > 0$ there exists an integer p such that $|\langle m,f\rangle - \langle m,f_p\rangle| < \varepsilon$ and:

$$\text{For all } n \in \mathbb{N} \qquad \frac{1}{n+1}\left|\sum_{k=0}^{n} f(u_k) - \frac{1}{n+1}\sum_{k=0}^{n} f_p(u_k)\right| < \varepsilon .$$

From this it follows that for all $\varepsilon > 0$:

$$\overline{\lim_{n\to\infty}}\left|\frac{1}{n+1}\sum_{k=0}^{n} f(u_k) - \langle m,f\rangle\right| < 2\varepsilon ,$$

therefore that:

$$\frac{1}{n+1}\sum_{k=0}^{n} f(u_k) \to \langle m,f\rangle ,$$

and consequently every element of A is m-distributed.

CHAPTER 6
Transcendental Numbers

INTRODUCTION

NOTATIONS

$\mathbb{Q}$ denotes the *field of rational numbers*, $\mathbb{C}$ the *complex numbers*, and $\bar{\mathbb{Q}}$ the *algebraic numbers* ($\bar{\mathbb{Q}}$ is the algebraic closure of $\mathbb{Q}$ in $\mathbb{C}$). We say that a complex number is a ***TRANSCENDENTAL NUMBER*** if it does not belong to $\bar{\mathbb{Q}}$.

If $P \in \mathbb{C}[X]$ is a non-zero polynomial in one variable with complex coefficients, we *denote* by $d(P)$ the ***DEGREE** of* P, we *denote* by $H(P)$ the ***HEIGHT** of* P (the maximum of the absolute values of the coefficients of P); lastly, $s(P) = d(P) + \log H(P)$ is called the ***SIZE*** of P.

The notation $[x]$ *denotes* the ***INTEGRAL PART** of a real number* x.

The first result about transcendence (Liouville, 1844) was based on an approximation theorem: an algebraic number α cannot be approximated too well by rational numbers. One can, using this, construct transcendental numbers (Exercises 6·1 and 6·2), but these numbers are "artificial"; it is much more important to show the transcendence of a naturally occurring number, like e (the base of Napierian logarithms) or the number π. Hermite (1873) shows that e is transcendental and Lindemann (1882) showed that π is.

THEOREM (1): (Hermite-Lindemann): *Let* $\alpha \in \mathbb{C}$ *be a non-zero complex number that is algebraic over* $\mathbb{Q}$. *Then the number* $\exp(\alpha)$ *is transcendental.*

Another, classical, statement, that had already been mentioned by Euler in 1748, and again by Hilbert in 1900 (Hilbert's Seventh Problem) was proved in 1934:

THEOREM (2): (Gel'fond-Schneider): *Let* $\ell \neq 0$ *and* $b \notin \mathbb{Q}$ *be two complex numbers. At least one of the three numbers*

$$a = e^{\ell}, \quad b, \quad a^{b} = e^{b\ell},$$

is transcendental over $\mathbb{Q}$.

This Theorem was generalized by A. Baker in 1966, and his work has had some important consequences in diverse areas of number theory (cf., [Bak]).

THEOREM (3): (Baker): *Let* $\ell_1,\ldots,\ell_n$ *be* $\mathbb{Q}$*-linearly independent complex numbers. If the numbers* $e^{\ell_1},\ldots,e^{\ell_n}$ *are all algebraic, then* $1,\ell_1,\ldots,\ell_n$ *are linearly independent over the algebraic closure* $\overline{\mathbb{Q}}$ *of* $\mathbb{Q}$ *in* $\mathbb{C}$.

The proofs of these three Theorems may be found in the books [Wal] and [Bak]. The first two are also proved in [Sch] and [Lang 2].

Many of the problems in transcendence theory are as yet unsolved.

A very general conjecture about the transcendence of the values of the exponential function has been given by S. Schanel ([Lang 2]).

CONJECTURE (S): *Let* $x_1,\ldots,x_n$ *be* $\mathbb{Q}$*-linearly independent complex numbers. The transcendence degree over* $\mathbb{Q}$ *of the field*

$$\mathbb{Q}(x_1,\ldots,x_n,e^{x_1},\ldots,e^{x_n})$$

is at least n.

Let K be an extension of the field k. We say that the n elements $\alpha_1,\ldots,\alpha_n$ of K are *ALGEBRAICALLY INDEPENDENT over* k if, for every polynomial $P \in k[X_1,\ldots,X_n]$, $P(\alpha_1,\ldots,\alpha_n) = 0 \Rightarrow P = 0$.

The *TRANSCENDENCE DEGREE of* K *over* k is the maximum number of algebraically independent elements over k that can be taken from K.

Modern proofs of transcendence have in common a preliminary construction resting upon the "Pigeonhole Principle": *If objects are put into pigeonholes, and if the number of objects is greater than the number of pigeonholes, then one of the pigeonholes contains at least two objects.* In other words a mapping of a set of n elements into a set of m elements is not injective if $m < n$. This result will be used in the following form:

PIGEONHOLE PRINCIPLE: Let $F_1,\ldots,F_m$ *be disjoint sets,* E *a set having at least* $m + 1$ *elements and* $\varphi : E \to F$ *a mapping of* E *into* $F = \bigcup_{1 \leqslant j \leqslant m} F_j$. *There exist two elements* $x \neq x'$ *of* E *whose images under* φ *are in the same subset* F_j *of* F.

BIBLIOGRAPHY (Chronologically Ordered)

[Sch], [Lang2], [Wal], [Bak].

PROBLEMS

EXERCISE 6·1: Let α be an *algebraic number*. The *DEGREE* $d(\alpha)$ *of* α, the *HEIGHT* (α) *of* α, and the *SIZE* $s(\alpha)$ *of* α are the degree, height, and size of its minimal polynomial over $\mathbb{Z}$.

Recall that for every non-zero algebraic number α there exists a smallest positive integer $q \geqslant 1$ such that $q.\alpha$ is an algebraic integer; we call q the *DENOMINATOR of* α. Lastly, $N(\alpha)$ *denotes* the *NORM of an algebraic number* α.

(1): Let α be a non-zero algebraic number of degree s.

(a): By using that:

$$|N(\beta)| \geqslant 1,$$

if β is a non-zero algebraic integer show that there exists a constant $c = c(\alpha) > 0$ such that for every polynomial $P \in \mathbb{Z}[X]$ of degree n and height $H \geqslant 1$ we have:

$$P(\alpha) = 0 \quad \text{or} \quad |P(\alpha)| \geqslant \frac{c^n}{H^{s-1}} .$$

(b): From this deduce that there exists a constant $a \geqslant 0$

depending only upon α such that for every non-zero polynomial $P \in \mathbb{Z}[X]$ we have:

$$P(\alpha) = 0 \quad \text{or} \quad |P(\alpha)| \geq e^{-as(P)}.$$

(2): Let $\vartheta \in \mathbb{C}$. Assume that there exists a sequence $(\vartheta_n)_{n\geq 1}$ of pairwise distinct algebraic numbers such that

$$|\vartheta - \vartheta_n| \leq e^{-k_n s(\vartheta_n)} \quad \text{for all } n \geq 1,$$

with $\lim_{n\to+\infty} k_n = +\infty$.

Show that ϑ is transcendental.

EXERCISE 6·2: (1): Let x be a real number; assume that there exists a sequence (p_n/q_n) of rational numbers and a sequence (A_n) of real numbers, with $\lim_{n\to+\infty} A_n = +\infty$, such that:

$$0 < \left| x - \frac{p_n}{q_n} \right| \leq A_n^{-q_n}.$$

Show that for every integer $a \geq 2$ the number a^x is transcendental. (Use part (1)(a) of the preceeding Exercise 6·1).

(2): From this deduce the existence of a constant $C > 0$ such that for every rational number p/q we have:

$$\left| \frac{\log 2}{\log 3} - \frac{p}{q} \right| > C^{-q}.$$

(3): Let $\xi \in \mathbb{Q}$. Show that there exists $C(\xi) > 0$ such that:

$$\left| \xi - \frac{p}{q} \right| > \frac{C(\xi)}{q} \quad \text{if } \frac{p}{q} \neq \xi.$$

APPLICATION: Let $a_1, \ldots, a_n, \ldots$ be a sequence of integers such that the sequence $|a_n|$ be increasing (in the large , but not stationary).

Show that the number:

$$\sum_{n=1}^{\infty} \prod_{i=1}^{n} \frac{1}{a_i}$$

is irrational.

EXAMPLE: For q an integer, the numbers:

$$e^{1/q}, \quad \cos\frac{1}{q}, \quad \sin\frac{1}{q}, \quad \cosh\frac{1}{q}, \quad \sinh\frac{1}{q},$$

are irrational.

EXERCISE 6·3: APPLICATIONS OF DIRICHLET'S PIGEONHOLE PRINCIPLE

(1): A LEMMA OF SIEGEL: (a): Let $u_{i,j}$ $(1 \leqslant i \leqslant \nu, 1 \leqslant j \leqslant \mu)$ be real numbers, and let $U_1,\ldots,U_\mu$ be positive rational integers such that:

$$\sum_{i=1}^{\nu} |u_{i,j}| \leqslant U_j \qquad (1 \leqslant j \leqslant \mu).$$

Let $X_1,\ldots,X_\nu,\ell_1,\ldots,\ell_\mu$ be positive integers such that:

$$\ell_1\cdots\ell_\mu \leqslant \prod_{i=1}^{\nu} (1 + X_i).$$

Show that there exist elements $\xi_1,\ldots,\xi_\nu \in \mathbb{Z}$, not all zero, such that:

$$|\xi_i| \leqslant X_i \qquad (1 \leqslant i \leqslant \nu),$$

and:

$$\left|\sum_{i=1}^{\nu} u_{i,j}\xi_i\right| \leqslant \frac{U_j}{\ell_j} \max_{1\leqslant i\leqslant\nu} X_i \qquad (1 \leqslant j \leqslant \mu).$$

(b) Let $a_{i,j}$ $(1 \leqslant i \leqslant n, 1 \leqslant j \leqslant m)$ be complex algebraic integers. For $1 \leqslant j \leqslant m$ let us denote by K_j the subfield of $\mathbb{C}$ obtained by adjoining to $\mathbb{Q}$ the n numbers:

$$a_{1,j},\ldots,a_{n,j},$$

and let:

$$\delta_j = [K_j:\mathbb{Q}]$$

be the degree of the number field K_j. Let $A_1,\ldots,A_m$ be positive integers satisfying:

$$\max_{1\leqslant h\leqslant\delta_j} \sum_{i=1}^{n} |\sigma_h^{(j)}(a_{i,j})| \leqslant A_j \qquad (1 \leqslant j \leqslant m),$$

where:

$$\sigma_1^{(j)},\ldots,\sigma_{\delta_j}^{(j)}$$

are the different embeddings of K_j in $\mathbb{C}$ $(1 \leqslant j \leqslant m)$.

Assume

$$n > \mu = \delta_1 + \cdots + \delta_m.$$

Show that there exist rational integers $x_1,\ldots,x_n$, not all zero, satisfying:

$$\sum_{i=1}^{n} a_{i,j}x_i = 0 \qquad (1 \leqslant j \leqslant m),$$

and:

$$\max_{1\leqslant i\leqslant n} |x_i| \leqslant (2^{\mu/2}A_1^{\delta_1}\cdots A_m^{\delta_m})^{1/(n-\mu)}$$

(2): CONSEQUENCES: (a): Let $u_0,\ldots,u_m$ be real numbers and

H a positive integer.

Show that there exist rational numbers $\xi_0,\ldots,\xi_m$, not all zero, such that:

$$\max_{0\leqslant j\leqslant m} |\xi_j| \leqslant H$$

and:

$$|u_0\xi_0 + \cdots + u_m\xi_m| \leqslant (|u_0| + \cdots + |u_m|)H^{-m}.$$

(b): Let $u_0,\ldots,u_m$ be complex numbers and H a positive integer.

Show that there exist rational integers $\xi_0,\ldots,\xi_m$, not all zero, such that:

$$\max_{0\leqslant j\leqslant m} |\xi_j| \leqslant H$$

and:

$$|u_0\xi_0 + \cdots + u_m\xi_m| < \sqrt{2}(|u_0| + \cdots + |u_m|)H^{-\frac{1}{2}(m-1)}$$

(c): Let $N_1,\ldots,N_q,H$ be positive integers, and $x_1,\ldots,x_q$ complex numbers.

Show that there exists a non-zero polynomial $P\in\mathbb{Z}[X_1,\ldots,X_q]$ of degree at most N_h with respect to X_h $(1 \leqslant h \leqslant q)$ and of height less than or equal to H, such that:

$$|P(x_1,\ldots,x_q)| \leqslant \sqrt{2}H^{-\frac{1}{2}M+1}e^{c(N_1+\cdots+N_q)}$$

where:

$$M = \prod_{i=1}^{q} (1 + N_k),$$

and:

$$c = 1 + \log\max(1, |x_1|, \ldots, |x_q|).$$

(d): Show that a complex number x is transcendental if and only if, for every real number $w > 0$, there exists an integer $n > 0$ such that the inequality:

$$0 < |a_0 + a_1 x + \cdots + a_n x^n| < (\max_{0 \leq i \leq n} |a_i|)^{-w}$$

has an infinite number of solutions:

$$(a_0, \ldots, a_n) \in \mathbb{Z}^{n+1}.$$

EXERCISE 6·4: CONSEQUENCES OF THE GEL'FOND-SCHNEIDER THEOREM

(1): Show that there does not exist a real number $\alpha \neq 0$ such that the mapping $f_\alpha : \mathbb{R} \to \mathbb{R}$ defined by $f_\alpha(x) = e^{\alpha x}$ sends every real algebraic number into an algebraic number.

REMARK: One can prove that the additive group of the field $\bar{\mathbb{Q}} \cap \mathbb{R}$ of real algebraic numbers is isomorphic to the multiplicative group $\bar{\mathbb{Q}} \cap \mathbb{R}_+^*$. The above is used to prove that there is no such isomorphism which is locally increasing, cf., J. Dieudonné: *Algèbre linéaire et géométrie élémentaire*, Enseignement des Sciences, (Hermann, Paris), (1964), p. 164.

(2): Let $P \in \mathbb{Z}[X,Y]$ be an irreducible polynomial such that:

$$P'_X \neq 0, \quad P'_Y \neq 0, \quad P(0,0) \neq 0, \quad P(1,1) \neq 0.$$

Let α be an irrational algebraic number.

Show that the equation in z:

$$P(z, z^\alpha) = 0$$

does not have roots in $\overline{\mathbb{Q}}$ (that is to say, if $\ell \in \mathbb{C}$ satisfies:

$$P(e^{\ell}, e^{\ell\alpha}) = 0,$$

then e^{ℓ} is transcendental).

(3): Let $M \in M_n(\mathbb{C})$ be an $n \times n$ square matrix with coefficients in $\mathbb{C}$, and α_1, α_2 two algebraic numbers such that the matrices:

$$\exp M\alpha_1 \quad \text{and} \quad \exp M\alpha_2$$

belong to the general linear group $GL_n(\overline{\mathbb{Q}})$ of invertible $n \times n$ matrices with algebraic coefficients.

Show that if M is not nilpotent, then α_1, α_2 are $\mathbb{Q}$-linearly dependent.

EXERCISE 6·5: CONSEQUENCES OF THE HERMITE-LINDEMANN AND GEL'FOND-SCHNEIDER THEOREMS

Let $M \in M_n(\overline{\mathbb{Q}})$ be an $n \times n$ matrix with algebraic coefficients. Assume that M is not nilpotent.

(1): Show that for all $\alpha \in \overline{\mathbb{Q}}$, $\alpha \neq 0$, the matrix:

$$\exp M\alpha$$

does not belong to $GL_n(\overline{\mathbb{Q}})$.

(2): Assume that there exists $u \in \mathbb{C}$ such that:

$$\exp Mu \in GL_n(\overline{\mathbb{Q}}).$$

Show that the $\mathbb{Q}$-vector subspace of $\mathbb{C}$ generated by the eigenvalues of M has dimension equal to one, and that M is diagonalizable.

EXERCISE 6·6: CONSEQUENCES OF BAKER'S THEOREM

(1): Let $\beta_{i,j}$ $(1 \leqslant i \leqslant h, 1 \leqslant j \leqslant k)$ be algebraic numbers, and let $\gamma_1,\ldots,\gamma_k$ be non-zero algebraic numbers. For $1 \leqslant j \leqslant k$ let $\log\gamma_k$ be a non-zero determination of the logarithm of γ_k. Assume:

$$\sum_{j=1}^{k} \beta_{i,j}\log\gamma_j = 0 \qquad (1 \leqslant i \leqslant h).$$

Show that there exist rational integers $c_1,\ldots,c_k$, not all zero, such that:

$$\sum_{j=1}^{k} \beta_{i,j}c_j = 0 \qquad (1 \leqslant i \leqslant h).$$

(2): Show that Baker's Theorem (3) is equivalent to the following:

Let $\alpha_1,\ldots,\alpha_n$ *be non-zero algebraic numbers; choose a determination of the logarithm at each of the points* $\alpha_1,\ldots,\alpha_n$, *and assume that the numbers* $\log\alpha_1,\ldots,\log\alpha_n$ *are* $\mathbb{Q}$*-linearly independent. If* $\beta_0,\ldots,\beta_{n-1}$ *are algebraic numbers, then the equality*

$$\alpha_0 = \alpha_1^{\beta_1}\cdots\alpha_{n-1}^{\beta_{n-1}}e^{\beta_0}$$

(where $\alpha_i^{\beta_i} = \exp(\beta_i\log\alpha_i)$, $1 \leqslant i \leqslant n-1$*) is possible only if* $\beta_1,\ldots,\beta_{n-1}$ *are all rational numbers, and* $\beta_0 = 0$.

(3): Show that the following four statements are equivalent, and that they follow from Baker's Theorem (3):

(i): If $\ell_1,\ldots,\ell_n$ are $\mathbb{Q}$-linearly independent logarithms of algebraic numbers, then $\ell_1,\ldots,\ell_n$ are $\overline{\mathbb{Q}}$-linearly independent;

(ii): Let $\log\alpha_1,\ldots,\log\alpha_n$ be non-zero logarithms of algebraic numbers, and $\beta_1,\ldots,\beta_n$ be algebraic numbers. If:

$$1, \beta_1, \ldots, \beta_n$$

are $\mathbb{Q}$-linearly independent, then the number:

$$\alpha_1^{\beta_1} \cdots \alpha_n^{\beta_n} = \exp(\beta_1 \log\alpha_1 + \cdots + \beta_n \log\alpha_n)$$

is transcendental;

(iii): Let $\log\alpha_1, \ldots, \log\alpha_n$ be $\mathbb{Q}$-linearly independent logarithms of algebraic numbers, and $\beta_1, \ldots, \beta_n$ algebraic numbers. Assume that at least one of the numbers $\beta_1, \ldots, \beta_n$ is irrational. Then the number:

$$\alpha_1^{\beta_1} \cdots \alpha_n^{\beta_n}$$

is transcendental.

(iv): Let $L = \{\ell \in \mathbb{C} : e^{\ell} \in \bar{\mathbb{Q}}\}$ be the set of logarithms of algebraic numbers. The injection of L into $\mathbb{C}$ can be extended to a $\bar{\mathbb{Q}}$-linear injective mapping of $\bar{\mathbb{Q}} \otimes_{\mathbb{Q}} L$ into $\mathbb{C}$.

(4): Show that the following two statements are equivalent, and are consequences of Baker's Theorem (3):

(v): Every non-zero element of the set:

$$\bar{\mathbb{Q}}.L = \{\beta_1 \ell_1 + \cdots + \beta_n \ell_n : \beta_i \in \bar{\mathbb{Q}}, e^{\ell_i} \in \bar{\mathbb{Q}}, n \geqslant 0\}$$

is transcendental;

(vi): Let $\log\alpha_1, \ldots, \log\alpha_n$ be logarithms of algebraic numbers, and $\beta_0, \ldots, \beta_n$ algebraic numbers with $\beta_0 \neq 0$. Then the number:

$$e^{\beta_0} \alpha_1^{\beta_1} \cdots \alpha_n^{\beta_n} = \exp(\beta_0 + \sum_{i=1}^{n} \beta_i \log\alpha_i)$$

is transcendental.

(5): Show that Baker's Theorem (3) is equivalent to (i) and (v). (Other pairs suffice).

(6): Let $M \in M_m(\mathbb{C})$ be a matrix that is not nilpotent. Let $t_1,\ldots,t_n$ be $\mathbb{Q}$-linearly independent complex numbers such that:

$$\exp(Mt_j) \in \mathrm{GL}_m(\overline{\mathbb{Q}}) \qquad (1 \leqslant j \leqslant n).$$

Show that the numbers $t_1,\ldots,t_n$ are $\overline{\mathbb{Q}}$-linearly independent.

Show that if $M \in M_m(\overline{\mathbb{Q}})$, then the numbers $1,t_1,\ldots,t_n$ are $\overline{\mathbb{Q}}$-linearly independent.

(7):(a): Let $\varphi:\mathbb{C}^n \to \mathrm{GL}_m(\mathbb{C})$ be a non-rational analytic homomorphism; assume $\varphi(\mathbb{Z}^n) \subset \mathrm{GL}_m(\overline{\mathbb{Q}})$, and let $\beta \in \overline{\mathbb{Q}}^n$, $\beta = (\beta_1,\ldots,\beta_n) \neq 0$, be such that $\varphi(\beta) \in \mathrm{GL}_m(\overline{\mathbb{Q}})$.

Show that $1,\beta_1,\ldots,\beta_n$ are $\mathbb{Q}$-linearly dependent.

(b): Let $\varphi:\mathbb{C}^n \to \mathrm{GL}_m(\mathbb{C})$ be an analytic homomorphism such that for all $\eta \in \overline{\mathbb{Q}}^n$, $\eta \neq 0$, the homomorphism $z \mapsto \varphi(\eta z)$ of $\mathbb{C}$ into $\mathrm{GL}_m(\mathbb{C})$ is non-rational. If $\alpha_1,\ldots,\alpha_\ell$ are $\mathbb{Q}$-linearly independent elements of $\overline{\mathbb{Q}}^n$ such that $\varphi(\alpha_j) \in \mathrm{GL}_m(\overline{\mathbb{Q}})$, then $\alpha_1,\ldots,\alpha_\ell$ are $\mathbb{C}$-linearly independent.

Show that the number:

$$\int_0^1 \frac{dx}{1+x^3}$$

is transcendental.

EXERCISE 6·7: LOWER BOUNDS OF LINEAR FORMS IN LOGARITHMS

Let $a_1,\ldots,a_n$ be positive rational integers and $b_1,\ldots,b_n$ rational integers. Assume that the number:

$$\Lambda = b_1\log a_1 + \cdots + b_n\log a_n$$

is non-zero. Denote:

$$A = \max\{a_1,\ldots,a_n\}, \qquad B = \max\{|b_1|,\ldots,|b_n|\}.$$

Then:

$$|\Lambda| > A^{-nB}.$$

REMARK: This inequality is very precise as a function of A, but not as a function of B. Using his transcendence method, Baker was able to prove a more precise inequality in terms of B:

$$|\Lambda| > \exp\{-(8n)^{400n}(\log A)^{n+1}\log B\}.$$

This kind of inequality is what makes this method applicable to many problems in number theory (see, A. Baker: 'The Theory of Linear Forms in Logarithms', *Transcendence Theory, Advances and Applications*, (Academic Press), (1977)).

EXERCISE 6·8: CONSEQUENCES OF SCHANUEL'S CONJECTURE (S)

Show that the following conjectures are consequences of Schanuel's conjecture.

(1): Let $\log\alpha_1,\ldots,\log\alpha_n$ be $\mathbb{Q}$-linearly independent logarithms of algebraic numbers. Let $\beta_1,\ldots,\beta_n$ be $\mathbb{Q}$-linearly independent algebraic numbers. Then the numbers:

$$\log\alpha_1,\ldots,\log\alpha_n, e^{\beta_1},\ldots,e^{\beta_n}$$

are algebraically independent (over $\mathbb{Q}$, or over $\bar{\mathbb{Q}}$, which amounts to the same thing).

(2): Let $\log\alpha$ be a non-zero logarithm of an algebraic number α, and let $1,\beta_1,\ldots,\beta_n$ be $\mathbb{Q}$-linearly independent algebraic numbers. Then the numbers:

$$\alpha^{\beta_1},\ldots,\alpha^{\beta_n}$$

are algebraically independent.

(3): Let $u_1,\ldots,u_n$ be $\mathbb{Q}$-linearly independent complex numbers, and v a transcendental complex number. Then the transcendence degree over $\mathbb{Q}$ of the field:

$$\mathbb{Q}(e^{u_1},\ldots,e^{u_n},e^{vu_1},\ldots,e^{vu_n})$$

is $\geqslant n - 1$.

(4): The sixteen numbers:

$$e,\pi,e^{\pi},\log\pi,e^{e},\pi^{e},\pi^{\pi},\log 2,2^{\pi},2^{e},2^{i},e^{i},\pi^{i},\log 3,(\log 2)^{\log 3},$$

$$2^{\sqrt{2}},$$

are algebraically independent over $\mathbb{Q}$.

(5): If x is a complex number different from 0 and 1, one of the two numbers:

$$x,x^{e}$$

is transcendental.

(6): If x is an irrational complex number, one of the two numbers x^{x},x^{x^2} is transcendental.

EXERCISE 6·9: Assume the following result: There exists an absolute constant $C_0 > 0$ such that for every non-zero polynomial $P \in \mathbb{Z}[X]$ of degree $\leqslant n$ and size $\leqslant s$, we have:

$$\log|P(\pi)| > - C_0 ns(1 + \log n).$$

(1): Show that if ξ is a complex number, algebraic over $\mathbb{Q}(\pi)$, there exists a constant $C_1 = C_1(\xi)$ such that for every non-zero polynomial $P \in \mathbb{Z}[X]$ of degree less than or equal to n and of size less than or equal to s, we have:

$$\log|P(\xi)| > - C_1 ns(1 + \log n).$$

(2): Let $(a_m)_{m\geq 1}$ be a sequence of rational numbers, an infinite number of which are not zero. Consider the function f defined by:

$$f(z) = \sum_{m\geq 0} a_m z^m \quad \text{with } a_0 = 1.$$

Let ξ be a non-zero complex number such that the sequence $\sum_{m\geq 0} a_m \xi^m$ is convergent.

(a): Show that there exists an infinite number of m's such that the number:

$$Q_m(\xi) = - f(\xi) + 1 + a_1\xi + \cdots + a_m\xi^m$$

is non-zero.

(b): Denote by D_m the lowest common denominator of the number $a_1,\ldots,a_m$, and assume:

$$\overline{\lim_{m\to\infty}} \frac{\log|a_{m+1}|}{m(m + \log D_m)(\log m)} = - \infty.$$

Show that if $f(\xi)$ is algebraic on $\mathbb{Q}(\xi)$ the two numbers ξ and π are algebraically independent on $\mathbb{Q}$.

(*Hint:* Consider the norm of $Q_m(\xi)$ over $\mathbb{Q}(\xi)$ and use Question (1) above).

SOLUTIONS

SOLUTION 6·1: (1):(a): Let us denote the denominator of α by q. Let $P \in \mathbb{Z}[X]$,

$$P(X) = \sum_{k=0}^{n} a_k X^k, \qquad a_n \neq 0,$$

be a non-zero polynomial of degree n and height H. If $P(\alpha) \neq 0$ the number $q^n P(\alpha)$ is then a non-zero algebraic integer, therefore:

$$|N[q^n P(\alpha)]| \geqslant 1. \tag{1}$$

Let $\alpha = \alpha^{(1)}, \alpha^{(2)}, \ldots, \alpha^{(s)}$ be the conjugates of α. This yields:

$$N[q^n P(\alpha)] = \prod_{i=1}^{s} q^n P(\alpha^{(i)}) = q^{ns} P(\alpha) \prod_{i=2}^{s} P(\alpha^{(i)}). \tag{2}$$

Now,

$$|P(\alpha^{(i)})| \leqslant \sum_{k=0}^{n} |a_k| |\alpha^{(i)}|^k \leqslant H(n+1)\max\{1, |\alpha^{(i)}|^n\} \leqslant$$

(Contd)

(Contd) $\leqslant H(n+1)\Delta^{n}$,

with:

$$\Delta = 1 + \sup_{1\leqslant i\leqslant s} |\alpha^{(i)}|.$$

And so:

$$\left|\prod_{i=2}^{s} P(\alpha^{(i)})\right| \leqslant [H(n+1)\Delta^{n}]^{s-1}. \quad (3)$$

From relations (1),(2),(3) we deduce:

$$|P(\alpha)| \geqslant [H(n+1)\Delta^{n}]^{1-s}q^{-ns} \geqslant [H2^{n}\Delta^{n}]^{1-s}q^{-ns} = \frac{c^{n}}{H^{s-1}},$$

where $c = (2\Delta)^{1-s}q^{-s} > 0$ is a constant depending only upon α.

(b): By the preceding, if $P \in \mathbb{Z}[X]$ is a non-zero polynomial of degree n and height H, we have:

$$P(\alpha) = 0 \quad \text{or} \quad |P(\alpha)| \geqslant \frac{c^{n}}{H^{s-1}} = e^{-[n\log 1/c + (s-1)\log H]}.$$

The result required is obtained by setting:

$$a = \max\{\log \frac{1}{c}, s-1\} \geqslant 0.$$

(2): Argue by *reductio ad absurdum* and assume ϑ to be algebraic. By Question (1) above there exists a constant $a \geqslant 0$ such that for every non-zero polynomial $P \in \mathbb{Z}[X]$ we have:

$$P(\vartheta) = 0 \quad \text{or} \quad |P(\vartheta)| \geqslant e^{-as(P)}.$$

Let P_n be the minimal polynomial of ϑ_n; let us denote its degree by d_n and its height by h_n, so that:

$$s(\vartheta_n) = d_n + \log h_n = s(P_n).$$

For large enough n, as the ϑ_n are distinct, the numbers ϑ and ϑ_n are not conjugate, hence we have $P_n(\vartheta) \neq 0$, and consequently:

$$|P_n(\vartheta)| \geqslant e^{-as(\vartheta_n)}. \tag{4}$$

If:

$$P_n(X) = \sum_{k=0}^{d_n} a_{nk} X^k,$$

this yields:

$$|P_n(\vartheta)| = |P_n(\vartheta) - P_n(\vartheta_n)| \leqslant \sum_{k=1}^{d} |a_{nk}||\vartheta^k - \vartheta_n^k|$$

$$\leqslant d_n h_n \sup_{1 \leqslant k \leqslant d_n} |\vartheta^k - \vartheta_n^k|. \tag{5}$$

By hypothesis we have:

$$|\vartheta - \vartheta_n| \leqslant e^{-k_n s(\vartheta_n)} \quad \text{with} \quad \lim_{n \to +\infty} k_n = +\infty.$$

For large enough n this yields $|\vartheta - \vartheta_n| \leqslant 1$, and consequently, for $1 \leqslant k \leqslant d_n$:

$$|\vartheta^k - \vartheta_n^k| = |\vartheta - \vartheta_n||\vartheta^{k-1} + \cdots + \vartheta_n^{k-1}|$$

$$\leqslant k(1 + |\vartheta|)^k |\vartheta - \vartheta_n| \leqslant d_n(1 + \vartheta)^{d_n} e^{-k_n s(\vartheta_n)}.$$

Hence, by relation (5):

$$|P_n(\vartheta)| \leqslant d_n^2 h_n (1 + |\vartheta|)^{d_n} e^{-k_n s(\vartheta_n)} \tag{6}$$

$$= e^{2\log d_n + \log h_n + d_n \log(1+|\vartheta|) - k_n s(\vartheta_n)}$$

$$\leqslant e^{-(k_n - d)s(\vartheta_n)},$$

with:

$$d = 2 + \log(1 + |\vartheta|).$$

From relations (4) and (6), for n large enough we conclude:

$$(k_n - d)s(\vartheta_n) \leqslant as(\vartheta_n),$$

or, again,

$$k_n \leqslant a + d,$$

which contradicts

$$\lim_{n\to+\infty} k_n = +\infty.$$

Consequently ϑ is transcendental.

REMARK: We can show that this condition, which is sufficient for a number ϑ to be transcendental, is also necessary, and similarly that we can choose $k_n = n/4$ ($n \geqslant 1$) and $d(\vartheta_n) \leqslant n$.

SOLUTION 6·2: (1): Let us assume $a \geqslant 2$ is an integer, with $a^x = \xi$ an algebraic number of degree d and height h. Let $P(X) = X^q - a^p$.

By Question (1)(a) of the preceding Exercise 6·1 there exists $B > 0$ (independent of p and q) such that for $\frac{p}{q}$ satisfying:

$$0 < \left|\frac{p}{q} - \xi\right| \leqslant 1, \qquad \frac{p}{q} \in \mathbb{Q},$$

we have:

$$|P(\xi)| = |\xi^q - a^p| > B^{-q}. \qquad (*)$$

For $p = p_n$ and $q = q_n$, by using the inequality:

$$|e^z - 1| \leqslant |z|e^{|z|}$$

we obtain:

$$0 \leqslant |\xi^{q_n} - a^{p_n}| \leqslant a^{|x|q_n}|xq_n - p_n|\log a\ a^{|xq_n - p_n|}$$

$$\leqslant a^{(|x|+3)q_n}\left|x - \frac{p_n}{q_n}\right|$$

whenever $\left|\frac{p_n}{q_n} - x\right| \leqslant 1$, and so:

$$0 < |P(\xi)| \leqslant \left(\frac{A_n}{a^{|x|+3}}\right) \quad < B^{-q_n}$$

for sufficiently large n, which contradicts (*).

REMARK: This proof extends to the case $a \in \mathbb{Q}$, $a > 0$, $a \neq 1$

(2): It suffices to remark that $\frac{\log 2}{\log 3}$ is irrational, and that $3^{\log 2/\log 3} = 2$ is not transcendental.

REMARK: This result extends to the case where α is a positive and algebraic number, $\beta \neq 1$ a positive rational number, with $\frac{\log\alpha}{\log\beta}$ irrational.

(3): Let $\xi = \frac{a}{b}$, with a,b integers, $b > 0$. If $\frac{p}{q} \neq \xi$ we have $|aq - bp| \geqslant 1$, therefore:

$$\left|\xi - \frac{p}{q}\right| \geqslant \frac{C(\xi)}{q} \quad \text{with } C(\xi) = \frac{1}{b}.$$

APPLICATION: Let $\xi = \sum_{n=1}^{\infty} \prod_{i=1}^{n} \frac{1}{a_i}$; for N a positive integer, we denote:

$$q_N = |a_1 \cdots a_N| \quad \text{and} \quad p_N = q_N \sum_{n=1}^{N} \prod_{i=1}^{h} \frac{1}{a_i} .$$

Whenever $a_{N+2} \geqslant 2$ we have:

$$\left|\xi - \frac{p_N}{q_N}\right| \leqslant \frac{1}{q_N} \sum_{h=N+1}^{\infty} \prod_{i=N+1}^{h} \frac{1}{|a_i|}$$

$$\leqslant \frac{2}{|a_{N+1}| q_N} .$$

Since $\frac{2}{|a_{N+1}|} \to 0$ when $N \to +\infty$, we deduce from this that ξ is irrational.

Using the classical expansions of the functions $\exp z$, $\cos z$, $\sin z$, $\cosh z$, $\sinh z$ as power series finishes off the numbers mentioned. For example, for $\exp(1/q)$ we will take $a_n = nq$.

SOLUTION 6·3: (1):(a): Let us consider the mapping φ of the set $\mathbb{N}(X_1,\ldots,X_\nu) = \{(\xi_1,\ldots,\xi_\nu) \in \mathbb{Z}^\nu : 0 \leqslant \xi_i \leqslant X_i, 1 \leqslant i \leqslant \nu\}$ into $\mathbb{R}^\mu$ which to $(\xi_1,\ldots,\xi_\nu)$ associates $(\eta_1,\ldots,\eta_\mu)$ with:

$$\eta_j = \sum_{i=1}^{\nu} u_{i,j}\xi_i \qquad (1 \leqslant j \leqslant \mu).$$

For $1 \leqslant j \leqslant \mu$ denote by $-v_j$ (*resp.* w_j) the sum of the negative (*resp.* positive) elements of the set:

$$\{u_{1,j},\ldots,u_{\nu,j}\}.$$

Hence we will have:

$$v_j + w_j \leqslant U_j \qquad (1 \leqslant j \leqslant \mu).$$

We notice that if $(\xi_1,\ldots,\xi_\nu)\in\mathbb{N}(X_1,\ldots,X_\nu)$ then the image $(\eta_1,\ldots,\eta_\mu) = \varphi(\xi_1,\ldots,\xi_\nu)$ belongs to the set:

$$E = \{(\eta_1,\ldots,\eta_\mu)\in\mathbb{R}^\mu : -v_jX \leqslant \eta_j \leqslant w_jX, 1 \leqslant j \leqslant \mu\},$$

with $X = \max_{1\leqslant i\leqslant\nu} X_i$. For $1 \leqslant j \leqslant \mu$ the interval $[-v_jX, w_jX]$ is divided into ρ_i equal intervals (of length less that or equal to U_jX/ℓ_j), which means that E is divided into $\ell_1\cdots\ell_\mu$ subsets E_k $(1 \leqslant k \leqslant \ell_1\cdots\ell_\mu)$. Since:

$$\ell_1\cdots\ell_\mu < \prod_{i=1}^{\nu}(1 + X_i) = \operatorname{card}\mathbb{N}(X_1,\ldots,X_\nu)$$

we can use the Pigeonhole Principle to show there exist two distinct elements ξ^* and ξ^{**} of $\mathbb{N}(X_1,\ldots,X_\nu)$ whose images under φ belong to the same subset E_k of E. Let us write ξ as the difference (componentwise) of ξ^* and ξ^{**}; we will have

$$\xi\in\mathbb{Z}^\nu,\quad \xi = (\xi_1,\ldots,\xi_\nu) \neq (0,\ldots,0) \quad\text{with } |\xi_i| \leqslant X_i \quad (1 \leqslant i \leqslant \nu),$$

and the image $\eta = (\eta_1,\ldots,\eta_\mu) = \varphi(\xi)$ of ξ under φ satisfies:

$$|\eta_j| \leqslant \frac{U_j}{\ell_j}X \qquad (1 \leqslant j \leqslant \mu),$$

hence the result.

(b): For $1 \leqslant j \leqslant m$ the different embeddings:

$$\sigma_1^{(j)},\ldots,\sigma_{\delta_j}^{(j)}$$

of K_j into $\mathbb{C}$ are numbered in such a way that we have:

$$\sigma_h^{(j)}(K_j)\subset\mathbb{R} \quad\text{for } 1 \leqslant h \leqslant r_j,$$

$$\sigma^{(j)}_{r_j+s_j+k} = \overline{\sigma^{(j)}_{r_j+k}} \quad \text{(complex conjugate of } \sigma^{(j)}_{r_j+k}\text{) for } 1 \leqslant k \leqslant s_j,$$

where r_j and s_j are two integers satifying:

$$\delta_j = r_j + 2s_j.$$

Define the mappings $\tau^{(j)}_h$, $1 \leqslant h \leqslant \delta_j$, of K_j into $\mathbb{R}$ by:

$$\tau^{(j)}_h = \begin{cases} \sigma^{(j)}_h & \text{for } 1 \leqslant h \leqslant r_j, \\ \mathrm{Re}(\sigma^{(j)}_h) & \text{for } r_j + 1 \leqslant h \leqslant r_j + s_j, \\ \mathrm{Im}(\sigma^{(j)}_h) & \text{for } r_j + s_j + 1 \leqslant h \leqslant \delta_j. \end{cases}$$

Define a positive integer X by:

$$X = [\sqrt{2}^{\mu/(n-\mu)} A_1^{\delta_1/(n-\mu)} \cdots A_m^{\delta_m/(n-\mu)}],$$

where $\mu = \delta_1 + \cdots + \delta_m$. Thus we have:

$$(1 + X)^{n-\mu} > \sqrt{2}^{\mu} A_1^{\delta_1} \cdots A_m^{\delta_m},$$

therefore, because $A_j \geqslant 1$ for $1 \leqslant j \leqslant m$,

$$(1 + X)^n > \prod_{j=1}^{m} (1 + \sqrt{2}A_j X)^{\delta_j}.$$

The preceding exercise (with $\nu = n$, $\mu = \delta_1 + \cdots + \delta_m$, and $\ell_j = [1 + \sqrt{2}A_j X]$, each ℓ_j being repeated δ_j times, $1 \leqslant j \leqslant m$) shows that there exist rational integers $x_1, \ldots, x_n$, not all zero, satisfying:

$$\max_{1\leqslant i\leqslant n} |x_i| \leqslant X,$$

and:

$$\max_{1\leqslant h\leqslant \delta_j} \left|\sum_{i=1}^{n} \tau_h(a_{i,j})x_i\right| \leqslant \frac{A_jX}{1 + [\sqrt{2}A_jX]} \qquad (1 \leqslant j \leqslant m).$$

From this we deduce:

$$\max_{1\leqslant h\leqslant r_j} \left|\sum_{i=1}^{n} \sigma_h(a_{i,j})x_i\right| \leqslant \frac{A_jX}{1 + [\sqrt{2}A_jX]} \qquad (1 \leqslant j \leqslant m),$$

and:

$$\max_{r_j+1\leqslant h\leqslant \delta_j} \left|\sum_{i=1}^{n} \sigma_h(a_{i,j})x_i\right| \leqslant \left(\frac{\sqrt{2}A_jX}{1 + [\sqrt{2}A_jX]}\right) \qquad (1 \leqslant j \leqslant m).$$

And so:

$$\left|N_{K_j/\mathbb{Q}}\left(\sum_{i=1}^{n} a_{i,j}x_i\right)\right| \leqslant 2^{s_j}\left(\frac{A_jX}{1 + [\sqrt{2}A_jX]}\right)^{\delta_j} \qquad (1 \leqslant j \leqslant m);$$

in this last inequality the left hand side is a rational integer (since $\sum_{i=1}^{n} a_{i,j}x_i$ are integers of K_j), and the right hand side is bounded , because $s_j \leqslant \frac{1}{2}\delta_j$, by:

$$\left(\frac{\sqrt{2}A_jX}{1 + [\sqrt{2}A_jX]}\right)^{\delta_j} < 1.$$

Hence:

$$\sum_{i=1}^{n} a_{i,j}x_i = 0 \qquad (1 \leqslant j \leqslant m),$$

with:

$$0 < \max_{1\leqslant i\leqslant n} |x_i| \leqslant X \leqslant \sqrt{2}^{\mu/(n-\mu)} (A_1^{\delta_1}\cdots A_m^{\delta_m})^{1/(n-\mu)}.$$

Hence the result.

Let us notice that if ℓ is an integer, $0 \leqslant \ell \leqslant m$, such that the fields $K_1,\ldots,K_\ell$ are totally real (that is to say $\sigma_h^{(j)}(K_j) \subset \mathbb{R}$ for $1 \leqslant h \leqslant \delta_j$, $1 \leqslant j \leqslant \ell$), then the coefficient

$$\sqrt{2}^{\mu/n - \mu}$$

can be replaced by

$$\sqrt{2}^{(\delta_{\ell+1}+\cdots+\delta_m)/(n-\mu)}$$

which is one if $\ell = m$).

(2):(a): The result required is an immediate consequence of Question (1)(a) (with $\mu = 1$, $\nu = m + 1$, $X_i = H$, and $\ell_1 = H^{m+1}$).

(2):(b): Question (1)(a) is used again, separating the real and imaginary parts. For $0 \leqslant j \leqslant m$ let:

$$v_j = \mathrm{Re} u_j \quad \text{and} \quad w_j = \mathrm{Im} u_j;$$

let us choose $\mu = 2$, $\mu = m + 1$, $X_i = H$, $\ell_1 = \ell_2 = H^{(m+1)/2}$; there exist rational integers $\xi_0,\ldots,\xi_m$, not all zero, such that:

$$\max_{0\leqslant j\leqslant m} |\xi_j| \leqslant H$$

and:

$$\begin{cases} |v_0\xi_0 + \cdots + v_m\xi_m| \leqslant \sum_{i=1}^{m} |v_i| \dfrac{H}{H^{(m+1)/2}}, \\ \\ |w_0\xi_0 + \cdots + w_m\xi_m| \leqslant \sum_{i=1}^{m} |w_i| \dfrac{H}{H^{(m+1)/2}}. \end{cases}$$

As:

$$|u_0\xi_0 + \cdots + u_m\xi_m|^2 = |v_0\xi_0 + \cdots + v_m\xi_m|^2 + |w_0\xi_0 + \cdots + w_m\xi_m|^2,$$

and as:

$$\sum_{i=0}^{m} |v_i| \leqslant \sum_{i=0}^{m} |u_i|, \qquad \sum_{i=0}^{m} |w_i| \leqslant \sum_{i=0}^{m} |u_i|,$$

the result is obtained.

(2):(c): We use the preceding Question with:

$$\{u_0,\ldots,u_m\} = \{x_1^{\lambda_1}\cdots x_q^{\lambda_q}, 0 \leqslant \lambda_h \leqslant N_h, 1 \leqslant h \leqslant q\};$$

therefore:

$$m + 1 = \prod_{h=1}^{q} (1 + N_h) = M,$$

and there exist integers $\xi_0,\ldots,\xi_h$, not all zero, such that:

$$|u_0\xi_0 + \cdots + u_m\xi_m| \leqslant \sqrt{2}\left(\sum_{i=0}^{m} |u_i|\right) H^{-\frac{1}{2}M+1};$$

now:

$$\sum_{i=0}^{m} |u_i| = \sum_{\lambda_1=0}^{N_1} \cdots \sum_{\lambda_q=0}^{N_q} |x_1^{\lambda_1}\cdots x_q^{\lambda_q}|$$

$$\leqslant (N_1 + 1)\cdots(N_q + 1) \prod_{h=1}^{q} [\max(1,|x_h|)]^{N_h};$$

we majorise $N_h + 1$ by e^{N_h} for $1 \leqslant h \leqslant q$, and the result follows:

(2):(d): Let x be a complex number and w a positive real number; an integer $N \geqslant 2w + 3$ (for example $N = [2w + 4]$) is chosen, and we define:

$$H_0 = \sqrt{2}e^N \max(1, |x|))^N.$$

By the preceding Question, for all $H > 0$ (therefore for all $H > H_0$) there exists a non-zero polynomial $P \in \mathbb{Z}[X]$ of degree less than or equal to N and height less than or equal to H, such that:

$$|P(x)| \leqslant H_0 H^{-\frac{1}{2}N+\frac{1}{2}};$$

if $H > H_0$ we will therefore have:

$$|P(x)| < H^{-w} \leqslant (H(P))^{-w},$$

where $H(P)$ denotes the height of the polynomial P.

Let us assume that x is transcendental, it is then easy to construct by induction on k a sequence $(P_k)_{k \geqslant 0}$ of polynomials of $\mathbb{Z}[X]$, of degree less than or equal to N and height $H(P_k)$, satisfying:

$$0 < |P_{k+1}(x)| < |P_k(x)| \quad \text{and} \quad |P_k(x)| < H(P_k)^{-w}$$

for all $k \geqslant 0$ (if $P_0, \ldots, P_k$ are constructed, choose an integer H such that $|P_k(x)| > H^{-w}$; this is possible since x is transcendental; then apply the preceding construction).

Therefore if x is transcendental, for every real $w > 0$ there exists an integer $N > 0$ such that the inequality:

$$0 < |P(x)| < (H(P))^{-w}$$

has an infinite number of solutions $P \in \mathbb{Z}[X]$, with the degree of $P \leqslant N$.

Conversely, let us assume x to be algebraic of degree δ over $\mathbb{Q}$; let d be a denominator of x. Let us select $w = \delta$, and let n be an arbitrary positive integer. Let us notice that the set of the $(n + 1)$-tuples $(a_0,\ldots,a_n) \in \mathbb{Z}^{n+1}$, with:

$$\max_{0\leq i\leq n} |a_i| \leq d^{\delta n}(n + 1)^{\delta-1} \max_{\sigma\neq 1}(1,|\sigma(x)|)^{n(\delta-1)},$$

(where $\{\sigma\}$ denotes the set of embeddings of K in $\mathbb{C}$) is finite. Let $(a_0,\ldots,a_n) \in \mathbb{Z}^{n+1}$ be such that:

$$\max_{0\leq i\leq n} |a_i| > d^{\delta n}(n + 1)^{\delta-1} \max_{\sigma\neq 1}(1,|\sigma(x)|)^{n(\delta-1)},$$

and:

$$a_0 + a_1 x + \cdots + a_n x^n \neq 0.$$

We are going to show that:

$$|a_0 + a_1 x + \cdots + a_n x^n| > (\max_{0\leq i\leq n} |a_i|)^{-w}.$$

In fact the number:

$$d^n \sum_{i=0}^{n} a_i x^i$$

is a non-zero algebraic integer of $\mathbb{Q}(x)$ over $\mathbb{Z}$, hence its norm satisfies:

$$|N_{\mathbb{Q}(x)/\mathbb{Q}}(d^n \sum_{i=0}^{n} a_i x^i)| \geq 1,$$

hence:

$$\left|\sum_{i=0}^{n} a_i x^i\right| d^{\delta n} \prod_{\sigma\neq 1} \left|\left(\sum_{i=0}^{n} a_i (\sigma(x))^i\right)\right| \geq 1.$$

For $\sigma \neq 1$:

$$\left|\sum_{i=0}^{n} a_i(\sigma(x))^i\right| \text{ is majorised by } (n+1)\max|a_i| \max_{\sigma\neq 1}(1,|\sigma(x)|)^n,$$

hence:

$$\left|\sum_{i=0}^{n} a_i x^i\right| \geq d^{-\delta n}(n+1)^{-\delta+1}(\max(1,|\sigma(x)|))^{-n(\delta-1)} \times \max_{0\leq i\leq n} |a_i|^{-(\delta-1)},$$

and the choice of a_i gives the desired result.

SOLUTION 6·4: (1): Let us assume that $\alpha \neq 0$ and $e^{\alpha x}$ is algebraic for all algebraic x. For an algebraic, but irrational, number x, one of the three numbers:

$$e^{\alpha}, \quad x, \quad e^{\alpha x}$$

is transcendental. It can only be e^{α}. But for $x = 1$, $e^{\alpha x} = e^{\alpha}$ would have to be algebraic.

(2): Let $\ell \in \mathbb{C}$ be such that:

$$P(e^{\ell}, e^{\ell\alpha}) = 0. \tag{1}$$

The field $\mathbb{Q}(e^{\ell}, e^{\ell\alpha})$ then has transcendence degree over $\mathbb{Q} \geq 1$ (by Theorem (2) and the hypothesis $P(1,1) \neq 0$). Finally, as P is irreducible (1) shows that if e^{ℓ} is algebraic, then $e^{\ell\alpha}$ is algebraic, hence that in this case the transcendence degree of $\mathbb{Q}(e^{\ell}, e^{\ell\alpha})$ over $\mathbb{Q}$ is zero.

(3): Let λ be a non-zero eigenvalue of M (since M is not nilpotent). Then for all $t \in \mathbb{C}$ $\exp(\lambda t)$ is an eigenvalue of $\exp(Mt)$.

If $\exp M\alpha \in GL_n(Q)$, $\exp(\lambda\alpha)$ is then algebraic. By the Gel'fond-Schneider Theorem, if:

$$a = \exp(\lambda\alpha_1), \quad b = \frac{\alpha_2}{\alpha_1}, \quad a^b = \exp(\lambda\alpha_2)$$

are algebraic numbers, $a \neq 0$, $\lambda\alpha_1 = \log a \neq 0$, then b is rational, hence α_1, α_2 are $\mathbb{Q}$-linearly dependent.

SOLUTION 6·5: (1): Let $\alpha \in \bar{\mathbb{Q}}$ be such that $\exp M\alpha \in GL_n(\bar{\mathbb{Q}})$. Let $P \in GL_n(\bar{\mathbb{Q}})$ be such that $P^{-1}MP$ is a diagonal matrix (if M is diagonalizable) or reduced to the form of diagonal blocks (Equation (2)). Let $\lambda_1,\ldots,\lambda_n$ be the eigenvalues of M, with $\lambda_1 \neq 0$. The diagonal of $\exp(P^{-1}MPz)$, for $z \in \mathbb{C}$, is:

$$(\exp\lambda_1 t,\ldots,\exp\lambda_n t).$$

Therefore $\exp\lambda_1\alpha \in \bar{\mathbb{Q}}$, whence $\alpha = 0$ by Theorem (1).

(2): Let us assume that M is not diagonalizable. Then, with the notations of Question (1), one of the blocks $A_1,\ldots,A_r$ of the matrix:

$$P^{-1}MP = \begin{bmatrix} A_1 & & 0 \\ & \ddots & \\ 0 & & A_r \end{bmatrix} \tag{2}$$

belongs to $M_\nu(\mathbb{Q})$ with $\nu \geq 2$; let A_i be one of these blocks:

$$A_i = \begin{bmatrix} \mu_1 & 1 & & 0 \\ & \ddots & \ddots & \\ & & \ddots & 1 \\ 0 & & & \mu_i \end{bmatrix},$$

and:

$$\exp A_i z = \exp\mu_i z \begin{bmatrix} 1 & z & \frac{z^2}{2!} & \cdots\cdots \\ 0 & 1 & z & \cdots\cdots \\ 0 & 0 & 0 & \cdots & 1 \end{bmatrix}.$$

From this we deduce that $u \in \mathbb{Q}$, which contradicts the preceding part (1). Hence M is diagonalizable:

$$P^{-1}MP = \begin{bmatrix} \lambda_1 & & 0 \\ & \ddots & \\ 0 & & \lambda_n \end{bmatrix},$$

and:

$$\exp P^{-1}MPz = \begin{bmatrix} e^{\lambda_1 z} & & 0 \\ & \ddots & \\ 0 & & e^{\lambda_n z} \end{bmatrix}.$$

Since all the number $e^{\lambda_i u}$ $(1 \leqslant i \leqslant n)$ are algebraic, Theorem (2) shows that the dimension of the $\mathbb{Q}$-vector subspace of $\mathbb{C}$ generated by $\lambda_1, \ldots, \lambda_n$ is equal to one.

SOLUTION 6·6: (1): Let us denote by $x_1, \ldots, x_r$ a basis over $\mathbb{Q}$ of the vector space generated by $\mathbb{Q}$ by $\log\gamma_1, \ldots, \log\gamma_k$. Let $b_{j,s} \in \mathbb{Q}$ $(1 \leqslant j \leqslant k, 1 \leqslant s \leqslant r)$ be defined by:

$$\log\gamma_j = \sum_{s=1}^{r} b_{j,s} x_s \qquad (1 \leqslant j \leqslant k).$$

By Baker's Theorem $x_1, \ldots, x_r$ are linearly independent over $\overline{\mathbb{Q}}$; now:

$$\sum_{j=1}^{k} \beta_{i,j} \log\gamma_j = \sum_{s=1}^{r} \left(\sum_{j=1}^{k} \beta_{i,j} b_{j,s} \right) x_s = 0.$$

Hence:

$$\sum_{j=1}^{k} \beta_{i,j} b_{j,s} = 0 \qquad (1 \leqslant i \leqslant h, 1 \leqslant s \leqslant r).$$

As the $b_{j,s}$ are not all zero, there exists s, $1 \leqslant s \leqslant r$, such that one of the numbers $b_{1,s},\dots,b_{k,s}$ is not zero. Multiplying by a common denominator gives the result.

(2): If Baker's Theorem were false, there would exist $\mathbb{Q}$-linearly independent logarithms of algebraic numbers such that $1,\log\alpha_1,\dots,\log\alpha_n$ were $\overline{\mathbb{Q}}$-linearly dependent:

$$\beta_0 + \beta_1\log\alpha_1 + \cdots + \beta_n\log\alpha_n = 0 \qquad (\beta_i \in \overline{\mathbb{Q}}),$$

with one of the β_i $(0 \leqslant i \leqslant n)$ being non-zero, for example $\beta_n \neq 0$. Start by dividing by $-\beta_n$, assume $\beta_n = -1$, and:

$$\log\alpha_n = \beta_0 + \beta_1\log\alpha_1 + \cdots + \beta_{n-1}\log\alpha_{n-1}.$$

If at least one of the numbers $\beta_0,\dots,\beta_{n-1}$ is irrational, the relation:

$$\alpha_n = e^{\beta_0}\alpha_1^{\beta_1}\cdots\alpha_{n-1}^{\beta_{n-1}} \tag{3}$$

gives the desired result. If $\beta_0,\dots,\beta_{n-1}$ are all rational, then $\beta_0 \neq 0$ because of the independence of $\log\alpha_1,\dots,\log\alpha_n$ over $\mathbb{Q}$, and relation (3) again gives a contradiction to the property stated in part (2).

Conversely, let us assume that Baker's Theorem is true and that:

$$\alpha_n = e^{\beta_0}\alpha_1^{\beta_1}\cdots\alpha_{n-1}^{\beta_{\ -1}}, \tag{4}$$

with $\log\alpha_1,\dots,\log\alpha_n$ $\mathbb{Q}$-linearly independent and $\beta_0,\dots,\beta_{n-1}$ algebraic. By taking the logarithms of two members of (4) we establish that there exists a rational integer k such that:

$$\log\alpha_n = \beta_1\log\alpha_1 + \cdots + \beta_{n-1}\log\alpha_{n-1} + \beta_0 + 2ik\pi. \tag{5}$$

Therefore $1, 2i\pi, \log\alpha_1,\dots,\log\alpha_n$ are $\overline{\mathbb{Q}}$-linearly dependent, therefore (by Baker's Theorem) $2i\pi, \log\alpha_1, \ldots, \log\alpha_n$ are $\bar{\mathbb{Q}}$-linearly dependent; from this we deduce the existence of rational numbers $b_1, \ldots, b_n$ such that:

$$2i\pi = b_1 \log\alpha_1 + \cdots + b_n \log\alpha_n.$$

By using Equation (5) we obtain:

$$\beta_0 + \sum_{i=1}^{n} (\beta_i + kb_i)\log\alpha_i = 0 \qquad \text{with } \beta_n = -1.$$

Since $1, \log\alpha_1, \ldots, \log\alpha_n$ are $\bar{\mathbb{Q}}$-linearly independent (Baker's Theorem) we find:

$$\beta_0 = 0 \quad \text{and} \quad \beta_i = -kb_i \in \mathbb{Q} \qquad (1 \leqslant i \leqslant n).$$

Notice that the two numbers $\ell_1 = \log 2$ (the Napierian logarithm of 2) and $\ell_2 = 2\ell_1 + 2i\pi$ are $\mathbb{Q}$-linearly independent logarithms of algebraic numbers, and yet:

$$e^{\ell_2} = e^{2\ell_1},$$

which shows that the relation $\alpha_n = \alpha_1^{\beta_1} \cdots \alpha_n^{\beta_{n-1}} e^{\beta_0}$ can hold without the β_i being all zero.

(3): As (i) is obviously a special case of Baker's Theorem, it suffices that we show the equivalence of the four statements (i),(ii),(iii),(iv). It is clear that (i) and (iv) are equivalent.

Let us show (i) => (ii) and (i) => (iii).

Let $\ell_1, \ldots, \ell_n$ be non-zero logarithms of algebraic numbers, and $\beta_1, \ldots, \beta_n$ algebraic numbers, such that the number:

$$\exp(\beta_1\ell_1 + \cdots + \beta_n\ell_n)$$

is algebraic, that is to say, such that:

$$\ell_0 = \beta_1 \ell_1 + \cdots + \beta_n \ell_n$$

is a logarithm of an algebraic number.

Let $\beta_0 = -1$; we have $\sum_{i=0}^{n} \beta_i \ell_i = 0$, therefore the numbers $\ell_0, \ldots, \ell_n$ are $\overline{\mathbb{Q}}$-linearly d pendent (hence $\mathbb{Q}$-linearly dependent, by (i)).

To prove (iii) we notice that if $\ell_1, \ldots, \ell_n$ are $\mathbb{Q}$-linearly independent, then ℓ_0 belongs to the vector space $\mathbb{Q}$ generated by $\ell_1, \ldots, \ell_n$:

$$\ell_0 = \sum_{i=1}^{n} k_i \ell_i ;$$

from this we deduce $\beta_i = k_i$ $(1 \leqslant i \leqslant n)$, hence $\beta_1, \ldots, \beta_n$ are rational numbers.

To prove (ii) we denote by $\lambda_1, \ldots, \lambda_q$ a basis of the $\mathbb{Q}$-vector space generated by $\ell_0, \ldots, \ell_n$. Thus:

$$\ell_i = \sum_{j=1}^{q} k_{i,j} \lambda_j \qquad (0 \leqslant i \leqslant n),$$

hence:

$$\sum_{i=0}^{n} \beta_i \ell_i = \sum_{j=1}^{q} \left(\sum_{i=0}^{n} \beta_i k_{i,j} \right) \lambda_j = 0;$$

by (i) $\lambda_1, \ldots, \lambda_q$ are $\overline{\mathbb{Q}}$-linearly independent:

$$\sum_{i=0}^{n} \beta_i k_{i,j} = 0 \quad \text{for } 1 \leqslant j \leqslant q.$$

If the $k_{i,j}$ were all zero we would have $\ell_i = 0$ for $0 \leqslant i \leqslant n$, which is contrary to the hypothesis that $\ell_1, \ldots, \ell_n$ are not zero.

Hence $\beta_0 = -1, \beta_1, \ldots, \beta_n$ are $\mathbb{Q}$-linearly dependent.

Let us show that (ii) => (i).

Let $\ell_1, \ldots, \ell_n$ be $\mathbb{Q}$-linearly independent logarithms of algebraic numbers, and $\bar{\mathbb{Q}}$-linearly dependent, then we have:

$$\sum_{i=1}^{n} \beta_i \ell_i = 0,$$

with $\beta_1, \ldots, \beta_n$ algebraic numbers not all zero. We can reduce to the case where the β_i are $\mathbb{Q}$-linearly independent, in fact, if:

$$\beta_n = \sum_{i=1}^{n-1} k_i \beta_i, \qquad k_i \in \mathbb{Q},$$

set:

$$\ell'_i = \ell_i + k_i \ell_n \qquad (1 \leqslant i \leqslant n - 1),$$

and $\ell'_1, \ldots, \ell'_{n-1}$ are $\mathbb{Q}$-linearly independent logarithms of algebraic numbers, but are $\bar{\mathbb{Q}}$-linearly dependent.

$\beta_1, \ldots, \beta_n$ can be assumed to be $\mathbb{Q}$-linearly independent; they are therefore non-zero; for $1 \leqslant i \leqslant n - 1$ let $b_i = -\beta_i/\beta_n$, and we have:

$$\ell_n = \sum_{i=1}^{n-1} b_i \ell_i$$

with $1, b_1, \ldots, b_{n-1}$ $\mathbb{Q}$-linearly independent and $\ell_1, \ldots, \ell_{n-1}$ non-zero; now,

$$\exp \sum_{i=1}^{n-1} b_i \ell_i = \exp \ell_n \in \bar{\mathbb{Q}},$$

which contradicts (ii). Hence (ii) => (i). It remains to verify that (iii) => (i).

Let $\ell_1,\ldots,\ell_n$ be $\mathbb{Q}$-linearly independent logarithms of algebraic numbers; if they are $\bar{\mathbb{Q}}$-linearly dependent we have:

$$\sum_{i=1}^{n} \beta_i\ell_i = 0, \quad \beta_i \in \bar{\mathbb{Q}}, \quad (\beta_1,\ldots,\beta_n) \neq (0,\ldots,0).$$

One of the β_i's is non-zero, for example $\beta_n \neq 0$; let $b_i = -\beta_i/\beta_n$ $(1 \leq i \leq n-1)$; thus:

$$\ell_n = \sum_{i=1}^{n-1} \beta_i\ell_i;$$

as $\ell_1,\ldots,\ell_n$ are $\mathbb{Q}$-linearly independent, at least one of the numbers $\beta_1,\ldots,\beta_{n-1}$ is irrational, which contradicts (iii).

(4): Let us show (vi) => (v). If (v) is false there exists a non-zero element:

$$\beta_0 = \beta_1\ell_1 + \cdots + \beta_n\ell_n \in \bar{\mathbb{Q}} \cap (\bar{\mathbb{Q}}.L);$$

since the numbers $\beta_1,\ldots,\beta_n$ are not all zero, we can assume $\beta_n \neq 0$, and then:

$$\exp\left[-\beta_0 + \frac{\beta_1}{\beta_n}\ell_1 + \cdots + \frac{\beta_{n-1}}{\beta_n}\ell_{n-1}\right] = e^h \in \bar{\mathbb{Q}},$$

contradicting (vi).

Let us show that (v) => (vi). Assume (v) to be true, and let $\beta_0,\beta_1,\ldots,\beta_n \in \bar{\mathbb{Q}}$, $\ell_1,\ldots,\ell_n \in L$ be such that:

$$\exp[\beta_0 + \beta_1\ell_1 + \cdots + \beta_n\ell_n] \in \bar{\mathbb{Q}}.$$

Let us show that β_0 is zero. Let $\ell_{n+1} = \beta_0 + \beta_1\ell_1 + \cdots + \beta_n\ell_n$; then $\ell_{n+1} \in L$ and:

$$\beta_0 = -\beta_1\ell_1 - \cdots - \beta_n\ell_n - \ell_{n+1} \in (\bar{\mathbb{Q}}.L) \cap \bar{\mathbb{Q}},$$

hence $\beta_0 = 0$.

Finally, (v) is obviously a consequence of Baker's Theorem.

(5): It remains to show that (i) together with (v) imply Baker's Theorem. Let $\ell_1,\ldots,\ell_n \in L$ be $\mathbb{Q}$-linearly independent; let $\beta_0,\beta_1,\ldots,\beta_n \in \bar{\mathbb{Q}}$ be such that $\beta_0 + \beta_1\ell_1 + \cdots + \beta_n\ell_n = 0$. We then have $\beta_0 = 0$ (by (v)), and then $\beta_1 = \cdots = \beta_n = 0$, by (i).

(6): The argument is exactly the same as for parts (1) and (2) of Exercise 6·5.

(7): First of all let us remark that an analytic homomorphism $\psi:\mathbb{C} \to \mathrm{GL}_m(\mathbb{C})$ is of the form $\psi(t) = \exp(Mt)$, with $M \in M_m(\mathbb{C})$, in fact, let $M = \psi'(0)$, then $\psi(t)\exp(-\psi'(0)t)$ is constant, and equal to I, the identity matrix, at 0.

Therefore a homomorphism $\varphi:\mathbb{C}^n \to \mathrm{GL}_m(\mathbb{C})$ is of the form:

$$\varphi(t_1,\ldots,t_n) = \exp(M_1t_1 + \cdots + M_nt_n),$$

since, as $(e_1,\ldots,e_n)$ is the canonical basis of $\mathbb{C}^n$, we have:

$$\varphi(t) = \varphi\left(\sum_{j=1}^{n} t_je_j\right) = \prod_{j=1}^{n} \varphi(t_je_j),$$

and $t_j \mapsto \varphi(t_je_j)$ is a homomorphism of $\mathbb{C}$ into $\mathrm{GL}_m(\mathbb{C})$. Furthermore, the matrices $M_1,\ldots,M_n$ commute with each other, because:

$$\varphi(t(e_j + e_k)) = \exp(t(M_j + M_k))$$

$$= I + (M_j + M_k)t + (M_j + M_k)^2\frac{t^2}{2} + \cdots,$$

and:

$$\varphi(te_j + te_k) = \varphi(te_j)\varphi(te_k) = (\exp(tM_j))(\exp(tM_k))$$

$$= \left(I + M_jt + M_j^2\frac{t^2}{2} + \cdots\right)\left(I + M_kt + M_k^2\frac{t^2}{2} + \cdots\right),$$

and identifying the terms in t^2 gives:

$$(M_j + M_k)^2 = M_j^2 + 2M_jM_k + M_k^2,$$

hence $M_jM_k = M_kM_j$.

(7):(a): As the matrices $M_1,\dots,M_n$ commute with each other and are not all nilpotent (otherwise $\varphi(t_1,\dots,t_n)$ would be a rational function), there exists a common eigenvector corresponding to the eigenvalues $\lambda_1,\dots,\lambda_n$ of $M_1,\dots,M_n$ respectively, with $(\lambda_1,\dots,\lambda_n) \neq (0,\dots,0)$. Since $\varphi(\mathbb{Z}^n) \subset \mathrm{GL}_n(\overline{\mathbb{Q}})$ we have:

$$\varphi(e_j) = \exp M_j \in \mathrm{GL}_n(\overline{\mathbb{Q}}) \qquad (1 \leqslant j \leqslant n),$$

therefore e^{λ_j}, which is an eigenvalue of $\exp M_j$, is algebraic. Similarly $\exp\left(\sum_{j=1}^{n} \lambda_j\beta_j\right)$, which is the characteristic value of $\varphi(\beta)$, is algebraic. Let us write $\gamma_j = \exp\lambda_j$, $\gamma_0 = \exp\left(\sum_{j=1}^{n} \lambda_j\beta_j\right)$; we have:

$$\log\gamma_0 = \sum_{j=1}^{n} \beta_j\log\gamma_j,$$

and part (1) above shows that there exist rational integers $c_0,c_1,\dots,c_n$, not all zero, such that:

$$c_0 = \sum_{j=1}^{n} \beta_jc_j.$$

Therefore $1,\beta_1,\dots,\beta_n$ are $\mathbb{Q}$-linearly dependent.

(7):(b): The result is proved by induction on n. For $n = 1$ we have:

$$\varphi(t) = \exp(Mt), \qquad M \in M_m(\mathbb{C}),$$

and M is not nilpotent. We now use Question (3) of Exercise 6·4.

Let us assume the result to have been proved for a homomorphism of $\mathbb{C}^\nu$ into $GL_m(\mathbb{C})$ for $\nu \leqslant n - 1$. If $\alpha_1,\ldots,\alpha_h$ are $\mathbb{C}$-linearly independent and:

$$\alpha_{h+1} = \sum_{j=1}^{h} \beta_j\alpha_j,$$

then the homomorphism $(t_1,\ldots,t_h) \mapsto \varphi(\alpha_1 t_1 + \cdots + \alpha_h t_h)$ of $\mathbb{C}^h$ into $GL_n(\mathbb{C})$ takes algebraic values at the points of $\mathbb{Z}^h$ and at the point $(\beta_1,\ldots,\beta_h)$. As $\alpha_1,\ldots,\alpha_h$ are $\mathbb{C}$-linearly independent, for $(\eta_1,\ldots,\eta_h) \in \bar{\mathbb{Q}}^h$, $\eta \neq 0$, $\varphi((\alpha_1\eta_1 + \cdots + \alpha_h\eta_h)z)$ is an irrational function of $z \in \mathbb{C}$. By the induction hypothesis, since $(\beta_1,\ldots,\beta_h) \notin \mathbb{Q}^h$, we have $h = n$; moreover, by (a) the numbers $1,\beta_1,\ldots,\beta_n$ are $\mathbb{Q}$-linearly dependent:

$$b_{n+1} = \sum_{j=1}^{n} b_j\beta_j,$$

with $b_1,\ldots,b_{n+1} \in \mathbb{Q}$, and let us say that $b_n \neq 0$. Set:

$$\alpha'_j = b_j\alpha_n - b_n\alpha_j \qquad (1 \leqslant j \leqslant n + 1),$$

so that:

$$\alpha'_{n+1} = \sum_{j=1}^{n-1} \beta_j\alpha'_j.$$

But $\alpha'_1,\ldots,\alpha'_{n-1}$ are $\mathbb{C}$-linearly independent, and the preceding argument applied to the homomorphism $(t_1,\ldots,t_{n-1}) \mapsto \varphi(\alpha'_1 t_1 + \cdots + \alpha'_{n-1}t_{n-1})$ leads to a contradiction of the induction hypothesis.

(8): As $\displaystyle\int_0^1 \frac{dx}{1+x^3} = \frac{1}{3}\left(\log 2 + \frac{\pi}{\sqrt{3}}\right)$, and as $i\pi$ is a logarithm of -1, by Baker's Theorem the integral has a transcendental number as its value.

REMARK: More generally, we can show the following result:

Let P,Q be two relatively prime polynomials of $\bar{\mathbb{Q}}[X]$. Assume that the degree of P is strictly less than the degree of Q and that the zeros of Q are all distinct. Let Γ be a path in the complex plane; assume that Γ is closed or that Γ has algebraic or infinite limits.

If the integral:

$$I = \int_\Gamma \frac{P(z)}{Q(z)}\, dz$$

exists, then I vanishes or is transcendental.

(*Cf.*, A.J. Van der Poorten: 'On the Arithmetic Nature of Definite Integrals of Rational Functions', *Proc. Amer. Math. Soc.*, **29**, (1971), 451-456).

SOLUTION 6·7: (a): In order to simplify matters, let us first show:

$$|\Lambda| \geqslant A^{-2nB}$$

By the inequality:

$$|e^x - 1| \leqslant |x| e^{|x|},$$

valid for all $x \in \mathbb{R}$; the number:

$$y = e^\Lambda - 1$$

satisfies:

$$|y| \leqslant |\Lambda| e^{|\Lambda|}$$

Now,

$$|\Lambda| \leqslant nB\log A.$$

On the other hand,

$$y = a_1^{b_1}\cdots a_n^{b_n} - 1$$

is rational, and has a denominator $d \geqslant 1$ satisfying:

$$1 \leqslant d \leqslant A^{nB}.$$

Therefore dy is an integer, and it is non-zero because Λ is a non-zero real number, therefore:

$$|dy| \geqslant 1.$$

We obtain:

$$A^{-nB} \leqslant \frac{1}{d} \leqslant |y| \leqslant |\Lambda|A^{nB},$$

and:

$$|\Lambda| \geqslant A^{-2nB}.$$

(b): The proof that:

$$|\Lambda| \geqslant A^{-nB}$$

is essentially the same. First of all $\Lambda \neq 0$ implies $A \geqslant 2$. For $n = 1$ we have:

$$|\Lambda| = |b\log a| \geqslant \log 2 > \frac{1}{2} \geqslant \frac{1}{A} \geqslant \frac{1}{A^B},$$

therefore the result is true in this case.

Let us assume that $n \geqslant 2$. We can also assume that $|\Lambda| < 1$ (otherwise the required inequality is true), and that at least

$[\frac{1}{2}n]$ of the b_j's are positive or zero (otherwise change Λ to $-\Lambda$). We can then choose:

$$1 \leqslant d \leqslant A^{[\frac{1}{2}n]B},$$

and since:

$$[\tfrac{1}{2}n]B\log A > 0,$$

we have:

$$|\Lambda| < [\tfrac{1}{2}n]B\log A.$$

Therefore:

$$|\Lambda| \geqslant \exp\{-2[\tfrac{1}{2}n]B\log A\} > A^{-nB}.$$

REMARK: Under the same hypothesis we have:

$$|\Lambda| > \exp\{-(|b_1|\log a_1 + \cdots + |b_n|\log a_n)\}.$$

On the other hand, an analogous proof shows that if $\alpha_1,\ldots,\alpha_n$ are non-zero algebraic numbers, if $\log\alpha_j$ is the value of the logarithm of α_j $(1 \leqslant j \leqslant n)$, if $b_1,\ldots,b_n$ are rational integers such that the number:

$$\Lambda = b_1\log\alpha_1 + \cdots + b_n\log\alpha_n$$

is not zero, then:

$$|\Lambda| > A^{-3nDB},$$

where $B = \max\{|b_j|\}$, $D = [\mathbb{Q}(\alpha_1,\ldots,\alpha_n):\mathbb{Q}]$, and:

$$A = \max\{e, \max_{1\leqslant j\leqslant n} \exp|\log\alpha_j|, \max_{1\leqslant j\leqslant n} H(\alpha_j)\}.$$

SOLUTION 6·8: (1): We observe that Schanuel's Conjecture is stronger than Baker's Theorem. In fact, if $\log\alpha_1,\dots,\log\alpha_n$ are $\mathbb{Q}$-linearly independent logarithms of algebraic numbers, then the transcendence degree over $\bar{\mathbb{Q}}$ of the field:

$$\bar{\mathbb{Q}}(\log\alpha_1,\dots,\log\alpha_n,\alpha_1,\dots,\alpha_n) = \bar{\mathbb{Q}}(\log\alpha_1,\dots,\log\alpha_n)$$

is equal to n, therefore $\log\alpha_1,\dots,\log\alpha_n$ are algebraically independent over $\bar{\mathbb{Q}}$, and in particular $1,\log\alpha_1,\dots,\log\alpha_n$ are $\bar{\mathbb{Q}} - L - I$.

Let us define $x_1,\dots,x_{m+n}$ by:

$$x_i = \log\alpha_i \quad (1 \leqslant i \leqslant n), \qquad x_{n+j} = \beta_j \quad (1 \leqslant j \leqslant m).$$

By Baker's Theorem the numbers $x_1,\dots,x_{m+n}$ are $\mathbb{Q}$-linearly independent, therefore, by using Schanuel's Conjecture (S), the transcendence degree over $\bar{\mathbb{Q}}$ of the field:

$$\bar{\mathbb{Q}}(x_1,\dots,x_{m+n},e^{x_1},\dots,e^{x_{m+n}}) = \bar{\mathbb{Q}}(\log\alpha_1,\dots,\log\alpha_n,e^{\beta_1},\dots,e^{\beta_m})$$

is equal to $m + n$.

(2): Let us choose:

$$x_i = \beta_i\log\alpha \qquad (1 \leqslant i \leqslant n),$$

and:

$$x_{n+1} = \log\alpha.$$

By hypothesis the numbers $x_1,\dots,x_{n+1}$ are $\mathbb{Q} - L - I$, therefore the transcendence degree over $\mathbb{Q}$ of the field:

$$\bar{\mathbb{Q}}(\beta_1\log\alpha_1,\dots,\beta_n\log\alpha_n,\log\alpha,\alpha^{\beta_1},\dots,\alpha^{\beta_n},\alpha)$$

$$= \bar{\mathbb{Q}}(\log\alpha,\alpha^{\beta_1},\dots,\alpha^{\beta_n})$$

is equal to $n + 1$, that is to say that the numbers $\log\alpha, \alpha^{\beta_1}, \ldots, \alpha^{\beta_n}$ are algebraically independent on $\bar{\mathbb{Q}}$.

REMARK: The number $2^{1-\sqrt{2}}2^{\sqrt{2}} = 2$ shows that it is not enough to assume $\beta_1, \ldots, \beta_n$ are irrational and $\mathbb{Q}$-linearly independent.

(3): Let $n + \ell$ $(0 \leqslant \ell \leqslant n)$ be the dimension of the $\mathbb{Q}$-vector subspace of $\mathbb{C}$ generated by the numbers;

$$u_1, \ldots, u_n, vu_1, \ldots, vu_n;$$

$u_1, \ldots, u_n$ are ordered in such a way that:

$$u_1, \ldots, u_n, vu_1, \ldots, vu_\ell$$

is a basis of this $\mathbb{Q}$-vector space.

Let us write $vu_{\ell+1}, \ldots, vu_n$ in this basis:

$$vu_{\ell+k} = \sum_{i=1}^{n} a_{k,i}u_i + \sum_{j=1}^{\ell} b_{k,j}vu_j \qquad (1 \leqslant k \leqslant n - \ell).$$

These relations may also be written:

$$-\sum_{h=\ell+1}^{n} (a_{k,h} - \delta_{h,k+\ell}v)u_h = \sum_{j=1}^{\ell} (a_{k,j} + b_{k,j}v)u_j \quad (1 \leqslant k \leqslant n - \ell),$$

where $\delta_{h,k+\ell}$ is the Kronecker δ (which has the value 1 if $h = k + \ell$ and 0 if $h \neq k + \ell$). Let us consider these relations as a system of equations in $u_{\ell+1}, \ldots, u_n$; the determinant of this system is non-zero, because of the transcendence of v (it would be enough, if v is not algebraic of degree $\leqslant n - \ell$). As a consequence, for $\ell + 1 \leqslant h \leqslant n$ we have:

$$u_h \in \mathbb{Q}(v, u_1, \ldots, u_\ell),$$

and the transcendence degree over $\mathbb{Q}$ of the field:

$$\mathbb{Q}(v,u_1,\ldots,u_n)$$

is $\leqslant \ell + 1$. Now, by Schanuel's Conjecture (S) the transcendence degree over $\mathbb{Q}$ of the field:

$$\mathbb{Q}(u_1,\ldots,u_n,vu_1,\ldots,vu_\ell,e^{u_1},\ldots,e^{u_n},e^{vu_1},\ldots,e^{vu_\ell})$$

is at least $n + \ell$. Hence the result.

(4): Let K be the field obtained by adjoining the sixteen given numbers to $\mathbb{Q}$, and let $L = K(\log\log 2)$. Let us consider the following seventeen numbers $x_1,\ldots,x_{17}$:

$$1,\ i\pi,\ \pi,\ \log\pi,\ e,\ e\log\pi,\ \pi\log\pi,\ \log 2,\ \pi\log 2,\ e\log 2,$$

$$i\log 2,\ i\log\pi,\ \log 3,\ \log\log 2,\ \log 3\log\log 2,\ \sqrt{2}\log 2.$$

Since the field $\mathbb{Q}(x_1,\ldots,x_{17},e^{x_1},\ldots,e^{x_{17}})$ has the same transcendence degree over $\mathbb{Q}$ as L, by Schanuel's Conjecture (S) it is sufficient for us to show that $x_1,\ldots,x_{17}$ are $\mathbb{Q}$-linearly independent. Now a non-trivial linear form in $x_1,\ldots,x_{17}$ is also a non-trivial polynomial in:

$$\pi,\ \log\pi,\ e, \log 2,\ \log 3,\ \log\log 2.$$

Thus we are led to showing that these six numbers are algebraically independent. The same argument starting from the numbers:

$$1,\ i\pi,\ \log\pi,\ \log 2,\ \log 3,\ \log\log 2$$

leads the problem to the algebraic independence of the five numbers:

$$\pi,\ \log\pi,\ \log 2,\ \log 3,\ \log\log 2,$$

therefore to the linear independence of the numbers:

$$i\pi,\ \log\pi,\ \log 2,\ \log 3,\ \log\log 2.$$

On considering the exponentials of the real parts of these numbers, we have to show that $\pi, 2, 3, \log 2$ are multiplicatively independent. Now, by Question (1) above, Schanuel's Conjecture (S) implies the algebraic independent of π and $\log 2$. Hence the result.

(5): If $\log x$ and e are algebraically independent, then 1, $\log x, e\log x$ are $\mathbb{Q}$-linearly independent, therefore one of the two numbers:

$$x,\ x^{e}$$

is transcendental over $\mathbb{Q}(e, \log x)$, hence over $\mathbb{Q}$.

If $\log x$ and e are algebraically independent, the transcendence degree over $\mathbb{Q}$ of the field:

$$\mathbb{Q}(\log x, e\log x, x, x^{e})$$

being $\geqslant 2$ (by Schanuel's Conjecture (S)), one of the two numbers x, x^{e} is transcendental over $\mathbb{Q}(e, \log x)$.

(6): If x is algebraically irrational, x^{x} is transcendental (for example, by Question (2) above, or more simply by the Gel'-fond-Schneider Theorem (2)).

If x is transcendental, the numbers:

$$\log x,\ x\log x,\ x^{2}\log x$$

are $\mathbb{Q}$-linearly independent, hence at least one of the two numbers $x^{x}, x^{x^{2}}$ is transcendental over $\mathbb{Q}(x, \log x)$.

SOLUTION 6·9: (1): If ξ is algebraic on $\mathbb{Q}(\pi)$ there exists a polynomial:

$$u_\delta X^\delta + \cdots + u_0, \qquad u_\delta \neq 0,$$

with coefficients u_j in $\mathbb{Z}[\pi]$, such that:

$$u_\delta \xi^\delta + \cdots + u_0 = 0.$$

By induction, for every integer $r \geqslant 1$ there exist elements $v_{0,r}, \ldots, v_{\delta-1,r}$ of $\mathbb{Z}[\pi]$ such that:

$$u_\delta^r \xi^{\delta+r-1} = v_{0,r} + v_{1,r}\xi + \cdots + v_{\delta-1,r}\xi^{\delta-1};$$

furthermore, the degrees of the $v_{j,r}$ (with respect to π) are majorised by:

$$\max_{0 \leqslant j \leqslant \delta-1} d(v_{j,r}) \leqslant r \max_{0 \leqslant i \leqslant \delta} d(u_i),$$

and their height by:

$$\max_{0 \leqslant j \leqslant \delta-1} H(v_{j,r}) \leqslant 2^r (1 + \max_{0 \leqslant i \leqslant \delta} d(u_i))^r (\max_{0 \leqslant i \leqslant \delta} H(u_i))^r.$$

Let $P(X) = a_0 + a_1 X + \cdots + a_n X^n$ be a polynomial of degree n and size s with coefficients in $\mathbb{Z}$. We have:

$$P(\xi) = \sum_{h=0}^{n} a_h \xi^h = \sum_{j=0}^{\delta-1} \frac{b_j}{u_\delta^{n-\delta}} \xi^j,$$

where $b_0, \ldots, b_{\delta-1}$ are elements of $\mathbb{Z}[\pi]$ defined by

$$b_j = a_j u_\delta^{n-\delta} + \sum_{r=1}^{n-\delta} u_\delta^{n-\delta-r} a_{r+\delta-1} v_{j,r}.$$

The norm $R(\pi)$ of $u_\delta^n P(\xi)$ on $\mathbb{Q}(\pi)$ can be written:

$$R(\pi) = u_\delta^{\delta^2} \prod_{\{\sigma\}} \left(\sum_{j=0}^{\delta-1} b_j \xi^j \right) ,$$

$\{\sigma\}$ being the set of $\mathbb{Q}(\pi)$-isomorphisms of $\mathbb{Q}(\pi,\xi)$ into $\mathbb{C}$. Consequently $R(\pi)$ is a polynomial in π, of degree $\leqslant k_1 n$, and of size $\leqslant k_2 s$ (where k_1, k_2 are independent of the polynomial P), with coefficients in $\mathbb{Z}$. Furthermore, we have:

$$R(\pi) = P(\xi) u_\delta^{\delta^2} \prod_{\sigma \neq 1} \left(\sum_{j=0}^{\delta-1} b_j \xi^j \right) .$$

Therefore:

$$\log|R(\pi)| \leqslant \log|P(\xi)| + k_3 s,$$

where k_3 depends neither upon s nor upon n. From this we easily deduce the existence of a constant $C_1 = C_1(\xi)$ such that:

$$\log|P(\xi)| \geqslant - C_1 ns(1 + \log n).$$

(2):(a): If m_0 is an integer such that $a_{m_0} \neq 0$, one of the numbers $Q_{m_0-1}(\xi), Q_{m_0}(\xi)$ is not zero (since $\xi \neq 0$), because:

$$Q_{m_0}(\xi) - Q_{m_0-1}(\xi) = a_{m_0} \xi^{m_0} \neq 0.$$

(2):(b): Since $f(\xi)$ is algebraic over $\mathbb{Q}(\xi)$, there exists an irreducible polynomial Φ, in one variable, with coefficients in $\mathbb{Z}[\xi]$:

$$\Phi = u_\delta X^\delta + \cdots + u_0, \qquad u_\delta \neq 0, \quad u_j \in \mathbb{Z}[\xi], \qquad 0 \leqslant j \leqslant \delta,$$

such that $\Phi(f(\xi)) = 0$. Let $\eta_1, \ldots, \eta_\delta$ be the roots of Φ (that is to say, the conjugates of $f(\xi)$ over $\mathbb{Q}(\xi)$). The norm $R_m(\xi)$ of $Q_m(\xi)$ on $\mathbb{Q}(\xi)$ is:

$$R_m(\xi) = \prod_{i=1}^{\delta} (-\eta_i + 1 + a_1\xi + \cdots + a_m\xi^m)$$

$$= \frac{1}{u_\delta}\Phi(1 + a_1\xi + \cdots + a_m\xi^m) = \frac{1}{u_\delta}\sum_{j=0}^{\delta} u_j\left(\sum_{i=0}^{m} a_i\xi^i\right)^j.$$

Therefore:

$$P_m(\xi) = D_m^\delta u_\delta R_m(\xi)$$

is a polynomial in ξ, with coefficients in $\mathbb{Z}$, of degree $\leqslant k_1'm$, and of height $\leqslant k_2'D_m^\delta$, where k_1', k_2' do not depend upon m. Furthermore, if m_ℓ is a sequence of integers for which $Q_{m_\ell}(\xi) \neq 0$, we have $P_{m_\ell}(\xi) \neq 0$; let us show that the hypothesis on the a_m's implies:

$$\overline{\lim_{\ell\to\infty}} \frac{|\log P_{m_\ell}(\xi)|}{(k_1'm_\ell)(k_1'm_\ell + \log k_2'D_{m_\ell}^\delta)\log(k_1'm_\ell)} = -\infty,$$

(this will show that ξ is not algebraic over $\mathbb{Q}(\xi)$).

Let $C > 0$; there exists $M = M(C) > 0$ such that, for $m \geqslant M$, we have:

$$|a_{m+k+1}| < e^{-2C\lambda_{m+k}},$$

with:

$$\lambda_{m+k} = (m + k)(m + k + \log D_{m+k})\log(m + k).$$

Clearly we have:

$$\lambda_{m+k} \geqslant \left(1 + \frac{k}{m}\right)\lambda_m;$$

as we have $\lambda_m > m^2$, we obtain:

$$\left|\sum_{k\geqslant 0} a_{m+k+1}\xi^{m+k+1}\right|$$

$$\leqslant e^{-2C\lambda_m}(|\xi| + 1)^{m+1} \sum_{k\geqslant 0} [e^{-2C\lambda_m/m}(|\xi| + 1)]^k \leqslant e^{-C\lambda_m}$$

whenever m is sufficiently large. Consequently:

$$\overline{\lim_{m\to\infty}} \frac{\log|Q_m(\xi)|}{m(m + \log D_m)\log m} = -\infty.$$

Finally, $\log|P_m(\xi)|$ is at most equal (up to a multiplicative constant) to $\log|Q_m(\xi)|$.

CHAPTER 7

Congruences Modulo p: Modular Forms

INTRODUCTION

7.1 MODULAR FORMS

Let $f(q) = \sum_{s \geqslant s_0} A_s q^s$ be a Laurent series that is convergent in the disc $0 < |q| < 1$. Let H be the Poincaré (upper) half-plane, that is to say, the set $z \in \mathbb{C}$ with $\text{Im} z > 0$. Set $q = e^{2i\pi z}$, and for $z \in H$ set $F(z) = f(q)$. We say that F is a *MODULAR FUNCTION OF WEIGHT* k, for k an even integer, if:

$$F\left(-\frac{1}{z}\right) = z^k F(z) \quad \text{for } z \in H.$$

Since we have $F(z + 1) = F(z)$, and because the group of linear fractional transformations $z \mapsto \dfrac{az + b}{cz + d}$, with coefficients in $\mathbb{Z}$ satisfying $ad - bc = 1$, is generated by $z \mapsto z + 1$ and by $z \mapsto -1/z$, it follows that we have:

$$F\left(\frac{az + b}{cz + d}\right) = (cz + d)^k F(z)$$

for $a,b,c,d \in \mathbb{Z}$, $ad - bc = 1$, and $z \in H$.

We say that F is a *MODULAR FORM* if the series f is holomorphic at the origin ($A_s = 0$ for $s < 0$), and that F is a *CUSP FORM* if, in addition, f vanishes at the origin ($A_s = 0$ for $s \leqslant 0$).

The set of modular forms of weight k is a vector space whose dimension, when k is even, is equal to:

$$\left[\frac{k}{12}\right] \qquad \text{if } k \equiv 2 \bmod 12,\ k > 0,$$

$$\left[\frac{k}{12}\right] + 1 \quad \text{if } k \not\equiv 2 \bmod 12,\ k \geqslant 0,$$

$$0 \qquad \text{if } k < 0.$$

The set of cusp forms of weight $k + 12$ is a vector space isomorphic to the vector space of modular forms of weight k.

EISENSTEIN SERIES: For even $k \geqslant 4$, the series:

$$G_k(z) = \sum_{\substack{(m,n)\in\mathbb{Z}\times\mathbb{Z} \\ (m,n)\neq(0,0)}} \frac{1}{(mz + n)^k}$$

is convergent for $z \in H$, and is a modular form of weight k. Its expansion at $q = e^{2i\pi z}$ is given by:

$$G_k(z) = 2\zeta(k) + \frac{2(2i\pi)^k}{(k - 1)} \sum_{n=1}^{\infty} \sigma_{k-1}(n)q^n,$$

where:

$$\zeta(k) = \sum_{n\geqslant 1} \frac{1}{n^k} \quad \text{and} \quad \sigma_{k-1}(n) = \sum_{d/n} d^{k-1}.$$

The vector space of modular forms of even weight k, has as a basis the monomials of the form $G_4^{\alpha}G_6^{\beta}$ with $\alpha,\beta \geqslant 0$ integers satisfying $4\alpha + 6\beta = k$.

If we set:

$$\Delta(z) = q \prod_{n=1} (1 - q^n)^{24} = \sum_{n=1}^{\infty} \tau(n)q^n,$$

the function $\Delta(z)$ is a cusp form of weight 12.

7.2 SUBGROUPS OF THE MODULAR GROUP

If A is an unitary ring, we write the group of 2×2 matrices with coefficients in A and determinant one as $SL_2(A)$. This group has the trivial subgroup $\{\pm I\}$. The group $\Gamma = SL_2(\mathbb{Z})/\{\pm I\}$ operates on the Poincaré upper half plane H by:

$$\begin{bmatrix} a & b \\ c & d \end{bmatrix} z \to \frac{az + b}{cz + d}$$

On the other hand Γ is generated by T and J with:

$$T = \begin{bmatrix} 1 & 1 \\ 0 & 1 \end{bmatrix}, \qquad J = \begin{bmatrix} 0 & -1 \\ 0 & -1 \end{bmatrix},$$

T and J satisfy the relations: $J^2 = I$, $(TJ)^3 = I$. From this we deduce that every point of H is transformed by an element of Γ to a point of the fundamental domain of Γ:

$$D = \{z = x + iy : |x| \leqslant \tfrac{1}{2}, |z| \geqslant 1\},$$

the transformations T and J, which generate Γ, permit one to identify the boundaries of this domain.

A subgroup of Γ, of index n, will have a fundamental domain formed by n copies of D, and the boundaries of this domain will be identified by elements of this subgroup.

The *genus of a Riemann surface* can be found by taking a triangulation, and we have the formula:

$$2 - 2g = \text{Number of Triangles} - \text{Number of Sides} + \text{Number of Vertices}.$$

If we want to apply this to the quotient of H by a subgroup of

Γ, we first have to compactify this quotient, that is to say, to look at what happens at the cusps: to do this we take a local uniformization—for example, for the point at infinity (denoted i) the uniformization is $q = e^{2i\pi x/h}$, where h is the smallest integer such that T^h belongs to the group.

If a Rieman surface has genus zero it is known that its field of functions is isomorphic to the field of rational functions over $\mathbb{C}$; in other words, if the quotient of H by a group gives a surface of genus zero there exists a meromorphic function f on H (including the cusps) invariant under the group such that every function having the same properties as f is a rational function of f. This is what happens with the group Γ, the function f then being the "modular invariant" function $\underline{j}(z)$, which we shall denote $\underline{j}(z)$ to avoid confusion with the number j such that $j^3 = 1$.

BIBLIOGRAPHY

[Serl], [Gun].

PROBLEMS

EXERCISE 7·1: IRREDUCIBLE POLYNOMIALS WITH COEFFICIENTS IN $\mathbb{F}_p$

(1): Let P be an irreducible polynomial of degree n with coefficients in $\mathbb{F}_p = \mathbb{Z}/p\mathbb{Z}$, where p is a fixed prime number.

Show, by studying the field generated by one of its roots, that P divides the polynomial $\Pi_n = X^{p^n} - X$.

(2): Show that an irreducible polynomial of $\mathbb{F}_p[X]$ divides Π_n if and only if its degree divides n.

(3): Deduce from this that the number $\psi_p(n)$ of monic irreducible polynomials of $\mathbb{F}_p[X]$, of degree n, is given by the formula:

$$\psi_p(n) = \frac{1}{n} \sum_{d/n} p^d \mu\left(\frac{n}{d}\right) ,$$

where μ denote the Möbius function.

EXERCISE 7·2: SOLUTION OF THE EQUATION $X^p - X = a$ IN A FINITE FIELD (OF CHARACTERISTIC p)

(1): Set $P(x) = x^p - x$, $Q_r(x) = x + x^p + \cdots + x^{p^{r-1}}$.

Show that the mapping of $\mathbb{F}_{p^r}$ to itself defined by:

$$u(a) = P(a) \quad \text{and} \quad v(a) = Q_r(a)$$

are $\mathbb{F}_p$-linear, and satisfy:

$$\operatorname{Ker}(v) = \operatorname{Im}(u).$$

(2): Establish the relation:

$$\prod_{a \in \mathbb{F}_p} [Q_r(x) - a] = x^{p^r} - x.$$

(3): Given an element a that is algebraic over $\mathbb{F}_p$, determine which field the roots of the polynomial $x^p - x - a$ lie.

EXERCISE 7·3: THE RING $\mathbb{Z}/p^h\mathbb{Z}$

Let $a \neq 1$ be an integer not divisible by p. Denote by $\omega(h)$ the order of the element a in the multiplicative group $(\mathbb{Z}/p^h\mathbb{Z})^*$.

(1) Show that ω is not constant and moreover satifies:

$$\omega(h) | \omega(h + 1) \quad \text{and} \quad \omega(h + 1) | p\omega(h)$$

(2): Assume $p \neq 2$. Show that $\omega(h) \neq \omega(h + 1)$ implies $\omega(h + 1) \neq \omega(h + 2)$. From this deduce that there exists h_0 such that:

$$\text{For all } h \geqslant h_0, \quad \omega(h) = \omega(h_0)p^{h-h_0}. \tag{1}$$

Define h_0. Study the case $p = 2$.

(3): p is again odd; show that $\omega(h)$ is equal to the order of $(\mathbb{Z}/p^h\mathbb{Z})^*$ for any h if and only if this holds for $h = 1$ and $h = 2$.

(4): Show that there exists a satisfying the conditions of Question (3), and deduce from this that $(\mathbb{Z}/p^h\mathbb{Z})^*$ is cyclic. Verify that the numbers a satisfying the conditions of Question (3) belong to distinct classes mod p^h, the number of these being equal to the number of generators of $(\mathbb{Z}/p^h\mathbb{Z})^*$. Determine the generators of $(\mathbb{Z}/5^h\mathbb{Z})^*$.

(5): Show that for $h \geqslant 3$ $(\mathbb{Z}/2^h\mathbb{Z})^*$ is not cyclic.

EXERCISE 7·4: BERNOULLI NUMBERS

We say that a series $\sum_{n=0}^{\infty} a_n \frac{z^n}{n}$ is *H-ENTIRE* if $a \in \mathbb{Z}$ for all n. Two H-entire series $\sum_{n=0}^{\infty} a_n \frac{z^n}{n}$ *and* $\sum_{n=0}^{\infty} b_n \frac{z^n}{n}$ are said to be CONGRUENT mod m if $a_n \equiv b_n \pmod{m}$.

(1): Show that if $f(z)$ and $g(z)$ are H-entire series, then the same is true of:

$$f'(z), \quad \int_0^z f(t)dt, \quad f(z)g(z), \quad \frac{f(z)^m}{m!} \quad \text{if } f(0) = 0.$$

(2): Show that for any non-prime $m > 4$:

$$(e^z - 1)^{m-1} \equiv 0 \pmod{m}.$$

$$(e^z - 1)^3 \equiv 2 \sum_{k=1}^{\infty} \frac{z^{2k+1}}{(2k+1)!} \pmod{4}.$$

For p prime, by using the periodicity mod p of the coefficients show that:

$$(e^z - 1)^{p-1} \equiv - \sum_{k=1}^{\infty} \frac{z^{k(p-1)}}{[k(p-1)]!} \pmod{p}.$$

(3): Set:

$$\frac{z}{e^z - 1} = 1 - \frac{z}{2} + \sum_{n=1}^{\infty} (-1)^{n-1} B_n \frac{z^{2n}}{(2n)!}$$

(the B_n are the Bernoulli numbers).

Show that if the p_i are the distinct prime numbers such that $p_i - 1$ divides $2n$, then $(-1)^{n-1}B_n + \sum_i \frac{1}{p_i}$ is an integer.

(4): Let q_n be the denominator of B_n (in lowest terms). Set:

$$\frac{q_1 q_2 \cdots q_n}{(2n)!} = \prod p^{a_p(n)}.$$

Calculate $a_p(n)$ (seperate the cases $p = 2$ and $p \neq 2$).

(5): Show that for $p > 2$ $a_p(n) \geqslant 0$, and that;

$$\text{If } 3 \leqslant p \leqslant \sqrt{2n}, \quad p^{a_p(n)} \leqslant (2n)^{3/2},$$

and lastly that:

$$\sum_{p > \sqrt{2n}} a_p(n) \leqslant \frac{2n}{\sqrt{2n} - 1}.$$

(6): Calculate:

$$\lim_{n \to \infty} \sqrt[n]{\frac{q_1 q_2 \cdots q_n}{(2n)!}}.$$

EXERCISE 7·5: RAMANUJAN'S IDENTITY FOR THE NUMBER OF PARTITIONS

Recall that the following identities between formal series (cf., [Har] Chapter XIX):

$$\varphi(x) = \prod_{1} (1 - x^m) = \sum_{n=-\infty}^{\infty} (-1)^n x^{\frac{1}{2}(3n^2+n)}, \tag{1}$$

$$\varphi^3(x) = \sum_{n=0}^{\infty} (-1)^n (2n + 1)x, \tag{2}$$

$$\frac{1}{\varphi(x)} = \sum_{n=0}^{\infty} p(n)x^n \quad \text{where } p(n) \text{ is the number of number of partitions of } n. \tag{3}$$

(1): For $s = 0,1,2,3,4$ set:

$$G_s(x) = \sum_{\frac{1}{2}(3n^2+n)\equiv s(\text{mod}5)} (-1)^n x^{\frac{1}{2}(3n^2+n)}.$$

Show that we have:

$$D = \begin{vmatrix} G_0 & G_4 & G_3 & G_2 & G_1 \\ G_1 & G_0 & G_4 & G_3 & G_2 \\ G_2 & G_1 & G_0 & G_4 & G_3 \\ G_3 & G_2 & G_1 & G_0 & G_4 \\ G_4 & G_3 & G_2 & G_1 & G_0 \end{vmatrix} = \frac{[\varphi(x^5)]^6}{\varphi(x^{25})}.$$

(2): Show that:

$$G_3 = G_4 = 0 \quad \text{and} \quad G_1 = -x\varphi(x^{25}).$$

By using Equation (2) show that:

$$G_0G_2 + G_1^2 = 0.$$

(3): Set:

$$P_s(x) = \sum_{n=0}^{\infty} p(5n + s)x^{5n+s}.$$

Show that:

$$P_4 \times D = G_1^4.$$

Conclusion?

EXERCISE 7·6: CALCULATION OF DETERMINANTS mod p: CHARACTERISATION OF LOGARITHMIC DIFFERENTIALS

Throughout this problem K is a field of characteristic p.

(1): With P a polynomial of $K[x]$, of (at most) degree $p - 1$, calculate (in K):

$$\sum_{n=0}^{p-1} P(n).$$

(2): With $a_1,\ldots,a_p$ elements of K, calculate the determinant:

$$\begin{vmatrix} a_1 & a_2 & \cdots & \cdots & a_p \\ a_p & a_1 & \cdots & \cdots & \vdots \\ \vdots & \vdots & \ddots & & \vdots \\ \vdots & \vdots & & \ddots & a_2 \\ a_2 & & & a_p & a_1 \end{vmatrix}.$$

(Begin by showing that if $a_i = b_i + c_i$, the determinant formed with the a_i's is the sum of that obtained by starting from the b_i's and that obtained by starting from the c_i's).

(3): Calculate the determinant:

$$\begin{vmatrix} a_1 & \cdots & \cdots & \cdots & a_p \\ a_p & a_1 - 1 & \cdots & \cdots & a_{p-1} \\ \vdots & \vdots & a_1 - i & \cdots & \vdots \\ a_2 & \cdots & \cdots & \cdots & a_1 - p + 1 \end{vmatrix}.$$

(4): Define an operator ϑ on the field $K((t))$ of formal series with coefficients in K by:

$$\vartheta\left(\sum_{n=-N}^{\infty} a_n t^n\right) = \sum_{n=-N}^{\infty} n a_n t^n.$$

Show that there exists an element b of $K((t))$ such that:

$$a = \sum_{n=-N}^{\infty} a_n t^n = \frac{\vartheta b}{b}$$

if and only if $a_{np} = (a_n)^p$ for all n.

EXERCISE 7·7: MULTIPLICATIVITY OF THE RAMANUJAN τ-FUNCTION

(1): Let p be a prime number, let m be an integer satisfying $1 \leqslant m \leqslant p - 1$.

Show that integers m', a, b, c, d can be found satisfying $1 \leqslant m' \leqslant p - 1$ and $ad - bc = 1$ such that for all z with $\mathrm{Im} z > 0$ we have:

$$\frac{-\frac{1}{z} + m}{p} = \frac{au + b}{cu + d} \quad \text{with } u = \frac{z + m'}{p}.$$

(2): Let F be a modular function of weight k. To it associate the function Φ_p defined for $\mathrm{Im} z > 0$ by:

$$\Phi_p(z) = p^{k-1} F(pz) + \frac{1}{p} \sum_{m=0}^{p-1} F\left(\frac{z + m}{p}\right).$$

Show that:

$$\Phi_p(z + 1) = \Phi_p(z) \quad \text{and} \quad \Phi_p\left(-\frac{1}{z}\right) = z^k \Phi_p(z).$$

(3): Assume that:

$$F(z) = \sum_{n \geqslant n_0} A_n q^n \quad \text{with } q = e^{2i\pi z}.$$

Show that $\Phi_p(z)$ may be written as $\sum_{n \geqslant n_0} a_n q^n$ and calculate a_n as a function of the coefficients A_n.

Show that if F is a modular form then Φ_p is a modular form of the same weight, and that if F is a cusp form then Φ_p also is a cusp form.

(4): Set:

$$F(z) = \Delta(z) = q \prod_{n=1}^{\infty} (1 - q^n)^{24} = \sum_{n=1}^{\infty} \tau(n) q^n.$$

We know that F is a cusp form of weight twelve.

Show that Φ_p and Δ are proportional. Show that for $\alpha \geqslant 1$ we have:

$$\tau(p^{\alpha+1}) = \tau(p)\tau(p^{\alpha}) - p^{11}\tau(p^{\alpha-1}).$$

Show by induction on α that for $\alpha \geqslant 0$ and p not dividing s we have:

$$\tau(sp^{\alpha}) = \tau(s)\tau(p^{\alpha}).$$

From this deduce that τ is a multiplicative function.

(5): Now choose $F(z) = G_k(z)$ = an Eisenstein series of weight k.

Show that Φ_p and G_k are proportional.

EXERCISE 7·8: STUDY OF A SUBGROUP OF THE MODULAR GROUP

(1): Show that the set:

$$\left\{ i = \begin{bmatrix} 1 & 0 \\ 0 & 1 \end{bmatrix}, \quad a = \begin{bmatrix} 0 & 3 \\ 1 & 0 \end{bmatrix}, \quad b = \begin{bmatrix} 1 & 2 \\ 1 & 3 \end{bmatrix}, \quad c = \begin{bmatrix} 1 & 1 \\ 2 & 3 \end{bmatrix}, \right.$$
$$\left. d = \begin{bmatrix} 3 & 1 \\ 1 & 2 \end{bmatrix}, \quad e = \begin{bmatrix} 2 & 1 \\ 1 & 1 \end{bmatrix} \right\}$$

forms a subgroup of the group $[SL_2(\mathbb{Z}/4\mathbb{Z})/\pm I]$.

(2): Denote by Γ_4 the subgroup of Γ generated by T^4, J, TJT^{-1}.

(a): Show that:

$$\Gamma = \bigoplus_{i=0}^{3} \Gamma_4 T^i.$$

(b): Find a fundamental domain for Γ_4 and the genus of the Riemann surface $\Gamma_4 \setminus H \cup \{i\infty\}$.

(3):(a): Set $q = e^{\frac{1}{2}i\pi z}$ and show that there exists an unique function $H(z)$, holomorphic on H, such that:

$$H(z) = \frac{e^{\frac{1}{4}i\pi}}{2\sqrt{2q}} + O(q) \quad \text{for } q \to 0,$$

$$H(\gamma z) = H(z) \qquad \text{for all } \gamma \in \Gamma_4.$$

(b): Calculate $\prod_{h=0}^{3} [x - H(z + h)] = \phi(x,z)$.

(4):(a): Let D be a positive integer congruent to 3 modulo 16, and set $\tau = \frac{1}{2}(1 + i\sqrt{D})$.

Show that there exists integers α, β, δ such that:

$$\alpha\delta = \frac{D + 1}{4}, \qquad \beta \equiv 0 \pmod 4, \qquad \alpha > 0,$$

which satisfy:

$$H(\tau + \ell) = H\left(\frac{\alpha(\tau + \ell) + \beta}{\delta}\right) \quad \text{if and only if } \ell \equiv 2 \pmod 4.$$

(b): Show that:

$$\prod_{\substack{\alpha,\beta,\delta \\ \alpha\delta=\frac{1}{4}(D+1) \\ 0\leqslant\beta<4\delta \\ \alpha>0,\,\beta\equiv 0 \bmod 4}} \left[x - H\left(\frac{\alpha z - \beta}{\delta}\right)\right]$$

is a polynomial in x and $H(z)$, we will denote it $\Phi(x,H(z))$.

(c): Show that $H(\tau + 2)$ is the unique root common to the two equations:

$$\begin{cases} (x - 3)(x + 1)^3 + \frac{1}{64}\, \underline{j}(\tau) = 0, \\ \Phi(x,x) = 0. \end{cases}$$

SOLUTIONS

SOLUTION 7·1:(1): Let ϑ be a root of P; since P is irreducible, $\mathbb{F}_p(\vartheta)$ is an n-dimensional vector space over $\mathbb{F}_p$, because it is a field it can only be $\mathbb{F}_{p^n}$. By the rules for calculating in characteristic p, it is clear that $\vartheta^p, \vartheta^{p^2}, \ldots, \vartheta^{p^{n-1}}$ are also roots of P; now, these numbers are distinct, for $\vartheta^{p^k} = \vartheta$ is equivalent to $\vartheta \in \mathbb{F}_{p^k}$, therefore P has all its roots in $\mathbb{F}_{p^n}$, that is to say $P|\Pi_n$.

(2): Let Q be an irreducible polynomial of degree m that divides Π_n. By Question (1) above Q splits completely in $\mathbb{F}_{p^m}$; $\mathbb{F}_{p^m}$ must therefore be included in $\mathbb{F}_{p^n}$, which implies that m divides n. Conversely, if m divides n it is clear that Π_m divides Π_n, hence that Q divides Π_n.

(3): Π_n is divisible by all the irreducible polynomials of degree m with m dividing n, and by them alone. It is not divisible by any power of these polynomials since it only has simple roots. By examining the degrees of these polynomials we see that:

$$p^n = \sum_{m/n} m\psi_p(m).$$

The Möbius Inversion Formula then gives the desired result (cf., chapter 1).

SOLUTION 7·2: (1): As we are in characteristic p, the mapping $a \to a^{p^h}$ is $\mathbb{F}_p$-linear, therefore u and v are linear also. We have:

$$v\circ u(a) = a^p - a + a^{p^2} - a^p + \cdots + a^{p^r} - a^{p^{r-1}} = a^{p^r} - a = 0,$$

that is to say, $\mathrm{Im}(u) \subset \mathrm{Ker}(v)$.

Ker(u) is made up of elements satisfying $a^p = a$, that is to say $\mathrm{Ker}(u) = \mathbb{F}_p$. Since $\mathbb{F}_{p^r}$ is r-dimensional over $\mathbb{F}_p$ and u has rank $r - 1$, Im(u) contains therefore p^{r-1} elements. On the other hand, as Ker(v) is made up of roots of a polynomial of degree p^{r-1}, it contains at most p^{r-1} elements, hence $\mathrm{Ker}(v) = \mathrm{Im}(u)$.

(2): We see that $u\,v = 0$, that is to say $\mathrm{Im}(v) \subset \mathrm{Ker}(u) = \mathbb{F}_p$, therefore $\mathrm{Im}(v) = \mathrm{Ker}(u)$. Therefore $\prod_{a\in\mathbb{F}_p} [Q_r(x) - a]$ is a polynomial of degree p^r which vanishes at all the elements of $\mathbb{F}_{p^r}$, therefore it is equal to $k(x^{p^r} - x)$; examining the coefficient of the term of degree p^r shows that $k = 1$.

(3): If b is a root of the polynomial $x^p - x - a$, the other roots of this are $b + \alpha$, with $\alpha \in \mathbb{F}_p$, they are all distinct and lie in the same field. From Question (1) it follows that the two conditions:

$$\begin{cases} \text{The roots of the polynomial } x^p - x - a \text{ lie in } \mathbb{F}_{p^r} & (a \in \mathrm{Im}(u)), \\ Q_r(a) = 0 & (a \in \mathrm{Ker}(v)), \end{cases}$$

are equivalent. Let then h be the smallest integer such that $a \in \mathbb{F}_{p^h}$. If $r = sh + k$, with $k < h$, we have:

$$Q_r(a) = a + \cdots + a^{p^{h-1}} + \cdots + a^{p^{2h-1}} + \cdots + a^{p^{sh-1}} + a^{p^{sh}} + \cdots + a^{p^{sh+k-1}}$$

$$= sQ_h(a) + Q_k(a),$$

with the convention $Q_0(a) = 0$. Now, if $k \neq 0$, $Q_k(a)$ is not in $\mathbb{F}_p$, for it follows from Question (2) above that the only numbers α such that $Q_k(\alpha)$ lies in $\mathbb{F}_p$ are those of $\mathbb{F}_{p^k}$. Since $Q_h(a) \in \mathbb{F}_p$, $Q_r(a)$ can vanish only if $sQ_h(a) = Q_k(a) = 0$. That is to say, if $k = 0$, either $Q_h(a) = 0$ or p divides s. From this we conclude:

$$\begin{cases} \text{If } Q_h(a) = 0, \text{ the roots of } x^p - x - a \text{ lie in } \mathbb{F}_{p^h}, \\ \text{If } Q_h(a) \neq 0, \text{ the roots of } x^p - x - a \text{ lie in } \mathbb{F}_{p^{ph}}. \end{cases}$$

SOLUTION 7·3: (1): $\omega(h) = k$ would imply $a^k \equiv 1 \pmod{p^h}$ for all h, that is to say, $a^k = 1$, which is impossible if $a \neq 1$. If $a^{\omega(h+1)} \equiv 1 \pmod{p^{h+1}}$, it is clear that $a^{\omega(h+1)} \equiv 1 \pmod{p^h}$, that is to say, $\omega(h)$ divides $\omega(h+1)$. Conversly, if $a^{\omega(h)} \equiv 1 \pmod{p^h}$, $a^{p\omega(h)} \equiv 1 \pmod{p^{h+1}}$, that is to say, $\omega(h+1)$ divides $p\omega(h)$. Therefore, either $\omega(h+1) = \omega(h)$, or $\omega(h+1) = p\omega(h)$.

(2): Assume $\omega(h+1) \neq \omega(h)$, that is to say, $a^{\omega(h)} = 1 + \alpha p^h$ with α not divisible by p. By Question (1) above we know that $\omega(h+1) = p\omega(h)$, therefore:

$$a^{\omega(h+1)} = a^{p\omega(h)} = (1 + \alpha p^h)^p$$

$$= 1 + p^{h+1}\alpha + \sum_{i=2}^{p-1} \binom{p}{i} \alpha^i p^{hi} + p^{ph}\alpha^p.$$

As p is different from two, ph is larger than $2h + 1$; on the other hand, as $\binom{p}{i}$ is divisible by p for $2 \leqslant i \leqslant p - 1$, it follows that:

$$a^{\omega(h+1)} = 1 + p^{h+1}\alpha + p^{2h+1}\beta,$$

which shows that $\omega(h + 1) \neq \omega(h + 2)$.

If we choose as h_0 the smallest number such that $\omega(h_0 + 1) \neq \omega(h_0)$ (h_0 exists, by (1) above), we see that $\omega(h + 1) = p\omega(h)$ for $h \geqslant h_0$, which gives the formula.

If $p = 2$, the same argument gives:

$$a^{\omega(h+1)} = 1 + 2^{h+1}\alpha + 2^{2h}\alpha^2,$$

which shows that if $h \neq 1$, $\omega(h + 1) \neq \omega(h + 2)$. Hence the same result is obtained by taking $h_0 \geqslant 2$.

(3): The condition is trivially necessary. It is immediately seen that the order of $(\mathbb{Z}/p\ \mathbb{Z})^*$ is $p^h - p^{h-1} = p^{h-1}(p - 1)$. Now, by Question (2) above, $\omega(1) = p - 1$ and $\omega(2) = p(p - 1)$ certainly implies that $\omega(h) = p^{h-1}(p - 1)$.

(4): We must choose a in such a way that $\omega(1) = p - 1$, that is to say, $a = \alpha + \lambda p$ with α a primitive $(p - 1)$-st root of 1 mod p, and that $\omega(2) = p(p - 1)$, that is to say that $a^{p-1} \not\equiv 1 \pmod{p^2}$. Now:

$$(\alpha + \lambda p)^{p-1} \equiv \alpha^{p-1} + \lambda p(p - 1)\alpha^{p-2} \pmod{p^2}$$

$$\equiv 1 + p[\nu + \lambda(p - 1)\alpha^{p-2}] \pmod{p^2},$$

where we have set $\alpha^{p-1} \equiv 1 + p\nu \pmod{p^2}$. Since $(p - 1)$ and α are non-zero (mod p), for each α there exists a single value of λ, say $\left(-\dfrac{\nu}{(p - 1)\alpha^{p-2}}\right)$, mod p, such that

$$(\alpha + \lambda p)^{p-1} \equiv 1 \pmod{p^2}.$$

For the $(p - 1)$ other values the resulting α's give us our result. It is clear that a generates $(\mathbb{Z}/p^h\mathbb{Z})^*$, therefore this group is cyclic. $(p - 1)\varphi(p - 1)$incongruent solutions mod p^2 have been found, that is to say, $p^{h-2}(p - 1)\varphi(p - 1) = \varphi[\varphi(p\ )]$ solutions modulo p^h; since $\varphi(p^h) = \mathrm{card}(\mathbb{Z}/p^h\mathbb{Z})^*$, $\varphi[\varphi(p^h)]$ is the number of its generators.

EXAMPLE: For $p = 5$ the primitive roots are 2 and 3. Now,

$$(2 + 5\lambda)^4 \equiv 1 + 5(2 + 4.8\lambda)$$

$$\equiv 1 + 5(3 + 2\lambda)$$

$$\Rightarrow \lambda \not\equiv 1 \bmod 5$$

$$(3 + 5\lambda)^4 \equiv 1 + 5(1 + 4.27\lambda)$$

$$\Rightarrow \lambda \not\equiv 3 \bmod 5;$$

the generators are therefore the numbers a congruent (mod 25) to 2,12,17,22;3,8,12,23.

(5): If $a = 4n + 1$ we find $\omega(1) = \omega(2) = 1$, hence $\omega(h) \leqslant 2^{h-2}$ for $h \geqslant 2$; if $a = 4n + 3$ we find $\omega(1) = 1$, $\omega(2) = \omega(3) = 2$, and therefore $\omega(h) \leqslant 2.2^{h-3}$ for $h \geqslant 3$. $\omega(h)$ can therefore never for $h \geqslant 3$ be equal to 2^{h-1} = the order of $(\mathbb{Z}/2^h\mathbb{Z})^*$.

SOLUTION 7·4: (1): Let us set:

$$f(z) = \sum_{n=0}^{\infty} a_n \frac{z^n}{n!} \quad \text{and} \quad g(z) = \sum_{n=0}^{\infty} b_n \frac{z^n}{n!},$$

we then find:

$$f'(z) = \sum_{n=0}^{\infty} a_{n+1} \frac{z^n}{n!}, \qquad \int_0^z f(t)dt = \sum_{n=1}^{\infty} a_{n-1} \frac{z^n}{n!},$$

$$f(z)g(z) = \sum_{n=0}^{\infty} \sum_{m=0}^{n} a_m b_{n-m} \binom{n}{m} \frac{z^n}{n!}.$$

These series are therefore clearly H-entire.

We now assume that $f(0) = 0$ and that $\frac{f(z)^{m-1}}{(m-1)!}$ is H-entire; since f and f' are H-entire, the same is true of $\frac{f(z)^{m-1}}{(m-1)!} f'(z)$, therefore also for:

$$\int_0^z \frac{f(t)^{m-1}}{(m-1)!} f'(t)dt = \frac{f^m(z)}{m!}.$$

Which proves the last result by induction.

(2): By part (1) we see that $(e^z - 1)^{m-1} = (m-1)!g(z)$, where $g(z)$ is H-entire, since for non-prime $m > 4$ $(m-1)! \equiv 0 \pmod{m}$, we find $(e^z - 1)^{m-1} \equiv 0 \pmod{m}$.

(To show that $(m-1)! \equiv 0 \pmod{m}$ set $m = pq$. If $p \neq q$, as $p,q < (m-1)$ the result is obvious. If $p = q$ we have the case $m = p^2$ with p prime; if $p \neq 2$ we then see that p and $2p$ are smaller than $(p^2 - 1)$, whence the result).

We have:

$$(e^z - 1)^m = \sum_{h=0}^{m} \binom{m}{h} e^{hz}(-1)^{m-h}$$

$$= \sum_{n=0}^{\infty} \left[\sum_{h=0}^{m} (-1)^{m-h} \binom{m}{h} h^n \right] \frac{z^n}{n!} \quad \text{(with } 0^0 = 1\text{)},$$

in particular,

$$(e^z - 1)^3 = \sum_{n=1}^{\infty} [3 - 3\times 2^n + 3^n] \frac{z^n}{n} \equiv \sum_{n=2}^{\infty} [3 + 3^n] \frac{z^n}{n} \pmod{4}.$$

Now $3^2 \equiv 1 \pmod 4$, hence $3 + 3^{2p+1} \equiv 2 \pmod 4$ and $3 + 3^{2p} \equiv 0 \pmod 4$, which yields:

$$(e^z - 1)^3 \equiv 2 \sum_{k=1}^{\infty} \frac{z^{2k+1}}{(2k+1)!} \pmod 4.$$

We now apply the formula with $m = p - 1$; if we set:

$$(e^z - 1)^{p-1} = \sum_{n=1}^{\infty} a_n \frac{z^n}{n!},$$

the formula $h^{p-1} \equiv 1 \pmod p$ implies that $a_{n+p-1} \equiv a_n \pmod p$, and the coefficients are periodic; on the other hand, we know that $(p-1)! \equiv -1 \pmod p$, hence:

$$(e^z - 1)^{p-1} = z^{p-1} + \cdots \equiv (-1) \frac{z^{p-1}}{(p-1)!} + \cdots \pmod p,$$

which certainly gives:

$$(e^z - 1)^{p-1} \equiv - \sum_{k=1}^{\infty} \frac{z^{k(p-1)}}{[k(p-1)]!} \pmod p.$$

(3): It is easy to verify that the function $\frac{z}{e^z - 1} + \frac{z}{2}$ is even, hence the given expansion. Now set $z = \log(1 + e^z - 1)$, and for $|e^z - 1| < 1$ this yields:

$$\frac{z}{e^z - 1} = \sum_{m=1}^{\infty} (-1)^{m+1} \frac{(e^z - 1)^{m-1}}{m}.$$

As $f(z) \equiv 0 \pmod m$ is equivalent to $\frac{f(z)}{m}$ being H-entire, we find:

$$\frac{z}{e^z - 1} = 1 - \frac{1}{2} \sum_{k=1}^{\infty} \frac{z^k}{k!} - \frac{2}{4} \sum_{k=1}^{\infty} \frac{z^{2k+1}}{(2k+1)!} - \qquad \text{(Contd)}$$

$$(\text{Contd}) \qquad - \sum_{\substack{p \text{prime} \\ p>2}} \frac{1}{p} \sum_{k=1}^{\infty} \frac{z^{k(p-1)}}{[k(p-1)]!} + f(z),$$

where $f(z)$ is H-entire, which finally gives:

$$(-1)^{n-1}B_n = -\frac{1}{2} - \sum_{\substack{p \text{prime} \\ p>2 \\ (p-1)/2n}} \frac{1}{p} + \text{integer},$$

which is the required result $(2 - 1 = 1$ always dividing $2n)$; this is the *von Staudt Clausen Theorem.*

(4): From Question (3) it follows that:

$$q_n = 2 \prod_{\substack{p>2 \\ (p-1)/2n}} p,$$

2 divides q_n with multiplicity 1 and $[2n]!$ with multiplicity $\left[\frac{2n}{2}\right] + \left[\frac{2n}{4}\right] + \cdots$, from this it follows that

$$a_2(n) = n - n - \left[\frac{n}{2}\right] - \cdots = -\left[\frac{n}{2}\right] - \left[\frac{n}{4}\right] - \cdots .$$

p divides q_n with multiplicity 1 if and only if $2n = k(p-1)$, that is to say $n = k\left(\frac{p-1}{2}\right)$, therefore p divides $(q_1,\ldots,q_n)$ with multiplicity $\left[\frac{2n}{p-1}\right]$, and consequently:

$$a_p(n) = \left[\frac{2n}{p-1}\right] - \left[\frac{2n}{p}\right] - \left[\frac{2n}{p^2}\right] - \cdots .$$

(5): We find:

$$a_p(n) \geqslant \left[\frac{2n}{p-1}\right] - \frac{2n}{p} - \frac{2n}{p^2} - \cdots = \left[\frac{2n}{p-1}\right] - \frac{2n}{p}\frac{1}{1-\frac{1}{p}} \geqslant$$

$$\geqslant \left[\frac{2n}{p-1}\right] - \frac{2n}{p-1} > -1.$$

As $a_p(n)$ is an integer, it certainly follows from this that $a_p(n) \geqslant 0$.

We define r by $p^r \leqslant 2n < p^{r+1}$, that is to say $r = [\log_p(2n)] \leqslant \log_p(2n)$; then:

$$a_p(n) = \left[\frac{2n}{p-1}\right] - \left[\frac{2n}{p}\right] - \cdots - \left[\frac{2n}{p^r}\right]$$

$$\leqslant \frac{2n}{p-1} - \left(\frac{2n}{p} - 1\right) - \cdots - \left(\frac{2n}{p^r} - 1\right)$$

$$\leqslant r + \frac{2n}{p-1} - \frac{2n}{p}\left(\frac{1 - \frac{1}{p^r}}{1 - \frac{1}{p}}\right)$$

$$= r + \frac{2n}{p^r(p-1)},$$

and by using the fact that $2n < p^{r+1}$, for $p \geqslant 3$ we find:

$$a_p(n) \leqslant r + \frac{p}{p-1} \leqslant r + \frac{3}{2}.$$

As $a_p(n)$ and r are integers, we deduce from this that:

$$a_p(n) \leqslant r + 1.$$

Therefore, for $3 \leqslant p \leqslant \sqrt{2n}$ we find:

$$p^{a_p(n)} \leqslant p^{\log_p(2n)+1} \leqslant 2np \leqslant 2n\sqrt{2n} = (2n)^{3/2}.$$

If $p > \sqrt{2n}$ then $r \leqslant 1$, therefore:

$$a_p(n) = \left[\frac{2n}{p-1}\right] - \left[\frac{2n}{p}\right].$$

Now, it is clear that $\left[\frac{2n}{k-1}\right] - \left[\frac{2n}{k}\right]$ is positive or zero for all k and is zero for $k > 2n + 1$, therefore:

$$\sum_{p>\sqrt{2n}} a_p(n) \leqslant \sum_{k>\sqrt{2n}} \left[\frac{2n}{k-1}\right] - \left[\frac{2n}{k}\right]$$

$$\leqslant \sum_{k\geqslant[\sqrt{2n}]+1} \left[\frac{2n}{k-1}\right] - \left[\frac{2n}{k}\right] = \left[\frac{2n}{[\sqrt{2n}]}\right]$$

$$\leqslant \frac{2n}{[\sqrt{2n}]} \leqslant \frac{2n}{\sqrt{2n}-1}$$

(6): We have:

$$\sqrt[n]{\frac{q_1\cdots q_n}{(2n)!}} = 2^{a_2(n)/n} \prod_{3\leqslant p\leqslant\sqrt{2n}} p^{a_p(n)/n} \prod_{p>\sqrt{2n}} p^{a_p(n)/n}.$$

We are going to calculate the limit of each of the three factors. For all $h \geqslant 1$:

$$-\frac{n}{2} - \frac{n}{4} \cdots \leqslant a_2(n) < -\frac{n}{2} + 1 - \frac{n}{4} + 1 \cdots - \frac{n}{2^n} + 1,$$

or:

$$-n \leqslant a_2(n) < -n + \frac{n}{2^{h+1}} + h,$$

that is to say, for all h we have:

$$2^{-1} \leqslant 2^{a_2(n)/n} \leqslant 2^{-1}\, 2^{(h/n)+(1/(2^h+1))}$$

that is to say,

$$\tfrac{1}{2} \leqslant \underline{\lim} 2^{a_2(n)/n} \leqslant \overline{\lim} 2^{a_2(n)/n} \leqslant \tfrac{1}{2}\times 2^{1/(2^h+1)}.$$

Since this holds for all h we find that

$$\lim 2^{a_2(n)/n} = \tfrac{1}{2}.$$

Also, by Question (5) above we find:

$$1 \leqslant \prod_{3 \leqslant p \leqslant \sqrt{2n}} p^{a_p(n)/n} \leqslant [(2n)^{3/2n}]^{\sqrt{2}n} = (2n)^{3/\sqrt{2}n},$$

and because:

$$\lim_{n \to \infty} (2n)^{3/\sqrt{2n}} = 1,$$

we have:

$$\lim_{3 \leqslant p \leqslant \sqrt{2n}} p^{a_p(n)/n} = 1.$$

It is known that if $p > 2n + 1$, $a_p(n) = 0$, hence:

$$1 \leqslant \prod_{p > \sqrt{2n}} p^{a_p(n)/n} \leqslant (2n + 1)^{\left(\sum_{p > \sqrt{2n}} \frac{a_p(n)}{n}\right)}$$

$$\leqslant (2n + 1)^{2/(\sqrt{2n}-1)},$$

and therefore:

$$\lim_{n \to \infty} \prod_{p > \sqrt{2n}} p^{a_p(n)/n} = 1.$$

Taking all the results together we find:

$$\lim \sqrt[n]{\frac{q_1 \cdots q_n}{(2n)!}} = \tfrac{1}{2}.$$

SOLUTION 7·5: (1): D is a circular determinant ("circulant"). Hence we find:

$$D = \prod_{\xi} (G_0 + \xi G_1 + \xi^2 G_2 + \xi^3 G_3 + \xi^4 G_4)$$

where ξ runs over the fifth roots of unity. This can be written:

$$D = \prod_{\xi} \varphi(\xi x) = \prod_{m=1}^{\infty} \prod_{\xi} (1 - \xi^m x^m).$$

Now it is clear that:

$$\prod_{\xi} (1 - \xi x) = 1 - x^5.$$

On the other hand, since for $m \not\equiv 0 \pmod 5$, ξ^m also runs over the fifth roots of unity, we find:

$$D = \prod_{m \not\equiv 0 (\mathrm{mod}\, 5)} (1 - x^{5m}) \prod_{m \equiv 0 (\mathrm{mod}\, 5)} (1 - x^m)^5$$

$$= \frac{\prod (1 - x^{5m})}{\prod_m (1 - x^{25m})} \prod_m (1 - x^{5m})^5 = \frac{[\varphi(x^5)]^6}{\varphi(x^{25})}.$$

(2): Let us try to solve the congruence $\frac{3n^2 + n}{2} \equiv s \pmod 5$. We multiply both sides by 24 (which is invertible mod 5), and we find:

$$36n^2 + 12n \equiv 24s = - s \pmod 5,$$

or:

$$(6n + 1)^2 \equiv 1 - s \pmod 5.$$

In order that this congruence have a solution it is necessary and sufficient that $1 - s$ be a quadratic residue mod 5. That is to say that $G_s = 0$ if $\left(\frac{1-s}{5}\right) = -1$, which gives $1 - s \equiv 2$ or 3, so $s = 3$ or 4. Hence $G_3 = G_4 = 0$. For $s = 1$ we find $(6n + 1)^2 \equiv 0$ (mod 5), which implies $6n + 1 \equiv 0$ (mod 5), that is to say $n \equiv -1$ (mod 5), therefore we find:

$$G_1(x) = \sum_{-\infty}^{+\infty} (-1)^{5k-1} x^{(3(5k-1)^2+5k-1)/2}$$

$$= - \sum_{-\infty}^{+\infty} (-1)^k x^{(75k^2-25k+2)/2} = - x\varphi(x^{25}).$$

It follows from what has gone before that:

$$(G_0 + G_1 + G_2)^3 = \sum_{0}^{\infty} (-1)^n (2n + 1) x^{(n^2+n)/2}.$$

Let us now try so solve $\frac{n^2 + n}{2} \equiv s$ (mod 5); by multiplying by 8 we find:

$$(2n + 1)^2 = 3s + 1 \pmod 5,$$

which is impossible if $3s + 1 = 2$ or 3, that is to say, if $s =$ 2 or 4.

Now, in $(G_0 + G_1 + G_2)^3$ the terms giving an exponent of x congruent to 2 (mod 5) arise from $3G_0G_1^2 + 3G_0^2G_2$. This term must therefore be zero, that is to say $G_1^2 = - G_0G_2$.

(3): We observe that:

$$(G_0 + G_1 + G_2 + G_3 + G_4)(P_0 + P_1 + P_2 + P_3 + P_4) = 1,$$

and by decomposing the series according to the congruences modulo 5 of the exponents, we find:

$$D \times \begin{bmatrix} P_0 \\ P_1 \\ P_2 \\ P_3 \\ P_4 \end{bmatrix} = \begin{bmatrix} 1 \\ 0 \\ 0 \\ 0 \\ 0 \end{bmatrix},$$

therefore if the fifth column of D is multiplied by P_4, and to it we add the first column multiplied by P_0, the second by P_1, the third by P_2, and the fourth by P_3, we find:

$$P_4 D = \begin{vmatrix} G_0 & G_4 & G_3 & G_2 & 1 \\ G_1 & G_0 & G_4 & G_3 & 0 \\ G_2 & G_1 & G_0 & G_4 & 0 \\ G_3 & G_2 & G_1 & G_0 & 0 \\ G_4 & G_3 & G_2 & G_1 & 0 \end{vmatrix}$$

and by making use of Question (2) we find:

$$P_4 D = \begin{vmatrix} G_1 & G_2 & 0 & 0 \\ G_2 & G_1 & G_0 & 0 \\ 0 & G_2 & G_1 & G_0 \\ 0 & 0 & G_2 & G_1 \end{vmatrix} = G_1^4 - G_1(2G_1G_2G_0) - G_2(G_0G_1^2 - G_0^2G_2)$$

$$= G_1^4 - 3G_1^2G_2G_0 + G_2^2G_0^2 = 5G_1^4.$$

From this we deduce:

$$P_4 = \frac{5G_1^4}{D} = 5\,\frac{x^4\varphi^4(x^{25})}{\varphi^6(x\ )/\varphi(x^{25})} = 5x^4\,\frac{\varphi^5(x^{25})}{\varphi^6(x^5)}\ ;$$

which shows, in particular, Ramanujan's Congruence:

$p(5n + 4) \equiv 0 \pmod 5$.

SOLUTION 7·6: (1): The numbers $n = 0,\dots,p - 1$ are the solutions of the equation $x(x - 1)\cdots(x - p + 1) = x^p - x$. As the sums of the powers of the roots of a polynomial can be expressed as symmetric functions of these roots, we see that $\sum_{n=0}^{p-1} n^h = 0$ for $h < p - 1$. On the other hand,

$$\sum_{n=0}^{p-1} n^{p-1} = \sum_{n=1}^{p-1} 1 = -1.$$

From this therefore, we deduce:

$$\sum_{n=0}^{p-1} P(n) = -\text{ (coefficient of degree } p-1 \text{ of the polynomial } P).$$

(2): The usual method of calculating this determinant gives nothing, because 1 does not have a p-th root in any extension of K.

By using the distributivity of determinants we find:

$$\begin{vmatrix} b_1 + c_1 & \cdots & b_p + c_p \\ b_p + c_p & \cdots & \vdots \\ \vdots & & \vdots \\ b_2 + c_2 & & b_1 + c_1 \end{vmatrix} = \sum_f \begin{vmatrix} f_1(1) & \cdots & f_p(p) \\ f_p(1) & \cdots & \vdots \\ \vdots & & \vdots \\ f_2(1) & \cdots & f_1(p) \end{vmatrix},$$

where f runs over the set of functions from $\{1,\dots,p\}$ to the set of two elements $\{b,c\}$. If f is considered as a function on $\mathbb{Z}/p\mathbb{Z}$, we see that for all r the determinant obtained by starting from the function f_r defined by $f_r(i) = f(i + r)$ is equal to that obtained by starting from f (it suffices to make a cyclic permutation of the columns and the same permuation on the rows).

Let us assume now that $f_r(i) = f(i)$ for $r \not\equiv 0$, then for every k we have $f(kr) = f(0)$; since $r \not\equiv 0$, is invertible in $\mathbb{Z}/p\mathbb{Z}$, and kr runs over the whole of $\mathbb{Z}/p\mathbb{Z}$ when k runs over $\mathbb{Z}/p\mathbb{Z}$. From this it follows that $f(i)$ = constant. Therefore, if $f(i)$ is not a constant there are p distinct elements, e.g. f_r, which give the same value to the determinant, that is to say, which give zero to the total in the summation. Hence we find:

$$\begin{vmatrix} b_1 + c_1 & \cdots & b_p + c_p \\ b_p + c_p & & \vdots \\ \vdots & & \vdots \\ b_2 + c_2 & \cdots & b_1 + c_1 \end{vmatrix} = \begin{vmatrix} b_1 & \cdots & b_p \\ \vdots & & \vdots \\ \vdots & & \vdots \\ b_2 & \cdots & b_1 \end{vmatrix} + \begin{vmatrix} c_1 & \cdots & c_p \\ \vdots & & \vdots \\ \vdots & & \vdots \\ c_2 & \cdots & c_1 \end{vmatrix}.$$

We notice then that the system $(a_1,\ldots,a_p)$ is the sum of the systems $(0,\ldots,a_i,\ldots,0)$, and that:

$$\begin{vmatrix} 0 & \cdots & a_i & 0 & \cdots & 0 \\ \vdots & & & \ddots & & \vdots \\ \vdots & & & & \ddots & 0 \\ 0 & \cdots & \cdots & \cdots & \cdots & a_i \\ a_i & & & & & 0 \\ \vdots & \ddots & & & & \vdots \\ 0 & & a_i & \cdots & \cdots & 0 \end{vmatrix} = a_i^p,$$

which shows that the determinant has the value:

$$\sum_{a=1}^{p} a_i^p.$$

(3): We use the same method, and we find:

$$D = \begin{vmatrix} a_1 & \cdots & \cdots & a_p \\ a_p & a_1 - 1 & & \vdots \\ & & \ddots & \vdots \\ a_2 & \cdots & \cdots & a_i - p + 1 \end{vmatrix} =$$

$$= \sum_{f} \begin{vmatrix} f_1(1) & \cdots & f_p(p) \\ \vdots & & \vdots \\ f_2(1) & \cdots & f_1(p) \end{vmatrix} \prod_{k\in K} (1 - k),$$

where, this time, $f(k)$ either has the value a or $f_h(k) = \delta_{hk}$ (the Kronecker symbol), and where $K = \{k : f(k) \neq a\}$. This then leads to:

$$D = \begin{vmatrix} a_1 & \cdots & a_p \\ \vdots & & \vdots \\ a_2 & \cdots & a_1 \end{vmatrix} + \prod_{k=1}^{p} (1 - k) \begin{vmatrix} 1 & \cdots & 0 \\ & \ddots & \\ 0 & \cdots & 1 \end{vmatrix}$$

$$\sum_{\text{class of } f} \sum_{r=0}^{p-1} \prod_{k\in K} (1 - k + r) \begin{vmatrix} f_1(1) & \cdots & f_p(p) \\ \vdots & & \vdots \\ f_2(1) & \cdots & f_1(p) \end{vmatrix},$$

where we sum only over a set of elements f such that $f(i) \neq f(r + i)$ for all i. But by part (1),

$$\sum_{r=0}^{p-1} \prod_{k\in K} (1 - k + r) = \begin{cases} 0 & \text{if } \operatorname{card} K < p - 1, \\ \sum_{r=0}^{p-1} \prod_{k=2}^{p} (1 - k + r) = -1 & \text{otherwise.} \end{cases}$$

Hence we find:

$$D = \begin{vmatrix} a_1 & \cdots & a_p \\ a_p & & \vdots \\ \vdots & & \vdots \\ a_2 & \cdots & a_1 \end{vmatrix} + \begin{vmatrix} a_1 & 0 & \cdots & 0 \\ a_p & 1 & \cdots & 0 \\ \vdots & & \ddots & \\ a_2 & 0 & \cdots & 1 \end{vmatrix} (-1) + \begin{vmatrix} 1 & & 0 \\ & \ddots & \\ 0 & & 1 \end{vmatrix} \prod_{k=1}^{p} (1 - k)$$

= (Continued)

(Contd) $$= \sum_{i=1}^{p} a_i^p - a_1.$$

REMARK: When, for $i \neq 1$, $a_i = 0$, we recover the well known equality:

$$a_1(a_1 - 1)\cdots(a_1 - p + 1) = a_1^p - a_1.$$

(4): If $a \in k((t))$, the field of "meromorphic" [that is having only finitely many terms of negative order] formal series, we can write:

$$a = \sum_{n=-N}^{\infty} a_n t^n = \sum_{k=0}^{p-1} a^{(k)} \quad \text{with } a^{(k)} = \sum_{n \equiv k \bmod p} a_n t^n.$$

If b is another element of $k((t))$ we see that:

$$(ab)^{(k)} = \sum_{h+h' \equiv k (\bmod p)} a^{(h)} b^{(h')},$$

which can be written:

$$\begin{bmatrix} a^{(0)} & \cdots & a^{(1)} \\ a^{(1)} & & \vdots \\ \vdots & & \vdots \\ a^{(p-1)} & \cdots & a^{(0)} \end{bmatrix} \begin{bmatrix} b^{(0)} \\ \vdots \\ b^{(p-1)} \end{bmatrix} = \begin{bmatrix} (ab)^{(0)} \\ \vdots \\ (ab)^{p-1} \end{bmatrix}.$$

On the other hand:

$$(\vartheta b)^{(k)} = \sum_{n \equiv k \bmod p} n b_n x^n = k \sum_{n \equiv k} b_n x^n = k b^{(k)}.$$

If we look for a non-zero solution of the equation:

$$ab - \vartheta b = 0, \qquad \text{in } k((t)),$$

we are led to solving the equation:

$$\begin{bmatrix} a^{(0)} & a^{(p-1)} & \cdots & a^{(1)} \\ a^{(1)} & a^{(0)} - 1 & & \vdots \\ \vdots & & \ddots & \vdots \\ a^{(p-1)} & \cdots & \cdots & a^{(0)} - (p-1) \end{bmatrix} \begin{bmatrix} b^{(0)} \\ \vdots \\ b^{(p-1)} \end{bmatrix} = 0.$$

Since $k((t))$ is a field, this system has a non-zero solution if and only if its determinant vanishes, which by Question (3) gives (with $a^{(0)} = a_1$, $a^{(p-1)} = a_2$, $a^{(p-2)} = a_3, \ldots, a^{(1)} = a^p$):

$$0 = \sum_{i=0}^{p-1} (a^{(i)})^p - a^{(0)} = a^p - a^{(0)},$$

that is to say, we finally obtain:

$$a^p = \sum a_n^p x^{np} = a^{(0)} = \sum a_{np} x^{np}.$$

REMARK: The coefficient a_0 satisfies $a_0^p = a_0$ and therefore is in $\mathbb{F}_p$, that is to say it can be considered as an integer; it corresponds to a factor x^n in b, for we see that:

$$\frac{\vartheta(x^n b)}{x^n b} = n + \frac{\vartheta b}{b};$$

in particular, the solution is not unique.

SOLUTION 7·7: (1): Using the homomorphism between the product of matrices and the composition of fractional transformations, the equality required:

$$\frac{-\frac{1}{z} + m}{p} = \frac{mz - 1}{pz} = \frac{au + b}{cu + d} \quad \text{with } u = \frac{z + m'}{p},$$

can be written:

$$\begin{bmatrix} m & -1 \\ p & 0 \end{bmatrix} = \begin{bmatrix} a & b \\ c & d \end{bmatrix} \begin{bmatrix} 1 & m' \\ 0 & p \end{bmatrix} .$$

Solving for $\begin{bmatrix} a & b \\ c & d \end{bmatrix}$ we obtain

$$\begin{bmatrix} a & b \\ c & d \end{bmatrix} = \begin{bmatrix} m & \dfrac{-mm' - 1}{p} \\ p & -m' \end{bmatrix}$$

In order that b be an integer it is necessary that m' be chosen in such a way that $mm' + 1$ is divisible by p. As $1 \leqslant m \leqslant p - 1$, $\dot{m}$, the class of m in $\mathbb{Z}/p\mathbb{Z}$, is invertible, and there exists an unique m', $1 \leqslant m' \leqslant p - 1$, which represents the class $\frac{-1}{\dot{m}}$ in $\mathbb{Z}/p\mathbb{Z}$.

(2): As F if a modular function, $F(z + 1) = F(z)$, and we obtain:

$$\Phi_p(z + 1) = p^{k-1}F(pz + p) + \frac{1}{p} \sum_{m=0}^{p-1} F\left(\frac{z + 1 + m}{p}\right) ,$$

and on setting $n = m + 1$,

$$\Phi_p(z + 1) = p^{k-1}F(pz) + \frac{1}{p} \sum_{n=1}^{p} F\left(\frac{z + n}{p}\right) .$$

As $F\left(\frac{z + p}{p}\right) = F\left(\frac{z}{p}\right)$, we find:

$$\Phi_p(z + 1) = p^{k-1}F(pz) + \frac{1}{p} \sum_{n=0}^{p-1} F\left(\frac{z + n}{p}\right) = \Phi_p(z).$$

As F is modular of weight k, we have $F\left(-\frac{1}{z}\right) = z^k F(z)$. This gives:

$$\Phi_p\left(-\frac{1}{z}\right) = p^{k-1}F\left(-\frac{p}{z}\right) + \frac{1}{p}F\left(-\frac{1}{zp}\right) + \frac{1}{p}\sum_{m=1}^{p-1} F\left(\frac{-\frac{1}{z}+m}{p}\right)$$

$$= p^{k-1}\frac{z^k}{p^k}F\left(\frac{z}{p}\right) + p^{k-1}z^kF(zp) + \frac{1}{p}\sum_{m=1}^{p-1} F\left(\frac{au+b}{cu+d}\right)$$

by using part (1) above. As $cu + d = z$,

$$F\left(\frac{au+b}{cu+d}\right) = z^kF(u) = z^kF\left(\frac{z+m'}{p}\right).$$

When m takes all the values between 1 and $p - 1$, m' takes all the values between 1 and $p - 1$, and we have:

$$\Phi_p\left(-\frac{1}{z}\right) = p^{k-1}z^kF(pz) + \frac{1}{p}z^kF\left(\frac{z}{p}\right) + \frac{1}{p}\sum_{m'=1}^{p-1} z^kF\left(\frac{z+m'}{p}\right)$$

$$= z^k\Phi_p(z).$$

(3): If:

$$F(z) = \sum_{n\geq n_0} A_n q^n = \sum_{n\geq n_0} A_n e^{2i\pi nz},$$

we have:

$$\Phi_p(z) = p^{k-1}\sum_{n\geq n_0} A_n e^{2i\pi npz} + \frac{1}{p}\sum_{n\geq n_0} A_n \sum_{m=0}^{p-1} e^{2i\pi n((z+m)/p)}$$

by the definition of the sum of a series. Setting $\zeta = e^{2i\pi n/p}$, we find:

$$\sum_{m=0}^{p-1} e^{2i\pi n(z+m)/p} = e^{2i\pi nz/p}\sum_{m=0}^{p-1}\zeta^m.$$

The sum $\sum_{m=0}^{p-1} \zeta^m$ has the value zero if $\zeta \neq 1$ (that is to say, if p does not divide n) and has the value p if $\zeta = 1$ (that is to say, if p divides n). We then obtain:

$$\Phi_p(z) = p^{k-1} \sum_{n \geqslant n_0} A_n q^{np} + \sum_{\substack{n \geqslant n_0 \\ p/n}} A_n q^{n/p} = \sum_{n \geqslant n_0} a_n q^n,$$

with:

$$n_0' = \min\left(n_0 p, \left[\frac{n_0 + p - 1}{p}\right]\right).$$

If p does not divide n, the second term alone gives:

$$a_n = A_{pn}. \tag{1}$$

If p does divide n, we obtain:

$$a_n = p^{k-1} A_{n/p} + A_{pn}. \tag{2}$$

Using part (2), if F is a modular form we have $n_0 = 0$ and Φ_p is also a modular form; if F is a cusp form we have $n_0 = 1$, and Φ_p is also a cusp form.

(4): We have seen that Φ is a cusp form of weight twelve. Since the vector space of cusp forms of weight twelve is one-dimensional, Φ_p and Δ are proportional.

In Formulae (1) and (2), by setting $A_n = \tau(n)$ we obtain:

$$a_1 = \tau(p),$$

and since $\tau(1) = 1$, we deduce from this:

$$\Phi_p = \tau(p)\Delta \quad \text{and} \quad a_n = \tau(p)\tau(n),$$

which gives us the relations:

$$\text{If } p \text{ does not divide } n: \quad \tau(pn) = \tau(p)\tau(n), \tag{3}$$

$$\text{If } p \text{ does divide } n: \quad p^{11}\tau\left(\frac{n}{p}\right) + \tau(pn) = \tau(p)\tau(n), \tag{4}$$

For $n = p^{\alpha}$, $\alpha \geqslant 1$, relation (4) gives:

$$\tau(p^{\alpha+1}) = \tau(p)\tau(p^{\alpha}) - p^{11}\tau(p^{\alpha-1}). \tag{5}$$

Let us now prove by induction on α the formula:

$$\tau(p^{\alpha}s) = \tau(p^{\alpha})\tau(s) \quad \text{if } p \text{ does not divide } s; \tag{6}$$

for $\alpha = 1$ this is given by relation (3), and for $\alpha = 0$ it is obvious. Let us assume it to be true for α and $\alpha - 1$ with $\alpha \geqslant 1$, and let us prove it for $\alpha + 1$. Formula (4) with $n = p^{\alpha}x$ gives:

$$\begin{aligned}\tau(p^{\alpha+1}s) &= \tau(p)\tau(p^{\alpha}s) - p^{11}\tau(p^{\alpha-1}s)\\ &= \tau(p)\tau(p^{\alpha})\tau(s) - p^{11}\tau(p^{\alpha-1})\tau(s)\\ &= \tau(p^{\alpha+1})\tau(s),\end{aligned}$$

when we use Formula (5).

To prove the multiplicativity of τ, let $n = \prod_i p_i^{\alpha_i}$. Let us calculate $\tau(n)$ by applying Formula (6) successively to all the prime factors of n, we find:

$$\tau(n) = \prod_i (\tau(p_i^{\alpha_i})),$$

which proves that τ is multiplicative.

(5): We know that:

$$G_k(z) = 2\zeta(k)E_k(z)$$

with:

$$E_k(z) = 1 + \gamma_k \sum_{n=1}^{\infty} \sigma_{k-1}(n)q \quad \text{and} \quad \gamma_k = \frac{(2i\pi)^k}{(k-1)!\zeta(k)}.$$

If we set:

$$F = \frac{1}{\gamma_k} E_k = \frac{1}{\gamma_k} + \sum_{n=1}^{\infty} \sigma_{k-1}(n)q^n,$$

Formula (1) gives for the first coefficient of Φ_p:

$$a_1 = A_p = \sigma_{k-1}(p).$$

To show that Φ_p and F are proportional we must show that:

$$\Phi_p = \sigma_{k-1}(p)F,$$

so for all n, set:

$$a_n = \sigma_{k-1}(p)A_n.$$

If $n = 0$, Formula (2) gives:

$$a_0 = p^{k-1}A_0 + A_0 = A_0(1 + p^{k-1}) = A_0\sigma_{k-1}(p);$$

If $n > 0$ and if $p \nmid n$, Formula (1) gives:

$$a_n = A_{pn} = \sigma_{k-1}(pn) = \sigma_{k-1}(p)\sigma_{k-1}(n) = \sigma_{k-1}(p)A_n,$$

taking into account the multiplicativity of the function

$n \mapsto \sigma_{k-1}(n)$.

If $n \geqslant 0$ and if p/n, Formula (2) gives, with $n = p^\alpha s$, $p \nmid s$:

$$a_n = p^{k-1}A_{n/p} + A_{pn} = p^{k-1}\sigma_{k-1}(p^{\alpha-1}s) + \sigma_{k-1}(p^{\alpha+1}s)$$

$$= [p^{k-1}\sigma_{k-1}(p^{\alpha-1}) + \sigma_{k-1}(p^{\alpha+1})]\sigma(s).$$

By the definition of the function σ_{k-1} we have:

$$\sigma_{k-1}(p^{\alpha-1}) = \sigma_{k-1}(p^\alpha) - p^{\alpha(k-1)},$$

$$\sigma_{k-1}(p^{\alpha+1}) = \sigma_{k-1}(p^\alpha) + p^{(\alpha+1)(k-1)},$$

and substituting these in the square brackets we find:

$$a_n = [(p^{k-1} + 1)]\sigma_{k-1}(p^\alpha)\sigma(s) = \sigma_{k-1}(p)\sigma_{k-1}(p^\alpha s)$$

$$= \sigma_{k-1}(p)A_n.$$

The functions G_k and Φ_p are therefore proportional.

SOLUTION 7·8: (1): Recall that $SL_2(\mathbb{Z}/4\mathbb{Z})$ is the set of 2×2 matrices with entries in $\mathbb{Z}/4\mathbb{Z}$ with determinant 1 (mod 4). In $SL_2(\mathbb{Z}/4\mathbb{Z})/\pm 1$ the two matrices $\begin{bmatrix} a & b \\ c & d \end{bmatrix}$ and $\begin{bmatrix} -a & -b \\ -c & -d \end{bmatrix}$ are identical. An elementary calculation allows us to establish the following multiplication table:

1	a	b	c	d	e
a	1	d	e	b	c
b	e	1	d	c	a
c	d	e	1	a	b
d	c	a	b	e	1
e	b	c	a	1	d

;

for example:

$$ed = \begin{bmatrix} 2 & 1 \\ 1 & 1 \end{bmatrix}\begin{bmatrix} 3 & 1 \\ 1 & 2 \end{bmatrix} = \begin{bmatrix} 7 & 4 \\ 4 & 3 \end{bmatrix} = \begin{bmatrix} -1 & 0 \\ 0 & -1 \end{bmatrix} = \begin{bmatrix} 1 & 0 \\ 0 & 1 \end{bmatrix} = i.$$

Therefore our group is isomorphic to S_3.

(2):(a): The relations $[JT]^3 = I$ and $J^2 = I$ with $I = \begin{bmatrix} 1 & 0 \\ 0 & 1 \end{bmatrix}$ are already known. From this we immediately deduce:

$$TJT = [JTJ]^{-1} = JT^{-1}J,$$

$$T^{-1}JT^{-1} = [TJT]^{-1} = JTJ,$$

and consequently:

$$TJT^3 = TJT^{-1}T^4,$$

$$T^2JT = TTJT = TJT^{-1}J,$$

$$T^3JT^{-2} = T^4T^{-1}JT^{-1}T^{-1} = T^4JTJT^{-1}.$$

From this it follows that for all n there exists $\varphi(n)$ such that:

$$T^nJT^{\varphi(n)} \in \Gamma_4.$$

Since every element of Γ (the modular group) can be written:

$$\gamma = T^{n_1}JT^{n_2}\cdots JT^{n_r},$$

that is to say:

$$\gamma = T^{n_1}JT^{\varphi(n_1)}T^{n_2-\varphi(n_1)}JT^{\varphi(n_2-\varphi(n_1))}T^{n_3-\varphi(n_2-\varphi(n_1))}J\cdots,$$

and therefore that:

$$\gamma \in \Gamma_4 T^{n_r - \varphi(n_{r-1} - \varphi(\cdots))},$$

and, since $T^4 \in \Gamma_4$, γ belongs to one of the $\Gamma_4 T^i$ $(i = 0,1,2,3)$.

Let us show that the sum is direct, that is to say that there do not exist $\gamma_1, \gamma_2 \in \Gamma_4$ and $n_1 \mid n_2$ (mod 4) such that $\gamma_1 T^{n_1} = \gamma_2 T^{n_2}$. In this case $T^{n_1 - n_2} = \gamma_1^{-1}\gamma_2$ would belong to Γ_4, it therefor is enough to show that T, T^2, T^3 are not in Γ_4. Now, if we reduce our matrices (mod 4) we see that J becomes the a of Question (1), T^4 the i and $TJT^{-1} = \begin{bmatrix} 1 & -2 \\ 1 & -1 \end{bmatrix}$ the b. By reduction (mod 4) Γ_4 therefore becomes a subgroup of the group in part (1) (since it contains a and b it is the same group as in part (1)). Now, it is easy to establish that the latter contains neither the reduction of T, nor of T^2, nor of T^3. [This would allows us to construct the 48 matrices of

$$SL_2(\mathbb{Z}/4\mathbb{Z}) = \bigoplus_{i=0}^{3} (\text{group of } 1) \times T^i \times \{\pm 1\}].$$

(2):(b): Let $\mathcal{D}$ be the "classical" fundamental domain of Γ $(-\frac{1}{2} \leqslant \mathrm{Re}(z) \leqslant \frac{1}{2}$, $|z| \geqslant 1)$. The fundamental domain of Γ_4 (a subgroup of Γ) is made up of a certain number of "copies" of $\mathcal{D}$. In particular, by the decomposition above, every point of H is congruent by Γ_4 to a point in $\mathcal{D} \cup T\mathcal{D} \cup T^2\mathcal{D} \cup T^3\mathcal{D}$.

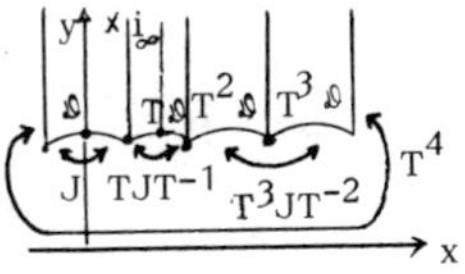

We now make the triangulation indicated in the Figure above, where we show how the sides are identified by the action of Γ_4 [the other points are not equivalent, for example the transformations of Γ which send j into $j + 3$ are T^3, $T^3(T^{-1}J)$ and

$T^3(T^{-1}J)^2$, which are, mod Γ_4, equivalent to T^3, those which send i into $i + 1$ are T and TJ which do not belong to Γ_4].

We then count 5 vertices: j ($\sim j + 1$, $j + 2$ and $j + 4$), $j + 3$, i, $i + 1$, i_∞, 6 vertical sides and 3 horizontal sides (j,i) $[\sim (i,j + 1)]$, $(j + 1,i + 1)$ $[\sim (i + 1),(j + 2)]$ and $(j + 2,j + 3)$ $[\sim (j + 3),(j + 4)]$, and finally 6 triangles. Therefore we have a surface of genus g with:

$$2 - 2g = 5 - 9 + 6 = 2,$$

that is to say, of genus zero.

(3):(a): q is the local uniformizer at the point i_∞ of the fundamental domain of Γ_4. Since $\Gamma_4 \backslash H$ is of genus zero, there exist functions having a simple pole at the point at infinity, and holomorphic elsewhere, moreover; they are of the form $\lambda H + \mu$ if H is any one of them. The two conditions correspond to the particular functions having a residue $e^{i\pi/4}/(2\sqrt{2})$ (which fixes λ) and zero constant term (which fixes μ), it is therefore uniquely defined.

(3):(b): By Question (2) it is clear that Γ, the full modular group, permutes the four functions $H(z + h) = H(T^h z)$ $[H(\gamma T^h(\gamma z)) = H(\gamma' T^{h'} z) = H(T^{h'} z)]$.

The polynomial in x, $\prod_{h=0}^{3} (x - H(z + h))$, therefore has coefficients which are functions of z invariant under Γ and having poles only at infinity, they are therefore polynomials in $\underline{j}(z)$.

On the other hand we have $e^{(i\pi/2)(z+h)} = q(e^{i\pi/2})^h = q(i)^h$, that is to say:

$$\Phi(x,z) = \prod_{h=0}^{3} \left(x - \frac{e^{i\pi/4}}{2\sqrt{2}\,i^h q} + O(q) \right) =$$

$$= x^4 + O(q)x^3 + O(1)x^2 + O\left(\frac{1}{q}\right)x + O\left(\frac{1}{q^2}\right) + \frac{1}{64q^4}$$

[by noticing that the $\frac{e^{i\pi/4}}{2\sqrt{2}q}\frac{1}{h}$ are the roots of $X^4 = -\frac{1}{64q^4}$].

Now, it is known that:

$$\underline{j}(z) = \frac{1}{q^4} + 744 + \sum c(n)q^4 = \frac{1}{q^4} + O(1),$$

therefore a non-constant polynomial in $\underline{j}(z)$ cannot be $o\left(\frac{1}{q^4}\right)$ and a constant could not be $o(1)$ without being zero, from which it follows that:

$$\Phi(x,z) = x^4 + ax^2 + bx + c + \frac{1}{64}\,j(z),$$

where a,b,c are constants.

By the construction of the domain we see that:

$$H(j) = H(j + 1) = H(j + 2),$$

that is to say that $\Phi(x,j)$ has a triple root; now, $\underline{j}(j) = 0$, therefore:

$$x^4 + ax^2 + bx + c = (x - \alpha)(x - \beta)^3,$$

and as the term in x^3 must be zero, $\alpha = -3\beta$.

On the other hand we see that $H(i + 2) = H(i + 3)$, that is to say, that $\Phi(x,i)$ has a double root; now, $\underline{j}(i) = 64\times 27 = 1{,}728$, hence $(x + 3\beta)(x - \beta)^3 + 27 = 0$ must have one double root in common with its derivative $[3x + 9\beta + x - \beta](x - \beta)^2$; this can not be β $(27 \neq 0)$, therefore it is $x = -2\beta$, hence it is necessary to have $(-2\beta + 3\beta)(2\beta - \beta)^3 + 27 = 0$, that is to say $\beta^4 = 1$; it remains to choose a suitable fourth root (the choice of β corresponds to the four functions $\pm H, \pm iH$).

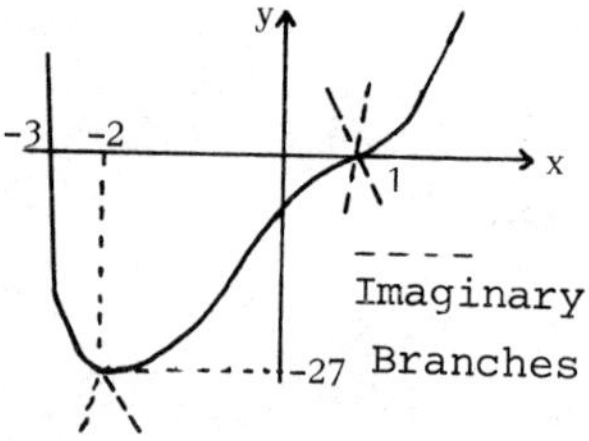

We draw the graph of the function $y = (x + 3)(x - 1)^3$ (see the Figure on the right). On the vertical half-line $j \to \infty$ the function $\underline{j}(z)$ takes real values decreasing from 0 to $-\infty$; the function $(\pm H, \pm iH)$, a root of $(H' + 3)(H' - 1)^3 = -\frac{j}{64}$, will be such that $(H'(z) + 3)(H'(z) - 1)^3$ will increase from 0 to $+\infty$. On the graph we see that the branch arising from the simple root is the branch where $x = H'(z)$ varies from -3 to $-\infty$; by the form of the fundamental domain this branch corresponds to the half-line $(j + 3) \to \infty$. The root of the equation:

$$(H' + 3)(H' - 1)^3 = -\frac{j}{64}$$

is therefore real and negative on this half-line; now, on the latter,

$$q = e^{(i\pi/2)((5/2)+ih)} = ae^{5i\pi/4} \quad \text{with } a \in \mathbb{R}^+,$$

therefore if $q \to 0$,

$$H(z) \sim \frac{e^{i\pi/4}}{2\sqrt{2}} \times e^{-5i\pi/4} \times \frac{1}{a}$$

and this number certainly belongs to $\mathbb{R}^-$, therefore $H' = H$ and

$$\Phi(x,z) = (x + 3)(x - 1)^3 + \frac{j(z)}{64} .$$

(4):(a): H is an univalent function on the fundamental domain of Γ_4, therefore the condition $H(\omega) = H\left(\frac{\alpha\omega + \beta}{\delta}\right)$ is equivalent to:

There exists $\begin{bmatrix} a & b \\ c & d \end{bmatrix} \in \Gamma_4$ which transforms $\frac{\alpha\omega + \beta}{\delta}$ into ω,

that is to say:

$$\omega = \frac{a\left(\frac{\alpha\omega + \beta}{\delta}\right) + b}{c\left(\frac{\alpha\omega + \beta}{\delta}\right) + d},$$

which leads to:

$$c\alpha\omega^2 + (c\beta + d\delta - a\alpha)\omega - \delta b - a\beta = 0.$$

The discriminant of this equation is:

$$(-c\beta - d\delta + a\alpha)^2 + 4c\alpha\delta b + 4c\alpha\beta a$$

$$= (a\alpha + c\beta + d\delta)^2 - 4a\alpha\delta d + 4\alpha\delta c$$

$$= (a\alpha + c\beta + d\delta)^2 - 4\alpha\delta,$$

because $ad - bc = 1$.

We can always assume $c \geqslant 0$ [otherwise take $\begin{bmatrix} -a & -b \\ -c & -d \end{bmatrix}$] and even $c > 0$, for otherwise ω would be real. The solution ω belonging to H is then:

$$\omega = \frac{a\alpha - c\beta - d\delta + i\sqrt{4\alpha\delta - (a\alpha + c\beta + d\delta)^2}}{2c\alpha};$$

in order that this be $\tau + \ell = \frac{1 + 2\ell + i\sqrt{D}}{2}$ it is therefore necessary that we have:

$$4\alpha\delta - (a\alpha + c\beta + d\delta)^2 = D(c\alpha)^2,$$

that is to say, with the conditions imposed:

$$D + 1 - (a\alpha + c\beta + d\delta)^2 = D(c\alpha)^2,$$

which implies:

$$c\alpha = 1,$$

$$a\alpha + c\beta + d\delta = \pm 1,$$

that is to say $c = \alpha = 1$, since $c,\alpha \geqslant 0$, and moreover it is necessary that:

$$1 + 2\ell = a\alpha - c\beta - d\delta \quad \text{and} \quad \delta = \frac{D+1}{4}.$$

Hence we must find a matrix $\begin{bmatrix} a & b \\ c & d \end{bmatrix}$ such that:

$$\begin{cases} c = 1, \\ a + B + d \times \dfrac{D+1}{4} = \pm 1, \qquad ad - b = 1, \\ a - \beta - d\,\dfrac{D+1}{4} = 1 + 2\ell, \end{cases}$$

which gives:

$$a = \begin{cases} 1 + \ell, \\ \ell, \end{cases} \qquad \beta + d\,\frac{D+1}{4} = \begin{cases} -\ell, \\ -1-\ell; \end{cases}$$

now, $\frac{D+1}{4} \equiv 1 \pmod 4$, therefore $\frac{D+1}{4}$ is prime to 4, therefore the second relation has a solution in d and β with β a multiple of 4 (and $0 \leqslant \beta < D + 1$), therefore for every ℓ the problem is solvable in Γ, it is enough to check that the solution is in Γ_4, therefore it is enough to check if, by reducing mod 4 the solution belongs to the group of 1); since $\beta \equiv 0 \pmod 4$ we have:

$$c = 1, \qquad a = \begin{cases} 1 + \ell, \\ \ell, \end{cases} \qquad d: \begin{cases} -\ell \\ -1-\ell \end{cases} \pmod 4 \quad b = ad - 1,$$

which gives for:

$\ell \equiv 0 \pmod 4$, the matrices $\begin{bmatrix} 1 & -1 \\ 1 & 0 \end{bmatrix}$ or $\begin{bmatrix} 0 & -1 \\ 1 & -1 \end{bmatrix}$,

$\ell \equiv 1 \pmod 4$, the matrices $\begin{bmatrix} 2 & -3 \\ 1 & -1 \end{bmatrix}$ or $\begin{bmatrix} 1 & -3 \\ 1 & -2 \end{bmatrix}$,

$\ell \equiv 2 \pmod 4$, the matrices $\begin{bmatrix} 3 & -7 \\ 1 & -2 \end{bmatrix} \equiv d$ or $\begin{bmatrix} 2 & -7 \\ 1 & -3 \end{bmatrix} \equiv e$,

$\ell \equiv 3 \pmod 4$, the matrices $\begin{bmatrix} 0 & -1 \\ 1 & -3 \end{bmatrix}$ or $\begin{bmatrix} 3 & -1 \\ 1 & 0 \end{bmatrix}$.

We see that only when $\ell \equiv 2$ do the solutions belong to Γ_4, in this case the solutions certainly belong to Γ_4, because by reducing (mod 4) we have a matrix equal to d or e.

(4):(b): We are going to prove that the coefficients of the given polynomial in x are invariant under Γ_4, as it is clear that these coefficients are holomorphic in H, from which it follows that these are polynomials in $H(z)$, which proves the result.

Let us then consider the set E of 2×2 matrices with integral entries and determinant $\frac{D+1}{4}$ (which is congruent to 1 modulo 4), and which are, modulo 4, congruent to an element of the group of 1). We shall say that two elements $A, B \in E$ are equivalent if there exists $\gamma \in \Gamma_4$ such that $A = \gamma B$; we are going to prove the each class contains an unique representative of the form:

$$\begin{bmatrix} \alpha & \beta \\ 0 & \delta \end{bmatrix} \text{ with } \begin{cases} \alpha > 0, \\ 0 \leqslant \beta < 4\delta, \\ \beta \equiv 0 \pmod 4), \end{cases}$$

so, in effect, $A = \begin{bmatrix} \alpha & \beta \\ \gamma & \delta \end{bmatrix} \in E$, then $\alpha = \lambda\gamma + \gamma'$ with $|\gamma'| < |\gamma|$.

Now in Γ_4 we have at our disposal four types of matrices:

$$\begin{bmatrix} 0 & -1 \\ 1 & 4p \end{bmatrix}, \quad \begin{bmatrix} 2 & 1 - 8p \\ 1 & 4p + 1 \end{bmatrix}, \quad \begin{bmatrix} -1 & -3 - 4p \\ 1 & 4p + 3 \end{bmatrix},$$

and:

$$\begin{bmatrix} 1 & 2 + 4p \\ 1 & 4p + 3 \end{bmatrix},$$

that is to say that for one of them we will have:

$$\begin{bmatrix} * & * \\ 1 & -\lambda \end{bmatrix}\begin{bmatrix} \alpha & \beta \\ \gamma & \delta \end{bmatrix} = \begin{bmatrix} * & * \\ \alpha - \lambda\gamma & * \end{bmatrix} = \begin{bmatrix} \alpha' & \beta' \\ \gamma' & \delta' \end{bmatrix},$$

which means every matrix of E is equivalent to a matrix having an entry γ whose absolute value is strictly less than the one you start with, and by iterating this procedure we will obtain an equivalent matrix having an entry $\gamma = 0$; multiplying by $\begin{bmatrix} -1 & 0 \\ 0 & -1 \end{bmatrix}$ if necessary we can always assume that $\alpha > 0$ (non-zero because otherwise $\alpha\delta = \dfrac{D + 1}{4}$); but since:

$$\begin{bmatrix} 1 & 4k \\ 0 & 2 \end{bmatrix}\begin{bmatrix} \alpha & \beta \\ 0 & \delta \end{bmatrix} = \begin{bmatrix} \alpha & \beta + 4k\delta \\ 0 & \delta \end{bmatrix},$$

we see that we could just as well choose $0 \leqslant \beta < 4\delta$; finally, the matrix obtained must be congruent modulo 4 to one of the elements $\pm i, \pm a, \pm b, \pm c, \pm d$, or $\pm e$, and it can only be $\pm i$, hence $\beta \equiv 0 \pmod 4$.

Let us show the uniqueness. Let us assume $\begin{bmatrix} \alpha & \beta \\ 0 & \delta \end{bmatrix} = \begin{bmatrix} \alpha' & \beta' \\ 0 & \delta' \end{bmatrix}$ with $\gamma \in \Gamma_4$; multiplying both sides by $\begin{bmatrix} \delta' & -\beta' \\ 0 & \alpha' \end{bmatrix}$ we find:

$$\gamma = \begin{bmatrix} \dfrac{\alpha}{\alpha'} & \dfrac{\alpha'\beta - \alpha\beta'}{\alpha'\delta'} \\ 0 & \dfrac{\delta}{\delta'} \end{bmatrix},$$

but since $\alpha\delta = \alpha'\delta'$, that is to say $\frac{\alpha}{\alpha'} = \frac{\delta}{\delta'}$, these two numbers, the product of which is one, must have value ±1. As α,α' are positive one finds $\alpha = \alpha'$, therefore $\frac{\beta - \beta'}{\delta'}$ must be integral, and by reduction (mod 4) one sees that it must be a multiple of 4. Since $0 \leqslant \beta < 4\delta$, and since $(4,\delta) = 1$, we see that $\beta = \beta'$.

It is clear that A is equivalent to B if and only if $H(Az) = H(Bz)$. Therefore the polynomial can be written in the form $\prod (x - H(Az))$, where the product runs over the equivalence classes; but if $A \sim B$ it is obvious that $A\gamma \sim B\gamma$ [in fact $A = \gamma'B$, hence $A\gamma = \gamma'B\gamma$], that means right multiplication by γ (an invertible operation) permutes the equivalence classes, which completes the proof.

(4):(c): The roots of the first equation are $H(\tau)$, $H(\tau + 1)$, $H(\tau + 2)$ and $H(\tau + 3)$, by Question (3).

Let us assume that x is a root of the equation $\Phi(x,x) = 0$. Since $H(z)$ is univalent, there exists $z \in H$ (unique up to the action of Γ_4) such that $x = H(z)$, therefore we have $\Phi(H(z),H(z)) = 0$, that is to say, there exist α,β,δ, satisfying the conditions, such that:

$$H(z) = H\left(\frac{\alpha z + \beta}{\delta}\right) = x.$$

Now, we have seen in Question (4)(a) above that this is true for $z = \tau + \ell$ if and only if $\ell \equiv 2 \pmod 4$, from which we obviously have that $H(\tau + 2)$ is the unique solution common to the two equations.

CHAPTER 8
Quadratic Forms

INTRODUCTION

The goal of this Chapter is to determine all the rational numbers which are represented by a quadratic form f with rational coefficients.

(A *QUADRATIC FORM* $f(x_1,x_2,\ldots,x_n)$ *with coefficients in a field* K *REPRESENTS* $\gamma \in K$ *IN* K if there exist $\alpha_1,\alpha_2,\ldots,\alpha_n \in K$, not all zero, such that $f(\alpha_1,\alpha_2,\ldots,\alpha_n) = \gamma$).

8.1 NOTATIONS

We denote by V the union of the set P of prime numbers and the symbol ∞, and the convention $\mathbb{Q}_\infty = \mathbb{R}$ is adopted.

Let $v \in V$ and $\alpha,\beta \in \mathbb{Q}_v^*$. The *HILBERT SYMBOL* $(\alpha,\beta)_v$ is defined by:

$$(\alpha,\beta)_v = 1 \quad \text{if } \alpha x^2 + \beta y^2 \text{ represents 1 in } \mathbb{Q}_v,$$

$$(\alpha,\beta)_v = -1 \quad \text{otherwise.}$$

Let $f = a_1x_1^2 + a_2x_2^2 + \cdots + a_nx_n^2$, where $a_i \in \mathbb{Q}_v$, be a non-singular diagonalised quadratic form in n variables. We *denote* by $d(f)$ the determinant of f and by $\varepsilon_v(f)$ the *HASSE SYMBOL*, defined by:

$$d(f) = \prod_{i=1,2,\ldots,n} a_i \quad \text{and} \quad \begin{cases} \varepsilon_v(f) = \prod_{i<j} (a_i,a_j)_v & \text{if } n > 2, \\ \varepsilon_v(f) = 1 & \text{if } n = 1. \end{cases}$$

8.2 PROPERTIES

The following two properties of the Hilbert symbol:

$$(\alpha,\beta\gamma)_v = (\alpha,\beta)_v(\alpha,\gamma)_v \quad \text{and} \quad (\alpha,\alpha)_v = (\alpha,-1)_v,$$

allow, in all cases, the calculation of the values of the Hilbert symbol, since we know that:

$$\text{if } p \in P - \{2\}, \quad (p,\varepsilon)_p = \left(\frac{\varepsilon_0}{p}\right) \quad \text{if } \varepsilon \text{ is an unit of } \mathbb{Q}_p, \ \varepsilon_0 \in \mathbb{Z}, \ \varepsilon \equiv \varepsilon_0(p);$$

$$\text{if } p = 2, \quad (p,\varepsilon)_2 = (-1)^{\frac{1}{8}(\varepsilon_0^2-1)} \quad \text{if } \varepsilon \text{ is an unit of } \mathbb{Q}_p, \ \varepsilon_0 \in \mathbb{Z}, \ \varepsilon \equiv \varepsilon_0(8);$$

$$\text{if } p \in P - \{2\}, \quad (\varepsilon,\eta)_p = 1 \quad \text{if } \varepsilon \text{ and } \eta \text{ are units of } \mathbb{Q}_p;$$

$$\text{if } p = 2, \quad (\varepsilon,\eta)_2 = (-1)^{\frac{(\varepsilon_0-1)(\eta_0-1)}{2}} \quad \text{if } \varepsilon \text{ and } \eta \text{ are units of } \mathbb{Q}_2, \ \varepsilon_0,\eta_0 \in \mathbb{Z}, \ \varepsilon = \varepsilon_0(4), \ \eta = \eta_0(4);$$

$$v = \infty \quad (\alpha,\beta) = \begin{cases} 1 & \text{if } \alpha \text{ or } \beta > 0, \\ -1 & \text{if } \alpha < 0, \beta < 0. \end{cases}$$

THEOREM (I): *Let f be a non-singular diagonalised quadratic form in n variables, with coefficients in $\mathbb{Q}_p$; in order that f represents zero in $\mathbb{Q}_p$ it is necessary and sufficient that one of the following* *conditions hold:*

(1): $n = 2$ *and* $-d(f)$ *is a square in* $\mathbb{Q}_p$;

(2): $n = 3$ *and* $(-1,-d(f))_p = \varepsilon_p(f)$;

(3) $n = 4$ *and* *either:* $d(f)$ *is not a square in* $\mathbb{Q}_p$,

or: $d(f)$ *is a square in* $\mathbb{Q}_p$ *and* $\varepsilon_p(f) = (-1,-1)_p$;

(4): $n \geq 5$.

THEOREM (II): *Let* f *be a non-singular diagonalised quadratic form in* n *variables, with coefficients in* $\mathbb{Q}_p$; *in order that* f *represents* $a \in \mathbb{Q}_p^*$ *in* $\mathbb{Q}_p$ *it is necessary and sufficient that one of the following conditions hold:*

(1) $n = 1$ *and* $ad(f)$ *is a square in* $\mathbb{Q}_p$;

(2) $n = 2$ *and* $(a,-d(f))_p = \varepsilon_p(f)$;

(3) $n = 3$ *and* — *either:* $ad(f)$ *is not a square in* $\mathbb{Q}_p$,

— *or:* $ad(f)$ *is a square in* $\mathbb{Q}_p$ *and* $\varepsilon_p(f) = (-1,-d(f))_p$;

(4): $n \geq 4$.

THEOREM (III): (Hasse-Minkowski): *A quadratic form with rational coefficients represents* $a \in \mathbb{Q}$ *in* $\mathbb{Q}$ *if and only if it represents* a *in* $\mathbb{Q}_v$ *for all* $v \in V$.

PRODUCT FORMULA: If $a,b \in \mathbb{Q}^*$, we have $(a,b)_v = 1$ for almost all $v \in V$ and:

$$\prod_{v \in V} (a,b)_v = 1.$$

The theory of quadratic forms may be found in the following works: [Bor], [Ser 1].

PROBLEMS

EXERCISE 8·1: Let $f = a_1x_1^2 + a_2x_2^2 + \cdots + a_nx_n^2$ and $g = b_1x_1^2 + b_2x_2^2 + \cdots + b_kx_k^2$ be two diagonal non-singular quadratic forms with p-adic coefficient. *Write* $f \dotplus g$ for the *form in* $n + k$ *variables* defined by:

$$f \dotplus g = a_1y_1^2 + a_2y_2^2 + \cdots + a_ny_n^2 + b_1y_{n+1}^2 + \cdots + b_ky_{n+k}^2.$$

Show that:

$$\varepsilon_p(f \dotplus g) = \varepsilon_p(f)\varepsilon_p(g)(d(f),d(g))_p.$$

EXERCISE 8·2: Determine all the elements of $\mathbb{Q}_7$ represented in $\mathbb{Q}_7$ by the form $3x^2 + 7y^2$.

EXERCISE 8·3: In this problem $a \in \mathbb{Z}$ is a non-zero integer that is written in the form:

$$a = 2^\alpha 5^\beta b,$$

where $\alpha,\beta \geqslant 0$ are integers and $b \in \mathbb{Z}$ is divisible neither by 2 nor by 5.

(1): Characterise by conditions on α, β, b those numbers a such that $30a$ is the square of a 2-adic number. Give the list of those numbers a whose absolute values are less than or equal to 50.

(2): The same questions, replacing the field $\mathbb{Q}_2$ of 2-adic numbers by the field $\mathbb{Q}_5$ of 5-adic numbers.

(3): In the remainder of the problem f denotes the quadratic form in three variables on $\mathbb{Q}$:

$$f = X_1^2 + 3X_2^2 - 10X_3^2,$$

and d denotes its discriminant.

Calculate ε_v and the Hilbert symbol $(-1,-d)_v$ for all $v \in V$.

(4): For which $v \in V$ does f represent every element of $\mathbb{Z}^*$ in $\mathbb{Q}_v$?

(5): Describe the elements of $\mathbb{Z}$ which are not represented by f in $\mathbb{Q}$, and give the list of thos whose absolute value is less than or equal to 50.

EXERCISE 8·4: This problem studies the rational integers represented in $\mathbb{Q}$ by the quadratic form:

$$f = 5X_1^2 - 7X_2^2.$$

(1): Does the form f represent 0 in $\mathbb{Q}$?

(2): Show that the form f represents the non-zero rational integer a in $\mathbb{Q}$ if and only if the two Hilbert symbols $(a,35)_p$ and $(5,-7)_p$ are equal for all odd prime numbers p.

(3): Assuming the rational integer a to be non-zero and to be square-free, characterise by conditions on Legendre symbols those a's that can be represented by f in $\mathbb{Q}$, distinguishing the following four cases:

a is divisible neither by 5 nor by 7;

a is divisible by 5 and not divisibly by 7;

a is divisible by 7 and not divisible by 5;

a is divisible by both 5 and 7.

(4): What is the list of rational integers between -14 and +14 which are represented by f in $\mathbb{Q}$?

SOLUTIONS

SOLUTION 8·1: We argue by induction on n. If $n = 1$ there certainly holds:

$$\varepsilon_p(a_1x_1^2 \dotplus g) = (a_1,b_1)_p(a_1,b_2)_p\cdots(a_1,b_k)_p\varepsilon_p(g)$$

$$= (a_1,d(g))_p\varepsilon\ (g), \qquad (1)$$

so:

$$\varepsilon_p(a_1x_1^2 \dotplus g) = (d(f),d(g))_p\varepsilon_p(g)\varepsilon_p(f).$$

Let us assume, therefore, the property to be true for forms in $n - 1$ variables. We then write $f = ax_1^2 \dotplus f'$, where f' is a diagonalized form in $n - 1$ variables. Then we have:

$$\varepsilon_p(a_1x_1^2 \dotplus f' \dotplus g) = \varepsilon_p(f' \dotplus g)(a_1,d(f' \dotplus g))_p,$$

by Equation (1). By the recurrence property

$$\varepsilon_p(f' \dotplus g) = \varepsilon_p(f')\varepsilon_p(g)(d(f'),d(g))_p.$$

On the other hand:

$$(a_1,d(f' \dotplus g))_p = (a_1,d(f'))_p(a_1,d(g))_p.$$

So:

$$\varepsilon_p(f \dotplus g) = \varepsilon_p(f')(a_1,d(f'))_p\varepsilon_p(g)(a_1d(f'),d(g))_p$$

$$= \varepsilon_p(f)\varepsilon_p(g)(d(f),d(g))_p.$$

SOLUTION 8·2: 0 is not represented, because $-\frac{7}{3}$ is not square in $\mathbb{Q}_7$ since $v_7\left(-\frac{7}{3}\right) = 1$.

By Theorem (II) γ is represented if and only if:

$$(3,7)_7 = (\gamma,-21)_7. \tag{1}$$

Or, in $\mathbb{Q}_7$ γ admits an unique expansion in the form:

$$\gamma = 7^n(a_0 + 7a_1 + \cdots + 7^i a_i + \cdots),$$

where:

$$n \in \mathbb{Z},\ a_i \in \{0,1,2,3,4,5,6\},\ a_0 \neq 0.$$

Let us set $\gamma = 7^n\gamma'$, where γ' is a 7-adic unit. If n is even:

$$(\gamma,-21)_7 = (\gamma',-21)_7 = (\gamma',-3)_7(\gamma',7)_7 = \left(\frac{a_0}{7}\right),$$

since γ' is a 7-adic unit congruent to a_0 modulo 7.

If n is odd, one obtains similarly:

$$(\gamma,-21)_7 = (7\gamma',-21)_7 = (7,-7)_7(7,3)_7(\gamma',-3)_7(\gamma',7)_7$$

$$= (7,3)_7\left(\frac{a_0}{7}\right).$$

But:

$$(3,7)_7 = \left(\frac{3}{7}\right) = \left(\frac{-4}{7}\right) = \left(\frac{-1}{7}\right) = (-1)^{\frac{1}{2}(7-1)} = -1,$$

and condition (1) becomes:

$$\begin{cases} \left(\frac{a_0}{7}\right) = -1 & \text{if } n \text{ is even,} \\ \left(\frac{a_0}{7}\right) = 1 & \text{if } n \text{ is odd,} \end{cases}$$

or:

$$\begin{cases} a_0 \in \{1,2,4\} & \text{if } n \text{ is odd,} \\ a_0 \in \{3,5,6\} & \text{if } n \text{ is even.} \end{cases}$$

The elements of $\mathbb{Q}_7$ represented by the form $3x^2 + 7y^2$ are therefore those whose expansion have the form:

$$\gamma = 7^{2k}(a_0 + 7a_1 + \cdots + 7^n a_n + \cdots),$$

where:

$$a_0 \in \{3,5,6\},\ a_n \in \{0,1,2,3,4,5,6\} \text{ for } n \geq 1,$$

or of the form:

$$\gamma = 7^{2k+1}(a_0 + 7a_1 + \cdots + 7^n a_n + \cdots),$$

where:

$$a_0 \in \{1,2,4\},\ a_n \in \{0,1,2,3,4,5,6\} \text{ for } n \geq 1.$$

SOLUTION 8·3: (1): In order that $30a = 2^{\alpha+1}5^{\beta+1}3b$ be a square in $\mathbb{Q}_2$ it is necessary and sufficient that we have:

α odd and $5^{\beta+1}3b$ square in $\mathbb{Q}_2$.

This second condition is equivalent to:

β odd and $3b$ square in $\mathbb{Q}_2$,

or:

β even and $15b$ square in $\mathbb{Q}_2$.

A necessary and sufficient condition for a unit x of $\mathbb{Q}_2$ to be a square in $\mathbb{Q}_2$ is that:

$x \equiv 1 \pmod 8$.

As $3.3 = 1 \pmod 8$ and $15(-1) \equiv 1 \pmod 8$, we finally obtain the following characterisation:

A necessary and sufficient condition for $30a$ to be a square in $\mathbb{Q}_2$ is that we have:

(i): α odd;

and:

(ii): either: β odd and $b \equiv 3 \pmod 8$,

or: β even and $b \equiv -1 \pmod 8$.

This allows us to obtain the (α,β,b) corresponding to such integers a, with $|a| \leqslant 50$:

if: $\alpha = 1$ $\beta = 0$, $b = -17,-9,-1,7,23$,

$\beta = 1$, $b = 3$,

$\beta = 2$, $b = -1$,

$\alpha = 3$ $\beta = 0$, $b = -1$,

$\alpha = 5$ $\beta = 0$, $b = -1$.

The set of numbers $a \neq 0$ such that $30a$ is a square in $\mathbb{Q}_2$, with $|a| \leqslant 50$, is:

$$\{-50,-34,-32,-18,-8,-2,14,30,46\}.$$

(2): In order that $30a = 2^{\alpha+1}5^{\beta+1}3b$ be a square in $\mathbb{Q}_5$ it is necessary and sufficient that we have:

β odd and $2^{\alpha+1}3b$ square in $\mathbb{Q}_5$.

This condition is equivalent to:

$2^{\alpha+1}3b$ square modulo 5, or $2^{\alpha}b$ square modulo 5.

Now the only squares of $\mathbb{F}_5$, the field with five elements, are +1 and -1; on the other hand, if α is even:

$$2^{\alpha} \equiv \pm 1 \text{ modulo } 5,$$

and this finally leads to the following characterisation:

A necessary and sufficient condition for $30a$ to be a square in $\mathbb{Q}_5$ is that we have:

(i): β odd;

and:

(ii): either: α odd and $b \equiv \pm 2 \bmod 5$,

or: α even and $b \equiv \pm 1 \bmod 5$.

This permits us to obtain the (α,β,b) corresponding to such integers a, with $|a| \leqslant 50$:

$\beta = 1$ $\quad \alpha = 0$, $b = -9,-1,1,9$,

$\alpha = 1$ $\quad b = -3,+3$,

$\alpha = 2$, $b = -1,+1$,

and the set of numbers a such that $30a$ is a square in $\mathbb{Q}_5$ with $|a| \leqslant 50$ is:

$$\{\pm 5, \pm 20, \pm 30, \pm 45\}.$$

$$(3): \quad \varepsilon_v = (1,3)_v(1,-10)_v(3,-10)_v$$

$$= (3,-10)_v ,$$

as 1 is square in $\mathbb{Q}_v$ for all v; on the other hand, in $\mathbb{R}$:

$$\varepsilon_\infty = (3,-10)_\infty = 1,$$

as 3 is positive.

For p an odd prime, if u and v are two p-adic units $(u,v)_p$ has the value 1, and from this it follows that:

$$\varepsilon_p = 1 \text{ for } p \neq 2,3,5.$$

If $p = 3$:

$$\varepsilon_3 = (3,-10)_3 = \left(\frac{-10}{3}\right) \quad \text{(Legendre symbol)}$$

$$= -1,$$

as -10 is not a square modulo 3.

If $p = 5$:

$$\varepsilon_5 = (3,-10)_5 = \left(\frac{3}{5}\right) = -1,$$

as 3 is not a square modulo 5.

By the Product Formula we obtain $\varepsilon_2 = 1$, and so we deduce that:

$$\varepsilon_v = 1 \text{ for all } v \text{ different from 3 or 5},$$

and:

$$\varepsilon_3 = \varepsilon_5 = -1.$$

The discriminant d of f is equal to -30:

$$(-1,-d)_v = (-1,30)_v,$$

in $\mathbb{R}$, $(-1,30)_\infty = 1$ as 30 is positive.

If $p \neq 2,3$ or 5 $\quad (-1,30)_p = 1,$

for -1 and 30 then are p-adic units.

If $p = 3$ $\qquad (-1,30)_3 = \left(\frac{-1}{3}\right) = -1,$

for -1 is a square modulo 5.

If $p = 5$ $\qquad (-1,30)_5 = \left(\frac{-1}{5}\right) = 1,$

for -1 is a square modulo 5.

We then obtain $(-1,30)_2 = -1$ by the Product Formula, which allows us to conclude that:

$$(-1,30)_v = 1 \quad \text{for all } v \text{ different from 2 Or 3},$$

and:

$$(-1,30)_2 = (-1,30)_3 = -1.$$

(4): *For* $v = \infty$: The coefficients of f do not all have the same sign, therefore f represents in $\mathbb{R}$ every element of $\mathbb{R}$.

For $p \in P$, $p \neq 2$, $p \neq 5$: We have:

$$\varepsilon_p(f) = (-1,d(f))_p ,$$

hence, by Theorems (I) and (II) f represents in $\mathbb{Q}_p$ every element of $\mathbb{Q}_p$, therefore every element of $\mathbb{Z}$.

For $p = 2$ *and* $p = 5$: We have:

$$\varepsilon_p(f) = -(-1,-d(f))_p,$$

hence by Theorems (I) and (II) f does not represent zero in $\mathbb{Q}_p$ nor the elements a of $\mathbb{Z}^*$ such that $30a$ be a square in $\mathbb{Q}_p$.

(5): f does not represent zero in $\mathbb{Q}$, for f does not represent zero in $\mathbb{Q}_2$ and $\mathbb{Q}_5$; and by Theorem (III) f represents the integer a non-zero in $\mathbb{Q}$ if and only if $30a$ is not square in $\mathbb{Q}_2$ and $\mathbb{Q}_5$. The integers lying between -50 and 50 not represented by f are therefore:

$$-50,-45,-34,-32,-30,-20,-18,-8,-5,-2,0,5,14,20,30,45,46.$$

SOLUTION 8·4: (1): f does not represent zero in $\mathbb{Q}$ since $\frac{5}{7}$ is not square in $\mathbb{Q}$.

(2): By Theorem (II) f represents a in $\mathbb{Q}_p$ if and only if $(a,35)_p = (5,-7)_p$. Now, $(a,35)_\infty = (5,-7)_\infty = 1$ for all $a \in \mathbb{Q}$. By the Product Formula:

$$\prod_{p\in V} (a,b)_p = 1 \text{ if } (a,35)_p = (5,-7)_p \text{ for all } p \in V - \{2\},$$

also we will have $(a,35)_2 = (5,-7)_2$. The condition:

$$(a,35)_p = (5,-7)_p \quad \text{for all } p \in P - \{2\} \tag{1}$$

is therefore necessary and sufficient for f to represent a in all the $\mathbb{Q}_v$ for $v \in V$, and therefore, by Theorem (III), for f to represent a in $\mathbb{Q}$.

(3): Let us spell out condition (1); we have:

$$(5,-7)_p = 1 \quad \text{if } p \neq 5 \text{ and } 7,$$

$$(5,-7)_5 = (\tfrac{3}{5}) = -1 \quad \text{and} \quad (5,-7)_7 = (5,7)_7 = (\tfrac{5}{7}) = -1.$$

Therefore the condition is equivalent to:

$$\begin{cases} (a,35)_p = 1 \quad \text{for all } p \in P - \{2\} - \{5\} - \{7\}, \\ (a,35)_p = -1 \text{ for } p \in \{5,7\}. \end{cases} \tag{1}$$

Now if $p \notin \{2,5,7\}$ and if p does not divide a, then $(a,35)_p = 1$. Therefore (1) is further equivalent to:

$$\begin{cases} (a,35)_p = 1 \quad \text{if } p \text{ divides } a \text{ and } p \notin \{2,5,7\}, \\ (a,35)_p = -1 \quad \text{if } p \in \{5,7\}. \end{cases} \tag{1}$$

But if $p|a$ and $p \notin \{2,5,7\}$, then, since a has no square factors,

$$(a,35)_p = (p,35)_p = \left(\frac{35}{p}\right),$$

and finally we obtain the condition:

$$\begin{cases} \text{(a):} \quad (a,35)_p = 1 \quad \text{for all } p \text{ dividing } a \text{ and } p \notin \{2,5,7\}, \\ \text{(b):} \quad (a,35)_p = -1 \quad \text{for } p \in \{5,7\}. \end{cases} \tag{1}$$

Let us make condition (1b) explicit in the different cases of the problem:

a is divisible by neither 5 nor 7:

$$(a,35)_5 = (a,5)_5 = \left(\frac{a}{5}\right) \quad \text{and} \quad (a,35)_7 = (a,7)_7 = \left(\frac{a}{7}\right),$$

and the conditions are therefore:

$$\begin{cases} \text{(a)}: \left(\dfrac{35}{p}\right) = 1 \quad \text{for all } p \text{ dividing } a \text{ and } p \notin \{2,5,7\}, \\ \text{(b)}: \left(\dfrac{a}{5}\right) = -1 \quad \text{and} \quad \left(\dfrac{a}{7}\right) = -1. \end{cases} \tag{2a}$$

a is divisible by 5 and not divisible by 7, i.e., $a = 5a'$ and $7 \nmid a'$, $5 \nmid a'$: then:

$$(a,35)_5 = (5a',7)_5(5a',5)_5 = \left(\tfrac{2}{5}\right)(5,-1)_5\left(\frac{a'}{5}\right)$$

$$= (-1)(1)\left(\frac{a'}{5}\right) = -\left(\frac{a'}{5}\right),$$

and:

$$(a,35)_7 = (5a',7)_7 = \left(\frac{5}{7}\right)\left(\frac{a'}{7}\right) = (-1)\left(\frac{a'}{7}\right),$$

the conditions are therefore:

$$\begin{cases} \text{(a)}: \left(\dfrac{35}{p}\right) = 1 \quad \text{for all } p \text{ dividing } a \text{ and } p \notin \{2,5,7\}, \\ \text{(b)}: \left(\dfrac{a'}{5}\right) = 1 \quad \text{and} \quad \left(\dfrac{a'}{7}\right) = 1. \end{cases} \tag{2b}$$

a is divisible by 7 and not divisible by 5, i.e. $a = 7a'$ and $7 \nmid a'$, $5 \nmid a'$: then:

$$(a,35)_7 = (7a',7)_7(7,5)_7 = \left(\frac{-1}{7}\right)\left(\frac{a'}{7}\right)\left(\frac{5}{7}\right) = (-1)\left(\frac{a'}{7}\right)(-1)$$

$$= \left(\frac{a'}{7}\right),$$

$$(a,35)_5 = (7a',5)_5 = \left(\frac{2}{5}\right)\left(\frac{a'}{5}\right) = -\left(\frac{a'}{5}\right),$$

the conditions are therefore:

$$\begin{cases} \text{(a)}: & \left(\frac{35}{p}\right) = 1 \quad \text{for all } p \text{ dividing } a \text{ and } p \notin \{2,5,7\}, \\ \text{(b)}: & \left(\frac{a'}{7}\right) = 1 \quad \text{and} \quad \left(\frac{a'}{5}\right) = -1. \end{cases} \tag{2c}$$

a is divisible by 5 *and* 7, *i.e.*, $a = 35a'$ *and* $5 \nmid a'$, $7 \nmid b'$: then:

$$(a,35)_7 = (a',35)_7(7,-1)_7(5,-1)_7 = \left(\frac{a'}{7}\right)\left(-\frac{1}{7}\right) = (-1)\left(\frac{a'}{7}\right)$$

$$= -\left(\frac{a'}{7}\right),$$

$$(a,35)_5 = (a',35)_5(5,-1)_5 = \left(\frac{a'}{5}\right)\left(\frac{4}{5}\right) = \left(\frac{a'}{5}\right),$$

and the conditions are therefore:

$$\begin{cases} \text{(a)}: & \left(\frac{35}{p}\right) = 1 \quad \text{for all } p \text{ dividing } a \text{ and } p \notin \{2,5,7\}, \\ \text{(b)}: & \left(\frac{a'}{7}\right) = 1 \quad \text{and} \quad \left(\frac{a'}{5}\right) = -1. \end{cases} \tag{2d}$$

(4): Let us look for the prime number p different from 2,5,7 and $\leqslant 13$ which satisfy the condition $\left(\frac{35}{p}\right) = 1$. We have:

$$\left(\frac{35}{p}\right) = \left(\frac{p}{5}\right)\left(\frac{p}{7}\right)(-1)^{\frac{1}{2}(p-1).2}(-1)^{\frac{1}{2}(p-1).3}.$$

Therefore if:

$$p = 3 \qquad \left(\frac{35}{p}\right) = \left(\frac{3}{5}\right)\left(\frac{3}{7}\right)(-1) = (-1)(-1)(-1) = -1,$$

$$p = 11 \qquad \left(\frac{35}{11}\right) = \left(\frac{1}{5}\right)\left(\frac{4}{7}\right)(-1) = -1,$$

$$p = 13 \qquad \left(\frac{35}{13}\right) = \left(\frac{3}{5}\right)\left(\frac{-1}{7}\right) = (-1)(-1) = 1.$$

Therefore the only prime number $p \leqslant 13$ different from 2,5,7 satisfying $\left(\frac{35}{p}\right) = 1$ is $p = 13$.

Hence amongst the integers with absolute value less than or equal to 14 not represented by f we find $\{\pm 3, \pm 6, \pm 11\}$, besides these there are numbers multiplied by a square from $\mathbb{Z}$, that is to say $\{\pm 3, \pm 6, \pm 11, \pm 12\}$.

After examining the other numbers we see that for them the conditions of (a) are realised. It remains, therefore to verify the conditions of (b). One then sees that ± 1 is not represented, therefore $\pm 4, \pm 9$ are not represented, $2, -5, 7, 8, \pm 10, -13, \pm 14$ are not represented, whilst $-2, 5, -7, -8, 13$ are represented.

In conclusion, the list of integers lying between -14 and +14 representable by f in $\mathbb{Q}$ is:

$$\{-8, -7, -2, 5, 13\}.$$

CHAPTER 9

Continued Fractions

INTRODUCTION

The problem is to determine the best approximations of a real number α by rational numbers.

DEFINITION: *The fraction* $\frac{p}{q}$ *, with* $(p,q) \in \mathbb{Z}\times\mathbb{N}^*$ *is called a* ***BEST APPROXIMATING FRACTION****, or a* ***CONVERGENT****, of the number* α *if* $q > 1$*, and if, furthermore:*

$$\textit{For all } (p',q') \in \mathbb{Z}\times\mathbb{N}^*, \quad q' < q \Rightarrow |q\alpha - p| < |q'\alpha - p'|.$$

It is easy to see that the definition allows us to classify the convergents of α as a sequence $\frac{p_n}{q_n}$ which tends to α, with $\frac{p_n}{q_n}$ is in lowest terms.

2.1 CALCULATION OF CONVERGENTS

We define the sequence (a_n) of partial quotients of α by the formulae:

$$\alpha = \alpha_0 = a_0 + \frac{1}{\alpha_1},$$

$$\alpha_1 = a_1 + \frac{1}{\alpha_1},$$

$$\cdot \quad \cdot \quad \cdot \quad \cdot \quad \cdot$$

$$\alpha_n = a_n + \frac{1}{\alpha_{n+1}},$$

with $a_n > 1$, $a_0 \in \mathbb{Z}$ and $a_n \in \mathbb{N}$ if $n \geqslant 1$, and the sequence a_n terminates at index n if $[\alpha_n] = \alpha_n$.

The convergents are then calculated by the following recurrence relations:

$$\begin{aligned} p_n &= a_n p_{n-1} + p_{n-2} \\ & \qquad\qquad \text{if } n \geqslant 0 \text{ with } \begin{cases} (p_{-2}, q_{-2}) = (0,1), \\ (p_{-1}, q_{-1}) = (1,0). \end{cases} \\ q_n &= a_n q_{n-1} + q_{n-2} \end{aligned}$$

2.2 FORMULAE

From the recurrence relations we deduce the formulae:

$$p_n q_{n-1} - q_n p_{n-1} = (-1)^{n+1},$$

$$\alpha = \frac{p_n \alpha_{n+1} + p_{n-1}}{q_n \alpha_{n+1} + q_{n-1}}, \qquad \frac{p_n}{q_n} = [a_0, \ldots, a_n],$$

and:

$$q_n \alpha - p_n = \frac{1}{q_n \alpha_{n+1} + q_{n-1}}.$$

THEOREM (I): *The sequence* (a_n) *is finite (resp. periodic starting from a certain index) if and only if* α *is rational (resp. quadratic).*

THEOREM (II): *Let* $\alpha = a + b\sqrt{d}$, *with* $a, b \in \mathbb{Q}$ *and* $d \in \mathbb{Q}^*$, *be a quad-*

ratic number, and let $\sigma(\alpha) = a - b\sqrt{d}$ *be its conjugate. The sequence* (a_n) *is periodic starting from index* 0 *if and only if* $\alpha > 1$ *and* $-1 < \sigma(\alpha) < 0$*, that is to say if* α *is reduced.*

THEOREM (III): (Lagrange's Criterion): *If the pair* $(p,q) \in \mathbb{Z} \times \mathbb{N}^*$ *and* α *satisfy* $\left|\alpha - \frac{p}{q}\right| \leqslant \frac{1}{2q^2}$, *then* $\frac{p}{q}$ *is a convergent of* α.

2.3 BIBLIOGRAPHY

[Har], [Niv].

PROBLEMS

EXERCISE 9·1: Let α be a quadratic irrational whose expansion as a continued fraction has period s. Let $\frac{p_n}{q_n}$ be the convergents of α and P_n the vector with components (p_n, q_n) in $\mathbb{R}^2$.

Show that there exists a linear transformation T and $n_0 \in \mathbb{N}$ such that:

$$\begin{cases} P_{n+s} = T(P_n) & \text{if } n \geqslant n_0, \\ \det T = (-1)^s. \end{cases}$$

EXERCISE 9·2: Let $\alpha = [\overline{a_0, a_1, \ldots, a_{k-1}}]$ be a continued fraction which is periodic starting from index $n = 0$ with period k. Define two fractions in lowest terms $\frac{u}{v}$ and $\frac{p}{q}$ by:

$$\frac{u}{v} = [a_0, a_1, \ldots, a_{k-1}] \quad \text{and} \quad \frac{p}{q} = [a_0, a_1, \ldots, a_{k-2}] \quad \text{if } k \geqslant 2,$$

and:

$$u = a_0, \quad q = 0 \quad \text{if } k = 1.$$

Let $\frac{p_n}{q_n}$ be the convergents of α, and P_n the vector in the plane $\mathbb{R}^2$ with components (p_n, q_n).

(1): Find a recurrence relation between $P_{(n+2)k}$, $P_{(n+1)k}$ P_{nk}.

(2): Show that there exist real numbers $\lambda, \mu, \rho, C, C'$ such that the following inequalities are satisfied for all $m \in \mathbb{N}^*$:

$$|p_{mk} - \lambda\rho^m| < \frac{C}{p_{mk}},$$

$$|q_{mk} - \mu\rho^m| < \frac{C'}{q_{mk}}.$$

EXERCISE 9·3: Let $d > 1$ be a rational number that is not a perfect square.

(1): Show that the development of $\sqrt{d}$ has the form:

$$\sqrt{d} = [a_0, \overline{a_1, \ldots, a_s}]$$

with:

$$a_s = 2a_0, \quad \text{and} \quad a_i = a_{s-i} \text{ for } i = 1, 2, \ldots, s-1.$$

(2):(a): Show that for $n \geqslant 1$ we have:

$$p_{ns-1}^2 - dq_{ns-1}^2 = (-1)^{ns}. \tag{1}$$

(2):(b): Show that for $1 \leqslant m \leqslant ns$ we have:

$$P_{ns-1} = q_{m-1}P_{ns-m} + q_{m-2}P_{ns-m-1}, \tag{2}$$

$$Q_{ns-1} = p_{m-1}P_{ns-m} + p_{m-2}P_{ns-m-1}, \tag{3}$$

where the components of P_n in $\mathbb{R}^2$ are (p_n,q_n) and those of Q_n are (dq_n,p_n).

(2):(c): From this deduce that:

$$\begin{cases} q_{2ns-1} = q_{ns-1}(q_{ns} + q_{ns-2}) = 2p_{ns-1}q_{ns-1}, \\ p_{2ns-1} = p_{ns-1}^2 + dq_{ns-1}^2. \end{cases} \tag{4}$$

(3): Apply Newton's method to the calculation of $\sqrt{d}$; that is to say, set:

$$x_{n+1} = \frac{1}{2}\left(x_n + \frac{d}{x_n}\right) .$$

Calculate x_n when we set $x_0 = \dfrac{p_{s-1}}{q_{s-1}}$ and give an upper bound for the error made by replacing $\sqrt{d}$ with x_n.

EXERCISE 9·4: DEVELOPMENT OF e AS A CONTINUED FRACTION

(1): With a,n,k being positive integers, $a \neq 0$, set:

$$u_{n+k} = \frac{2^n(n+k)!}{k!(2n+2k)!} \cdot \frac{1}{a^{2k}},$$

$$v_n = \sum_{k=0}^{\infty} u_{n,k}, \qquad w_n = \frac{av_n}{v_{n+1}} .$$

Show that w_n satisfies the recurrence relation:

$$w_n = (2n+1)a + \frac{1}{w_{n+1}} .$$

From this deduce the expansion as a continued fraction of

coth $\frac{1}{\alpha}$ as well as that of $\frac{e+1}{e-1}$.

(2): Taking b,c as strictly positive integers, g_n as a sequence of strictly positive integers (with the exception of g_0 which may be negative) we set:

$$\alpha = [g_0, b, c, g_1, b, c, \ldots],$$

$$\alpha' = [g_0(bc+1) + b + c, g_1(bc+1) + b + c, \ldots].$$

Let $\frac{p_n}{q_n}$ (*resp.* $\frac{p'_n}{q'_n}$) be the n-th convergent of α (*resp.* α').

Show that:

$$p_{3n+2} = p'_n - bq'_n,$$

$$q_{3n+2} = (bc+1)q'_n.$$

Calculate α' as a function of α. From the expansion of $\frac{e+1}{e-1}$ as a continued fraction deduce the one for e.

EXERCISE 9·5: THE POINT-OF-INTERROGATION FUNCTION

The function we are going to study was introduced by Minkowski under the name of "point of interrogation", here we shall denote it by φ.

φ is a numeric function defined on the real numbers in $[0,1]$ and satisfying the conditions:

$$\varphi(0) = 0, \qquad \varphi(1) = 1; \tag{1}$$

$$\varphi\left(\frac{p+p'}{q+q'}\right) = \frac{1}{2}\left(\varphi\left(\frac{p}{q}\right) + \varphi\left(\frac{p'}{q'}\right)\right) \tag{2}$$

for any adjacent fractions $\frac{p}{q}$, $\frac{p'}{q'}$ (i.e., $p'q - q'p = \pm 1$);

φ is continuous. (3)

(1): Show that every fraction in lowest terms $\frac{P}{Q}$ in $(0,1)$ can be written uniquely in the form $\frac{p + p'}{q + q'}$, where $\frac{p}{q}$, $\frac{p'}{q'}$ are two adjacent fractions in $[0,1]$. Deduce from this that conditions (1) and (2) completely determine $\varphi\left(\frac{P}{Q}\right)$.

(2): Set $\frac{P}{Q} = [0,a_1,\ldots,a_N]$. Let $\frac{p_n}{q_n}$ be the n-th convergent of $\frac{P}{Q}$ and set $\varphi\left(\frac{p_n}{q_n}\right) = u_n$.

Show that:

$$u_n = \left(1 - \frac{1}{2^{a_n}}\right)u_{n-1} + \frac{1}{2^{a_n}}u_{n-2}.$$

Deduce from this an expression for $\varphi\left(\frac{P}{Q}\right)$ as a function of the quotients a_n and write $\varphi\left(\frac{P}{Q}\right)$ in binary.

(3): Prove the existence and uniqueness of φ defined by the conditions (1), (2), and (3).

Calculate $\varphi(x)$ as a function of the partial quotients a_n of the expansion of x as a continued fraction, and write $\varphi(x)$ in binary. Determine the set of x's such that $\varphi(x)$ is rational.

(4): Prove that for $x \in [0,1]$, φ satisfies the identities:

$$\varphi(x) + \varphi(1 - x) = 1,$$

$$\varphi\left(\frac{ax + b}{cx + d}\right) = A\varphi(x) + B,$$

where $\frac{a}{b}$, $\frac{c}{d}$ are two adjacent fractions in $[0,1]$, and A,B are independent of x.

EXERCISE 9·6: SYMMETRIC DEVELOPMENT

(1): With $a_0, a_1, \ldots, a_n, \ldots$ indeterminates, define two sequences of polynomials with coefficients in $\mathbb{Z}$, $P_n(a_0, a_1, \ldots, a_n)$ and $Q_n(a_0, a_1, \ldots, a\ )$, by the relations:

$$\begin{cases} P_n(a_0, \ldots, a_n) = a_n P_{n-1}(a_0, \ldots, a_{n-1}) + P_{n-2}(a_0, \ldots, a_{n-2}) \\ \qquad\qquad\qquad\qquad\qquad\qquad\qquad (n = 2,3,\ldots), \\ P_0(a_0) = a_0, \qquad P_1(a_0, a_1) = a_0 a_1 + 1; \end{cases}$$

$$\begin{cases} Q_n(a_0, \ldots, a_n) = a_n Q_{n-1}(a_0, \ldots, a_{n-1}) + Q_{n-2}(a_0, \ldots, a_{n-2}) \\ \qquad\qquad\qquad\qquad\qquad\qquad\qquad (n = 2,3,\ldots), \\ Q_0(a_0) = 1, \qquad Q_1(a_0, a_1) = a_1. \end{cases}$$

Denote by M_i the matrix:

$$\begin{bmatrix} a_i & 1 \\ 1 & 0 \end{bmatrix}.$$

Calculate the product $M_0 M_1 \cdots M_n$. Show that:

$$P_n(a_n, a_{n-1}, \ldots, a_0) = P_n(a_0, a_1, \ldots, a_n),$$

$$P_n(a_0, a_1, \ldots, a_n) = P_{n-1}(a_1, \ldots, a_n).$$

Prove the following identity, called the *FUNDAMENTAL IDENTITY*, valid for $m \geqslant 2$, $n \geqslant 1$:

$$P_{m+n}(a_0, \ldots, a_{m+n}) = P_{m-1}(a_0, \ldots, a_{m-1}) P_n(a_m, \ldots, a_{m+n})$$
$$+ P_{m-2}(a_0, \ldots, a_{m-2}) P_{n-1}(a_{m+1}, \ldots, a_{m+n}).$$

(2): Let $\frac{A}{B} > 1$ be a fraction in lowest terms, $[a_0,\dots,a_N]$ *one* of its expansions as a continued fraction. Denote by $\frac{p_n}{q_n}$ the n-th convergent of this expansion.

Show that:

$$[a_N,\dots,a_0] = \frac{p_N}{p_{N-1}} .$$

Show that if the expansion of $\frac{A}{B}$ is symmetric (i.e., $a_n = a_{N-n}$, $n = 0,1,\dots,$) then A divides $B^2 + (-1)^{N-1}$.

Conversely, show that if A divides $B^2 \pm 1$, one of the expansions of $\frac{A}{B}$ as a continued fraction is symmetric. In the case where A divides $B^2 + 1$ show, with the help of the fundamental identity, that A is a sum of two square. Is the result still valid if $\frac{A}{B} < 1$?

EXERCISE 9·7: CONVERGENTS OF THE SECOND KIND

Let α be an irrational number. Set $\alpha = [a_0,a_1,\dots,a_n,\dots]$. We call the *SECONDARY CONVERGENTS of* α *associated with the convergent form* $\frac{p_n}{q_n}$ the fractions:

$$\frac{p'_n}{q'_n} = \frac{p_n + p_{n-1}}{q_n + q_{n-1}} \quad \text{and} \quad \frac{p''_n}{q''_n} = \frac{p_n - p_{n-1}}{q_n - q_{n-1}} \quad \text{if } n \geqslant 2.$$

Define λ',λ'' by the equalities:

$$\left|\alpha - \frac{p'_n}{q'_n}\right| = \frac{1}{\lambda'_n q'^2_n} , \qquad \left|\alpha - \frac{p''_n}{q''_n}\right| = \frac{1}{\lambda''_n q''^2_n} .$$

(1):(a): Show that $\frac{p'_n}{q'_n}$, $\frac{p''_n}{q''_n}$ are irreducible.

(1):(b): Calculate λ'_n, λ''_n as functions of $\alpha_n, \alpha_{n+1}, \beta_n, \beta_{n+1}$, with $\beta_n = [a_n, a_{n-1}, \ldots, a_1] = \frac{q_n}{q_{n-1}}$.

(1):(c): Deduce from (1)(b) that every secondary convergent satisfies $\left|\alpha - \frac{p}{q}\right| < \frac{2}{q^2}$.

(1):(d): By calculating $\lambda'_n + \lambda''_n$ show that at least one of the secondary convergents associated with the same convergent satisfy:

$$\left|\alpha - \frac{p}{q}\right| < \frac{1}{q^2}.$$

(2):(a): Let $\frac{p}{q}$ be a rational number whose expansion as a continued fraction is chosen in such a way that:

$$\frac{p}{q} = [b_0, b_1, \ldots, b_n] \quad \text{with } b_n \geqslant 2.$$

Show that if α lies between the rational numbers:

$$[b_0, b_1, \ldots, b_n + 2] \quad \text{and} \quad [b_0, b_1, \ldots, b_n - 1],$$

then α admits $\frac{p}{q}$ as a convergent or secondary convergent.

(2):(b): From this deduce that if $\frac{p}{q}$ satisfies $\left|\alpha - \frac{p}{q}\right| \leqslant \frac{1}{q^2}$, then $\frac{p}{q}$ is a convergent or a secondary convergent of α.

EXERCISE 9·8: GOLDEN NUMBERS

Let $\alpha_0 = \frac{\sqrt{5} - 1}{2}$, $\frac{P_n}{Q_n}$ its n-th convergent, $m \geqslant 1$ an integer fixed once and for all. Set:

$$C_m = \frac{\sqrt{5} + 1}{2} + \frac{P_{2m-1}}{Q_{2m-1}}.$$

With α a given irrational number, we want to study the number of solutions of the inequality:

$$\left|\alpha - \frac{p}{q}\right| \leqslant \frac{1}{C_m q^2} \qquad (1)$$

(p,q relatively prime integers, $q > 0$); for linguistic convenience a solution of (1) will be called the fraction $\frac{p}{q}$.

(1): Let $[a_0,a_1,\ldots,a_n,\ldots]$ be the expansion of α as a continued fraction, $\frac{p_n}{q_n}$ its n-th convergent, and define λ_n by the equality:

$$\left|\alpha - \frac{p_n}{q_n}\right| = \frac{1}{\lambda_n q_n^2}.$$

Show that:

$$\lambda_n = [a_{n+1},a_{n+2},\ldots] + [0,a_n,\ldots,a_1].$$

(2):(a): Verify that:

$$2 < \sqrt{5} < C_m \leqslant \frac{\sqrt{5}+3}{2} < \frac{8}{3}.$$

(2):(b): Show that every solution of (1) is a convergent of α. From this deduce that (1) is equivalent to:

$$\lambda_n \geqslant C_m. \qquad (2)$$

(3): Take $\alpha = \alpha_0$. Show that (2) has exactly m solutions and that the inequality:

$$\left|\alpha - \frac{p}{q}\right| < \frac{1}{Cq^2}, \quad \text{with } C > C_m,$$

has fewer than m solutions.

(4): Take $\alpha \neq \alpha_0$.

(4):(a): Show that if $a_n \geqslant 3$ for an infinite number of n's then (2) has an infinite number of solutions.

(4):(b): Show that the same holds if we have $a_n \leqslant 2$ from a certain index on but $a_n = 2$ for an infinite number of n's (we will show that if $a_{n+1} = 2$, $a_n \leqslant 2$, $a_{n+2} \leqslant 2$, we have $\lambda_n > \frac{8}{3}$).

(4):(c): Assume that $a_n = 1$ from a certain index on, let r be the largest integer such that $a_r > 1$.

Show that:

$$\frac{p_{r-1}}{q_{r-1}}, \quad \frac{p_{r+1}}{q_{r+1}}, \quad \frac{p_{r+3}}{q_{r+3}}, \quad \ldots, \quad \frac{p_{r+2m-3}}{q_{r+2m-3}}$$

are m solutions of (2).

(5): State a theorem on the minimum number of solutions of (1).

SOLUTIONS

SOLUTION 9·1: Let $T \in M_2(\mathbb{R})$ be the matrix of the mapping T with respect to the canonical basis. Let us assume that the development of α is periodic starting from the index n_0. Let us determine $T = \begin{bmatrix} a & b \\ c & d \end{bmatrix}$ by the relations:

$$p_{n_0+s-2} = ap_{n_0-2} + bq_{n_0-2}, \qquad q_{n_0+s-2} = cp_{n_0-2} + dq_{n_0-2},$$

$$p_{n_0+s-1} = ap_{n_0-1} + bq_{n_0-1}, \qquad q_{n_0+s-1} = cp_{n_0-1} + dq_{n_0-1},$$

These two systems have a solution $(a,b,c,d) \in \mathbb{Z}^4$ since the determinant of these two systems is equal to:

$$p_{n_0-2}q_{n_0-1} - q_{n_0-2}p_{n_0-1} = (-1)^{n_0-1}.$$

Let T be the transformation determined in this way. Let us show by induction that $P_{n+s} = T(P_n)$ if $n \geqslant n_0 - 2$. We have:

$$P_{n_0+s-1} = T(P_{n_0-1}) \quad \text{and} \quad P_{n_0+s-2} = T(P_{n_0-2}),$$

and if we assume that:

$$P_{n+s-2} = T(P_{n-2}) \quad \text{and} \quad P_{n+s-1} = T(P_{n-1}),$$

we then have:

$$T(P_n) = T(a_n P_{n-1} + P_{n-2}) = a_n P_{n+s-1} + P_{n+s-2};$$

and so $P_{n+s} = T(P_n)$, since $a_{n+s} = a_n$ if $n \geqslant n_0$.

On the other hand, from the equality:

$$\begin{vmatrix} p_{n+s+1} & q_{n+s+1} \\ p_{n+s} & q_{n+s} \end{vmatrix} = \begin{vmatrix} a & b \\ c & d \end{vmatrix} \begin{vmatrix} p_{n+1} & q_{n+1} \\ p_n & q_n \end{vmatrix} \quad \text{if } n \geqslant n_0 - 2,$$

we can show that:

$$\det T = (-1)^s.$$

SOLUTION 9·2: (1): Since the development is purely periodic, there exists a linear transformation T such that $P_{n+k} = T(P_n)$ for $n \geqslant -2$; we determine T with the help of the equalities $P_{k-2} = T(P_{-2})$ and $P_{k-1} = T(P_{-1})$. If T is the matrix of T with respect to the canonical basis (cf., Exercise 9·1), we then obtain:

$$T = \begin{bmatrix} p_{k-1} & p_{k-2} \\ q_{k-1} & q_{k-2} \end{bmatrix} \quad \text{and} \quad T^{-1} = (-1)^k \begin{bmatrix} q_{k-2} & -p_{k-2} \\ -q_{k-1} & p_{k-1} \end{bmatrix},$$

hence:

$$T + (-1)^k T \quad = (p_{k-1} + q_{k-2})I = (u + q)I.$$

From the relations:

$$P_{(n+2)k} = T(P_{(n+1)k}), \qquad P_{nk} = T^{-1}(P_{(n+1)k}),$$

and from the preceding inequality we deduce that:

$$P_{(n+2)k} - (u + q)P_{(n+1)k} + (-1)^k P_{nk} = 0. \qquad (1)$$

(2): Suppose we have the equation:

$$x^2 - (u + q)x + (-1)^k = 0$$

If $k \geqslant 2$ we have:

$$u + q = p_{k-1} + q_{k-2} \geqslant p_1 + q_0 \geqslant 3,$$

and, if $k = 1$:

$$u + q = a_0 \geqslant 1.$$

Therefore this equation has a real root ρ such that $\rho > 1$ and a real root ρ' such that $|\rho'| < 1$.

From Equation (1) we deduce that:

$$p_{mk} = \lambda\rho^m + \lambda'\rho'^m, \qquad q_{mk} = \mu\rho^m + \mu'\rho'^m,$$

with:

$$\lambda + \lambda' = a_0, \qquad \mu + \mu' = 1$$
$$\text{and } m \in \mathbb{N}.$$
$$\lambda\rho + \lambda'\rho' = p_k, \qquad \mu\rho + \mu'\rho' = q_k,$$

$\lambda, \lambda', \mu, \mu'$ can be calculated with the aid of the two systems, because the determinant has the value $\rho - \rho' \neq 0$

We then have:

$$0 < p_{mk} < (|\lambda| + |\lambda'|)\rho^m,$$

$$0 < q_{mk} < (|\mu| + |\mu'|)\rho^m.$$

Hence we obtain:

$$|p_{mk} - \lambda\rho^m| = |\lambda'\rho'^m| = \frac{|\lambda'|}{\rho^m} < \frac{|\lambda'|(|\lambda| + |\lambda'|)}{p_{mk}} = \frac{C}{p_{mk}},$$

and similarly:

$$|q_{mk} - \mu\rho^m| = |\mu'\rho'^m| = \frac{|\mu'|}{\rho^m} \quad \frac{|\mu'|(|\mu| + |\mu'|)}{q_{mk}} = \frac{C'}{q_{mk}},$$

SOLUTION 9·3: (1): If we set $\alpha = \sqrt{d}$, with the usual notations, α_1 is a convergent. In fact:

$$\sqrt{d} = \alpha = a_0 + \frac{1}{\alpha_1},$$

therefore:

$$\alpha' = -\sqrt{d} = a_0 + \frac{1}{\alpha'_1},$$

and therefore:

$$\frac{1}{\alpha_1} - \frac{1}{\alpha'_1} = 2\sqrt{d} > 2,$$

and since $\frac{1}{\alpha_1} < 1$, we deduce from this that $-\frac{1}{\alpha'_1} > 1$, and α_1 is certainly the convergent. The expansion of α is (cf., Theorem (II)) therefore of the form:

$$\alpha = [\overline{a_0, a_1, a_2, \ldots, a_s}],$$

and we have $\alpha_{s+1} = \alpha_1$.

We can obtain an equation of the second degree satisfied by α by using the formula given in the introduction:

$$\alpha = \frac{p_s\alpha_{s+1} + p_{s-1}}{q_s\alpha_{s+1} + q_{s-1}} = \frac{\dfrac{p_s}{(\alpha - a_0)} + p_{s-1}}{\dfrac{q_s}{(\alpha - a_0)} + q_{s-1}},$$

which gives:

$$\alpha^2 q_{s-1} - \alpha(a_0 q_{s-1} - q_s + p_{s-1}) + a_0 p_{s-1} - p_s = 0.$$

This polynomial must be proportional to the polynomial $\alpha^2 - d$. Hence:

$$a_0 q_{s-1} - q_s + p_{s-1} = 0,$$

so:

$$\frac{q_s}{q_{s-1}} = a_0 + \frac{p_{s-1}}{q_{s-1}},$$

or, again:

$$[a_s, a_{s-1}, \ldots, a_1] = a_0 + [a_0, a_1, \ldots, a_{s-1}],$$

so:

$$[a_s, a_{s-1}, \ldots, a_1] = [2a_0, a_1, \ldots, a_{s-1}],$$

and by the uniqueness of the expansion as a continued fraction we obtain:

$$a_s = 2a_0, \quad a_{s-1} = a_1, \quad a_2 = a_{s-2}, \quad \ldots, \quad a_1 = a_{s-1}.$$

(2):(a): Since $\alpha_{ns+1} = \alpha_1$ for all $n \in \mathbb{N}$ we can also write:

$$\alpha = \frac{p_{ns} + p_{ns-1}(\alpha - a_0)}{q_{ns} + q_{ns-1}(\alpha - a_0)},$$

so:

$$\alpha^2 q_{ns-1} - \alpha(a_0 q_{ns-1} - q_{ns} + p_{ns-1}) + a_0 p_{ns-1} - p_{ns} = 0.$$

Using that the polynomial is proportional to $\alpha^2 - d$ we obtain:

$$\begin{cases} a_0 q_{ns-1} - q_{ns} + p_{ns-1} = 0, \\ p_{ns} + a_0 p_{ns-1} = d q_{ns-1}. \end{cases} \tag{1}$$

Multiplying the first equation by p_{ns-1} and the second by q_{ns-1}, and then adding them, we obtain:

$$\boxed{p_{ns-1}^2 - d q_{ns-1}^2 = - p_{ns} q_{ns-1} + p_{ns-1} q_{ns} = (-1)^{ns}.}$$

(2):(b): Let us prove the following relation by induction on m.

$$P_{ns-1} = q_{m-1} P_{ns-m} + q_{m-2} P_{ns-m-1}, \tag{2}$$

using the equality $a_m = a_{ns-m}$ if $1 \leqslant m \leqslant ms$.

If $m = 1$, since $q_0 = 1$, $q_{-1} = 0$, and (2) is certainly satisfied. Therefore it suffices to prove that:

$$q_{m-1} P_{ns-m} + q_{m-2} P_{ns-m-1} = q_m P_{ns-m-1} + q_{m-1} P_{ns-m-2},$$

or again:

$$q_{m-1}(a_{ns-m} P_{ns-m-1} + P_{ns-m-2}) + q_{m-2} P_{ns-m-1}$$

$$= (a_m q_{m-1} + q_{m-2}) P_{ns-m-1} + q_{m-1} P_{ns-m-2},$$

which obviously holds.

Similarly, let us prove (3) by induction; if $m = 1$ we must therefore show that:

$$Q_{ns-1} = a_0 P_{ns-1} + P_{ns-2}.$$

Now the formulae (1) are equivalent to:

$$Q_{ns-1} = P_{ns} - a_0 P_{ns-1}.$$

Now:

$$P_{ns} = 2a_0 P_{ns-1} + P_{ns-2},$$

since $a_{ns} = 2a_0$, and (3) is certainly satisfied if $m = 1$.

Next, it suffices to verify the equality:

$$p_{m-1} P_{ns-m} + p_{m-2} P_{ns-m-1} = p_m P_{ns-m-1} + p_{m-1} P_{ns-m-2},$$

which can be shown in the same way as formula (2).

(2):(c): From (2) we deduce (by replacing n by $2n$ and m by ns) that:

$$\begin{cases} q_{2ns-1} = q_{ns-1}(q_{ns} + q_{ns-2}), \\ p_{2ns-1} = q_{ns-1} p_{ns} + q_{ns-2} p_{ns-1}, \end{cases} \tag{5}$$

but:

$$\begin{aligned} q_{ns} + q_{ns-2} &= 2q_{ns} + (q_{ns-2} - q_{ns}) \\ &= 2a_0 q_{ns-1} + 2p_{ns-1} - 2a_0 q_{ns} \end{aligned}$$

by formula (1), and therefore:

$$q_{2ns-1} = 2p_{ns-1} q_{ns-1} \quad \text{since} \quad 2p_{ns-1} = q_{ns} + q_{ns-2},$$

similarly:

$$p_{ns} + p_{ns-2} = 2p_{ns} + (p_{ns-2} - p_{ns})$$

$$= 2a_0p_{ns-1} + 2dq_{ns-1} - 2a_0p_{ns-1}$$

by ormulae (1), so:

$$2dq_{ns-1} = p_{ns} + p_{ns-2}.$$

Therefore we have:

$$p_{2ns-1} = q_{ns-1}(2dq_{ns-1} - p_{ns-2}) + p_{ns-1}q_{ns-2}$$

$$= 2dq^2_{ns-1} + (-1)^{ns},$$

so, with Question (2)(a) above:

$$p_{2ns-1} = p^2_{ns-1} + dq^2_{ns-1}.$$

(3): By induction let us show that:

$$x_n = \frac{p_{2^n s-1}}{q_{2^n s-1}}.$$

If $n = 0$ the formula holds; and if $x_{n-1} = \dfrac{p_{2^n s-1}}{q_{2^n s-1}}$ we deduce from this that:

$$x_n = \frac{1}{2}\left(\frac{p^2_{2^{n-1}s-1} + dq^2_{2^{n-1}s-1}}{q_{2^{n-1}s-1}\,p_{2^{n-1}s-1}}\right),$$

so:

$$x_n = \frac{p_{2^n s-1}}{q_{2^n s-1}}$$

by formulae (4).

Therefore we deduce from this that:

$$\lim x_n = \sqrt{d},$$

and that:

$$q_{2^n s-1}|x_n - \sqrt{d}| = \left|\frac{1}{q_{2^n s-1}\alpha_{2^n s} + q_{2^n s-2}}\right| < \frac{1}{q_{2^n s-1}a_{2^n s}}$$

$$\leqslant \frac{1}{2q_{2^n s-1}} .$$

But by formulae (5):

$$q_{2ns-1} > (q_{ns-1})^2,$$

so:

$$q_{2^n s-1} > (q_{2^n s-1})^{2^{n-1}} .$$

But $q_{2s-1} \geqslant 2$, for if $s \geqslant 2$ then $q_{2s-1} > q_3 > 2$, and if $s = 1$ then $a_1 = 2a_0$ and $q_{2s-1} = q_1 = a_1 \geqslant 2$. Therefore the error in replacing $\sqrt{d}$ by x_n is less that $\dfrac{1}{2^{2^n} + 1}$.

SOLUTION 9·4: (1): The convergence of the series giving v_n is shown using d'Alembert's Criterion:

$$\frac{u_{n,k+1}}{u_{n,k}} = \frac{1}{(k + 1)(2n + 2k + 1)}\frac{1}{2a^2},$$

whence:

$$\lim_{k\to+\infty} \frac{u_{n,k+1}}{u_{n,k}} = 0.$$

For positive n and k we have:

$$u_{n,k} = (2n + 1)u_{n+1,k} + \frac{1}{a^2} u_{n+2,k-1} \quad (\text{with } u_{n+2,-1} = 0).$$

On summing over k we obtain:

$$v_n = (2n + 1)v_{n+1} + \frac{1}{a^2} v_{n+1},$$

and by multiplying by $\frac{a}{v_{n+1}}$,

$$w_n = (2n + 1)a + \frac{1}{w_{n+1}} .$$

We can apply the continued fractions algorithm, for since w_n is positive we have:

$$w_n \geqslant (2n + 1)a \geqslant 1 \quad \text{and} \quad 0 < \frac{1}{w_{n+1}} < 1.$$

From this algorithm we deduce:

$$w_0 = [a,3a,5a,\dots].$$

The expansions of the hyperbolic functions as power series give:

$$v_0 = \cosh \frac{1}{a} , \qquad v_1 = a\sinh \frac{1}{a} ,$$

whence:

$$w_0 = \coth \frac{1}{a} ;$$

by taking $a = 2$ we obtain:

$$\frac{e+1}{e-1} = [2,6,10,14,\dots].$$

(2): For the indices $3n - 2, 3n - 1, 3n, 3n + 1, 3n + 2$ write the recurrence relation satisfied by p :

$$\begin{aligned}
p_{3n-2} &= bp_{3n-3} + p_{3n-4}, & 1,\\
p_{3n-1} &= cp_{3n-2} + p_{3n-3}, & -b,\\
p_{3n} &= g_n p_{3n-1} + p_{3n-2}, & bc+1,\\
p_{3n+1} &= bp_{3n} + p_{3n-1}, & c,\\
p_{3n+2} &= cp_{3n+1} + p_{3n}, & 1.
\end{aligned}$$

Multiplying by the integers indicated above, and adding, we obtain:

$$p_{3n+2} = (g_n(bc + 1) + b + c)p_{3n-1} + p_{3n-4}.$$

The sequence p_{3n+2} satisfies the recurrence relation:

$$u_n = (g_n(bc + 1) + b + c)u_{n-1} + u_{n-2}.$$

We know that the set of solutions of this equation in u_n forms a two-dimensional vector space, a basis of which is p'_n and q'_n. Hence there exist constants λ,μ such that:

$$p_{3n+2} = \lambda p'_n + \mu q'_n.$$

λ and μ are calculated by taking $n = -1$ and $n = 0$; we obtain:

$$p_{3n+2} = p'_n - bq'_n.$$

An analogous calculation gives:

$$q_{3n+2} = (bc + 1)q'_n.$$

By division we obtain the relation:

$$(bc + 1)\,\frac{p_{3n+2}}{q_{3n+2}} = \frac{p'_n}{q'_n} - b,$$

and by passing to the limit we obtain:

$$\alpha' = (bc + 1)\alpha + b;$$

taking $b = c = 1$, $g_n = 2n$, by Question (1) above we have $\alpha' = \frac{e + 1}{e - 1}$, whence:

$$\alpha = \frac{\alpha' - 1}{2} = \frac{1}{e - 1} = [0,1,1,2,1,1,4,1,1,\ldots\ ],$$

and:

$$e = [2,1,2,1,1,4,1,1,\ldots\ ] = [2,\overline{1,2n,1}]_{n=1}^{\infty},$$

because $\frac{1}{e - 2} = \alpha_2$.

SOLUTION 9·5: (1): We must find two fractions $\frac{p}{q}$ and $\frac{p'}{q'}$ such that:

$$p'q - pq' = 1, \quad p + p' = P, \quad q + q' = Q.$$

Eliminating p' and q' we obtain:

$$Pq - Qp = 1, \quad \text{with} \quad 1 \leqslant q \leqslant Q - 1.$$

Using Bezout's Theorem we can verify the existence and uniqueness of p and q. Next we have the existence and uniquess of p' and q'.

An induction on Q then shows the existence and uniqueness of $\varphi\left(\frac{P}{Q}\right)$:

$$\varphi\left(\frac{1}{2}\right) = \frac{1}{2}, \quad \varphi\left(\frac{1}{3}\right) = \frac{1}{4}, \quad \varphi\left(\frac{2}{3}\right) = \frac{3}{4}, \quad \dots .$$

(2): For fixed n, set:

$$\alpha_h = \varphi\left(\frac{hp_{n-1} + p_{n-2}}{hq_{n-1} + q_{n-2}}\right), \quad \text{for } h \in \mathbb{N}.$$

The fractions:

$$\frac{p_{n-1}}{q_{n-1}} \quad \text{and} \quad \frac{(h-1)p_{n-1} + p_{n-2}}{(h-1)q_{n-1} + q_{n-2}}$$

are adjacent, as is shown by the relation:

$$p_{n-2}q_{n-1} - p_{n-1}q_{n-2} = (-1)^{n-1}.$$

Hence we have:

$$\alpha_h = \tfrac{1}{2}(\alpha_{h-1} + \alpha_\infty) \quad \left(\text{with } \alpha_\infty = \varphi\left(\frac{p_{n-1}}{q_{n-1}}\right)\right).$$

From this we deduce:

$$\alpha_h = \left(1 - \frac{1}{2^h}\right)\alpha_\infty + \frac{1}{2^h}\alpha_0,$$

which certainly gives:

$$u_n = \left(1 - \frac{1}{2^{a_n}}\right)u_{n-1} + \frac{1}{2^{a_n}}(u_{n-2} - u_{n-2})$$

Writing this relation in the form:

$$u_n - u_{n-1} = -\frac{1}{2^{a_n}}(u_{n-1} - u_{n-2}),$$

we have:

$$u_n - u_{n-1} = \frac{(-1)^{n-1}}{2^{a_1+\cdots+a_n-1}},$$

hence:

$$\varphi\left(\frac{P}{Q}\right) = \frac{1}{2^{a_1-1}} - \frac{1}{2^{a_1+a_2-1}} + \cdots + \frac{(-1)^{N-1}}{2^{a_1+a_2+\cdots+a_N-1}}.$$

REMARK: $\frac{P}{Q}$ has two expansions as a continued fraction, of which only one ends in a 1; we could verify that this would require that $\varphi\left(\frac{P}{Q}\right)$ not depend upon the expansion chosen.

Taking N even and grouping the terms in pairs, we have in binary:

$$\varphi\left(\frac{P}{Q}\right) = 0.\underbrace{00\cdots0}_{a_1\text{ digits}}\underbrace{11\cdots1}_{a_2\text{ digits}}\underbrace{00\cdots0}_{a_3\text{ digits}}\cdots\underbrace{11\cdots1}_{a_N\text{ digits}}.$$

(3): It follows from the preceding that the restriction of φ to the rational numbers in [0,1] is continuous and increasing, this restriction can therefore be extended in an unique way to a map on the real numbers in [0,1].

If $x = [0,a_1,a_2,\ldots,a_n,\ldots]$ we have:

$$\varphi(x) = \sum_{n=1}^{\infty} \frac{(-1)^{n-1}}{2^{a_1+a_2+\cdots+a_n-1}}$$

and the binary development of $\varphi(x)$ is written as before.

$\varphi(x)$ is rational if and only if its binary development is finite or periodic, therefore if and only if the development of x as a continued fraction is finite or periodic. The set sought is formed by the union of the rationals and quadratic irrationals in $[0,1]$.

(4): In order to prove the identity $\varphi(x) + \varphi(1 - x) = 1$ we can assume that $0 < x < \frac{1}{2}$, and we have:

$$x = [0,a_1,a_2,\ldots] \quad \text{with } a_1 > 1,$$

and we can verify that:

$$1 - x = [0,1,a_1 - 1,a_2,\ldots].$$

The identity clearly follows from this.

Let us set $y = \frac{ax + b}{cx + d}$; first of all we show that $x \in [0,1]$ implies $y \in [0,1]$. In order to prove that $\varphi(y) = A\varphi(x) + B$ we can use the following result (cf., [Har] Theorem 172]: if $\frac{a}{c}$ and $\frac{b}{d}$ are two adjacent fractions, then they are two consecutive convergents in the expansion of y, and x is a complete quotient of them. More precisely, if $y = [0,b_0,b_1,\ldots]$ then there exists an index k such that:

$$\frac{a}{c} = [0,b_1,b_2,\ldots,b_k], \qquad \frac{b}{d} = [0,b_1,b_2,\ldots,b_{k-1}],$$

and:

$$x = [0,b_{k+1},b_{k+2},\ldots].$$

We then have:

$$\varphi(y) = \sum_{n=1}^{\infty} \frac{(-1)^{n-1}}{2^{b_1+b_2+\cdots+b_n-1}},$$

$$\varphi(x) = \sum_{n=1}^{\infty} \frac{(-1)^{n-1}}{2^{b_{k+1}+b_{k+2}+\cdots+b_{k+n}-1}},$$

where the summations may be finite.

By writing:

$$\varphi(y) = \sum_{n=1}^{k} \frac{(-1)^{n-1}}{2^{b_1+\cdots+b_n-1}} + \frac{(-1)^{k}}{2^{b_1+\cdots+b_k}} \sum_{n=1}^{\infty} \frac{(-1)^{n-1}}{2^{b_{k+1}+b_{k+2}+\cdots+b_{k+n}-1}},$$

we find:

$$A = \frac{(-1)^{k}}{2^{b_1+\cdots+b_k}} = \frac{1}{2}\left(\varphi\left(\frac{a}{c}\right) - \varphi\left(\frac{b}{d}\right)\right),$$

$$B = \sum_{n=1}^{k} \frac{(-1)^{n-1}}{2^{b_1+\cdots+b_n-1}} = \varphi\left(\frac{a}{c}\right).$$

SOLUTION 9·6: (1): We show by induction that:

$$M_0M_1\cdots M_n = \begin{bmatrix} P_n(a_0,\ldots,a_n) & P_{n-1}(a_0,\ldots,a_{n-1}) \\ Q_n(a_0,\ldots,a_n) & Q_{n-1}(a_0,\ldots,a_{n-1}) \end{bmatrix},$$

whence:

$$M_n M_{n-1} \cdots M_0 = \begin{bmatrix} P_n(a_n,\dots,a_0) & P_{n-1}(a_n,\dots,a_0) \\ Q_n(a_n,\dots,a_0) & Q_{n-1}(a_n,\dots,a_0) \end{bmatrix},$$

and by transposition:

$$M_0 M_1 \cdots M_n = \begin{bmatrix} P_n(a_n,\dots,a_0) & Q_n(a_n,\dots,a_0) \\ P_{n-1}(a_n,\dots,a_1) & Q_{n-1}(a_n,\dots,a_1) \end{bmatrix}.$$

Comparing the two expressions for the product $M_0 M_1 \cdots M_n$ we obtain the two identities required.

The fundamental identity expresses the associativity of the matrix product:

$$M_0 \cdots M_{m+n} = [M_0 \cdots M_{m-1}][M_m \cdots M_{m+n}].$$

Taking the entry in the first row and first column we obtain:

$$P_{m+n}(a_0,\dots,a_{m+n}) = P_{m-1}(a_0,\dots,a_{m-1})P_n(a_m,\dots,a_{m+n})$$
$$+ P_{m-2}(a_0,\dots,a_{m-2})Q_n(a_m,\dots,a_{m+n}).$$

Since:

$$Q_n(a_m,\dots,a_{m+n}) = P_{n-1}(a_{m+1},\dots,a_{m+n}),$$

we obtain the fundamental identity.

(2): $\frac{A}{B} > 1$ implies $a_0 \geqslant 1$, therefore the continued fraction $[a_N,\dots,a_0]$ exists. By the results of Question (1) above we have:

$$[a_N,\dots,a_0] = \frac{P_N(a_N,\dots,a_0)}{Q_N(a_N,\dots,a_0)} = \frac{P_N(a_N,\dots,a_0)}{P_{N-1}(a_0,\dots,a_{N-1})} = \frac{P_N}{P_{N-1}};$$

if the development of $\frac{A}{B}$ is symmetric we have:

$$[a_0,\dots,a_N] = [a_N,\dots,a_0],$$

hence:

$$p_{N-1} = q_N = B.$$

The identity:

$$p_N q_{N-1} - p_{N-1} q_N = (-1)^{N-1}$$

can be written:

$$A q_{N-1} - B^2 = (-1)^{N-1},$$

so A certainly divides $B^2 + (-1)^{N-1}$.

Conversely, if A divides $B^2 \pm 1$ we choose the parity of N (choosing $a_N = 1$ if need be) in such a way that A divides $B^2 + (-1)^{N-1}$ and we set:

$$B^2 + (-1)^{N-1} = AB' \quad (B' \text{ an integer}).$$

On the other hand we know that:

$$A q_{N-1} - p_{N-1} B = (-1)^{N-1}.$$

Hence by addition:

$$B(B - p_{N-1}) = A(B' - q_{N-1}),$$

and since $(A,B) = 1$, A divides $B - p_{N-1}$.

However, $A > B$ and $A > p_{N-1}$; whence $A > |B - p_{N-1}|$. From this we deduce $B - p_{N-1} = 0$, therefore:

$$\frac{A}{B} = \frac{p_N}{p_{N-1}} = [a_N,\dots,a_0],$$

and the uniqueness of the expansion as a continued fraction shows that this expansion is symmetric.

EXAMPLE: $\frac{10}{7} = [1,2,3]$, the expansion is not symmetric; however, $10|(7^2 + 1)$. We can write $\frac{10}{7} = [1,2,2,1]$, and the development is symmetric.

We now assume that A divides $B^2 + 1$ and we choose N odd, say $N = 2M + 1$. Apply the fundamental identity with $m = M + 1$, $n = M$:

$$A = P_M(a_0,\dots,a_M)P_M(a_{M+1},\dots,a_{2M+1})$$

$$+ P_{M-1}(a_0,\dots,a_{M-1})P_{M-1}(a_{M+2},\dots,a_{2M+1}),$$

and since $a_n = a_{2M+1-n}$ for $n = 0,\dots,2M + 1$ we have:

$$A = p_M^2 + p_{M-1}^2;$$

thus S is a sum of two squares.

The result is still valid if $\frac{A}{B} < 1$. Dividing B by A, there exist two integers Q,R such that $B = AQ + R$ with $\frac{A}{R} > 1$. Since A divides $R^2 + 1$, A is a sum of two squares.

SOLUTION 9·7: (1):(a): $\frac{p'_n}{q'_n}$ and $\frac{p''_n}{q''_n}$ are well defined because $q_n > q_{n-1}$, and they are irreducible because:

$$|p'_n q_n - p_n q'_n| = |p_{n-1}q_n - q_{n-1}p_n| = 1,$$

and:

$$|p''_n q_n - p_n q''_n| = |p_{n-1}q_n - q_{n-1}p_n| = 1.$$

(1):(b): We have:

$$\left|\alpha - \frac{p'_n}{q'_n}\right| = \left|\frac{p_n\alpha_{n+1} + p_{n-1}}{q_n\alpha_{n+1} + q_{n-1}} - \frac{p_n + p_{n-1}}{q_n + q_{n-1}}\right| = \frac{\alpha_{n+1} - 1}{q'_n(\alpha_{n+1}q_n + q_{n-1})},$$

so:

$$\lambda'_n = \frac{\alpha_{n+1}q_n + q_{n-1}}{(\alpha_{n+1} - 1)(q_n + q_{n-1})} = \frac{q_n}{q_n + q_{n-1}} + \frac{1}{\alpha_{n+1} - 1},$$

and so:

$$\lambda' = 1 + \frac{1}{\alpha_{n+1} - 1} - \frac{1}{\beta_n + 1}$$

since $\frac{q_n}{q_{n-1}} = \beta_n$; similarly,

$$\left|\alpha - \frac{p''_n}{q''_n}\right| = \left|\frac{p_n\alpha_{n+1} + p_{n-1}}{q_n\alpha_{n+1} + q_{n-1}} - \frac{p_n - p_{n-1}}{q_n - q_{n-1}}\right| = \frac{\alpha_{n+1} + 1}{q''_n(q_n\alpha_{n+1} + q_{n-1})},$$

so:

$$\lambda''_n = \frac{q_n\alpha_{n+1} + q_{n-1}}{(\alpha_{n+1} + 1)(q_n - q_{n-1})} = -\frac{1}{\alpha_{n+1} + 1} + \frac{q_n}{q_n - q_{n-1}},$$

and so:

$$\lambda'' = 1 + \frac{1}{\beta_n - 1} - \frac{1}{\alpha_{n+1} + 1}.$$

(1):(c): We have:

$$\lambda'_n > 1 - \frac{1}{\beta_n + 1} > 1 - \tfrac{1}{2} = \tfrac{1}{2},$$

since $\alpha_{n+1} > 1$ and $\beta_n > 1$, and similarly:

$$\lambda''_n > 1 - \frac{1}{\alpha_{n+1} + 1} > \tfrac{1}{2},$$

and therefore every secondary convergent $\frac{p}{q}$ satisfies:

$$\left|\alpha - \frac{p}{q}\right| < \frac{2}{q^2}.$$

(1):(d): We have:

$$\lambda'_n + \lambda''_n = 2 + \frac{2}{\beta_n^2 - 1} + \frac{2}{\alpha_{n+1}^2 - 1} > 2;$$

therefore λ'_n or λ''_n is strictly greater than one; and therefore we have:

$$\text{Either: } \left|\alpha - \frac{p'_n}{q'_n}\right| < \frac{1}{q'^2_n}, \quad \text{or} \quad \left|\alpha - \frac{p''_n}{q''_n}\right| < \frac{1}{q''^2_n}.$$

(2):(a): If α lies between the rational numbers $[b_0,\ldots,b_{n-1},b_n + 2]$ and $[b_0,\ldots,b_{n-1},b_n - 1]$, then α can be written as:

$$\alpha = [b_0,b_1,\ldots,b_{n-1},\alpha_n],$$

with:

$$b_n - 1 \leqslant \alpha_n = a_n + \frac{1}{\alpha_{n+1}} \leqslant bn + 2,$$

but $a_n \in \mathbb{N}^*$ and $\alpha_{n+1} > 1$, therefore:

$$b_n = a_n + \varepsilon \quad \text{with } \varepsilon \in \{-1,0,1\}.$$

But if $\frac{p_i}{q_i}$ denotes the convergents of α which are also the convergents of $\frac{p}{q}$ for $i \leqslant n - 1$, we have:

$$\frac{p}{q} = \frac{b_n p_{n-1} + p_{n-2}}{b_n q_{n-1} + q_{n-2}} = \frac{a_n p_{n-1} + p_{n-2} + \varepsilon p_{n-1}}{a_n q_{n-1} + q_{n-2} + \varepsilon q_{n-1}},$$

so:

$$\frac{p}{q} = \frac{p_n + \varepsilon p_{n-1}}{q_n + \varepsilon q_{n-1}}$$

with $\varepsilon \in \{-1,0,1\}$, that is to say, that $\frac{p}{q}$ is either a convergent or a secondary convergent of α.

(2):(b): By (2)(a) above it will be enough to prove that if $\frac{p}{q} = [b_0,b_1,\ldots,b_n]$, with $b_n \geqslant 2$, then α lies between the numbers B_1 and B_2, where:

$$B_1 = [b_0,b_1,\ldots,b_n + 2] \quad \text{and} \quad B_2 = [b_0,b_1,\ldots,b_n - 1].$$

Because $\frac{p}{q}$ lies between B_1 and B_2 it suffices to show that:

$$\left|\alpha - \frac{p}{q}\right| \leqslant \inf_{i=1,2} \left|B_i - \frac{p}{q}\right| .$$

Now,

$$B_1 = \frac{(b_n + 2)p_{n-1} + p_{n-2}}{(b_n + 2)q_{n-1} + q_{n-2}} \quad \text{and} \quad B_2 = \frac{(b_n - 1)p_{n-1} + p_{n-2}}{(b_n - 1)q_{n-1} + q_{n-2}},$$

where $\frac{p_i}{q_i}$ denotes the convergents of $\frac{p}{q}$, or again,

$$B_1 = \frac{p + 2p_{n-1}}{q + 2q_{n-1}} \quad \text{and} \quad B_2 = \frac{p - p_{n-1}}{q - q_{n-1}} .$$

We obtain:

$$\left|B_1 - \frac{p}{q}\right| = \frac{2}{(q + 2q_{n-1})q} \geqslant \frac{1}{q^2},$$

because $2q_{n-1} \leqslant q_n = q$, since $q = b_n q_{n-1} + q_{n-2}$ and $b_n \geqslant 2$, and similarly:

$$\left|B_2 - \frac{p}{q}\right| = \frac{1}{q(q_n - q_{n-1})} \geqslant \frac{1}{q^2}.$$

Therefore, if $\left|\alpha - \frac{p}{q}\right| \leqslant \frac{1}{q^2}$ we have:

$$\left|\alpha - \frac{p}{q}\right| \leqslant \inf_{i=1,2} \left|B_i - \frac{p}{q}\right|,$$

and therefore $\frac{p}{q}$ is consequently either a convergent of a secondary convergent of α.

SOLUTION 9·8: (1): From the equality:

$$\left|\frac{p_n}{q_n} - \alpha\right| = \frac{1}{q_n(q_n\alpha_{n+1} + q_{n-1})}$$

we deduce that:

$$\lambda_n = \alpha_{n+1} + \frac{q_{n-1}}{q_n},$$

so:

$$\lambda_n = [a_{n+1}, a_{n+2}, \dots] + [0, a_n, a_{n-1}, \dots, a_1].$$

(2):(a): The convergents of odd order of a number form a decreasing seqeunce tending towards the number. Therefore:

$$\sqrt{5} = \frac{\sqrt{5}+1}{2} + \frac{\sqrt{5}-1}{2} < C_m < \frac{\sqrt{5}+1}{2} + \frac{P_1}{Q_1} = \frac{\sqrt{5}+3}{2},$$

because the development of $\frac{\sqrt{5}-1}{2}$ is $[0,1,1,\ldots]$.

Now $3\sqrt{5} < 7$, therefore:

$$\frac{\sqrt{5}+3}{2} < \frac{8}{3}.$$

(2):(b): By Lagrange's Criterion, because $C_m > 2$, every solution of (1) is a convergent of α, and $\frac{p_n}{q_n}$ will be a solution of (1) if and only if $\lambda_n \geqslant C_m$.

(3): If $\alpha = \alpha_0$ let us evaluate λ_n, where:

$$\lambda_n = [1,1,\ldots] + [0,\underbrace{1,1,\ldots,1}_{n\text{ times}}],$$

so:

$$\lambda_n = \frac{1}{\alpha} + \frac{P_n}{Q_n}.$$

Inequality (2) can therefore be written:

$$\frac{P_n}{Q_n} \geqslant \frac{P_{2m-1}}{Q_{2m-1}},$$

which is equivalent to $n \in \{1,2,\ldots,2k-1,\ldots,2m-1\}$. Therefore in this case Inequality (2) has exactly m solutions, and there fore the inequality:

$$\left|\alpha - \frac{p}{q}\right| < \frac{1}{Cq^2} \quad \text{with} \quad C > C_m$$

has fewer than m solutions.

(4):(a): From Question (1) above we deduce that if $a_{n+1} \geqslant 3$, then $\lambda_n > 3$, and that if $a_n \geqslant 3$ for an infinite number of n's, then $\lambda_n \geqslant 3 \geqslant C_m$ for an infinite number of n's, and therefore that the inequality:

$$\left|\alpha - \frac{p}{q}\right| \leqslant \frac{1}{Cq^2}$$

has an infinite number of solutions if $C \leqslant 3$.

(4):(b): From Question (1) above, we deduce that if $n \geqslant 1$:

$$\lambda_n = a_{n+1} + \frac{1}{\alpha_{n+2}} + \frac{1}{[a_n, a_{n-1}, \ldots\ ]}$$

$$> a_{n+1} + \frac{1}{a_{n+2} + 1} + \frac{1}{a_n + 1},$$

and if $a_{n+2} \leqslant 2$, $a_n \leqslant 2$, $a_{n+1} = 2$, then:

$$\lambda_n > 2 + \frac{2}{3} = \frac{8}{3}.$$

Since these conditions are satisfied for an infinite number of n's, Inequality (1) has an infinite number of solutions for arbitrary m, and the inequality:

$$\left|\alpha - \frac{p}{q}\right| \leqslant \frac{1}{Cq^2}$$

has an infinite number of solutions for arbitrary $C \leqslant \frac{8}{3}$.

(4):(c): We then have:

$$\lambda_{r-1} > [a_r, 1, 1, \ldots\ ],$$

so:

$$\lambda_{r-1} > a_r + \frac{1}{[1,1,\ldots]} = a_r + \frac{\sqrt{5}-1}{2}$$

$$\geqslant \frac{\sqrt{5}+3}{2} \geqslant C_m.$$

Therefore $\frac{p_{r-1}}{q_{r-1}}$ is certainly a solution of (1).

— If $n \geqslant r$, in order that $\frac{p_n}{q_n}$ be a solution of (1) it is necessary and sufficient to show that:

$$\lambda_n = [1,1,\ldots] + [0,\underbrace{1,\ldots,1}_{(n-r)\text{ times}},a_r,\ldots,a_1]$$

$$\geqslant \frac{\sqrt{5}+1}{2} + [0,\underbrace{1,1,\ldots,1}_{(2m-1)\text{ times}}],$$

or again,

$$[\underbrace{1,\ldots,1}_{(n-r)\text{ times}},a_r,\ldots,a_1] \leqslant [\underbrace{1,1,\ldots,1}_{(2m-1)\text{ times}}].$$

— If $n - r$ is odd and $2m - 3 \geqslant n - r$, this last inequality is equivalent to:

$$[a_r,\ldots,a_1] \geqslant [\underbrace{1,1,\ldots,1}_{(2m-n+r-1)\text{ times}}],$$

which is realised because $a_r \geqslant 2$ and:

$$[a_r,\ldots,a_1] \geqslant a_r \geqslant 2 > [1,\ldots,1].$$

Therefore for $j \in \{0,1,\ldots,m-1\}$ the $\frac{p_{r-1+2j}}{q_{r-1+2j}}$ are solutions of Inequality (1).

(5): The results of Questions (3) and (4) therefore allow us to deduce that for an arbitrary irrational number α Inequality (1) has at least m solutions.

CHAPTER 10
p-Adic Analysis

INTRODUCTION

Denote a prime number by p; $\mathbb{N}$, $\mathbb{Z}$, $\mathbb{Q}$, $\mathbb{R}$, $\mathbb{R}$ have their usual meanings.

We know that for every prime number p there exists on $\mathbb{Q}$ an non-Archimedean (ultrametric) absolute value (called the *p-ADIC ABSOLUTE VALUE*) *denoted* $|\cdot|_p$ (or $|\cdot|$ if no confusion can arise) defined in the following way: let $x \in \mathbb{Q}$, $x = \frac{m}{n} p^h$ be an expression for x in lowest terms, with $h \in \mathbb{Z}$, $m,n \in \mathbb{Z}$, and set $|x|_p = p^{-h}$. We also define the *p-ADIC VALUATION of* x, *denoted* $v_p(x)$ (or $v(x)$ if no confusion can arise), by $v_p(x) = h$. $\mathbb{Q}_p$ (*resp.* $\mathbb{Z}_p$) is the completion of $\mathbb{Q}$ (*resp.* $\mathbb{Z}$) in this metric.

PROPOSITION 1: *Let* $x \in \mathbb{Q}$, *and let us denote the usual absolute value on* $\mathbb{Q}$ *by* $|x|_\infty$. *Then we have the product formula:*

$$|x|_\infty \prod_{p\,\text{prime}} |x|_p = 1.$$

PROPOSITION 2: *Let* $x \in \mathbb{Q}_p$, *there exists an unique sequence* $(b_n)_{n \geq n_x}$ *of integers* ($0 \leq b_n < p$, $n_x \in \mathbb{Z}$) *such that the series* $\sum_{n \geq n_x} b_n p^n$

converges p-adically to x. This sequence is called the **HENSEL EXPANSION** *of x.*

We note that if K is a complete non-Archimedean field, whose valuation is denoted $v(\cdot)$, associated with it is an *absolute value, denoted*:

$$|x| = \rho^{-v(x)} \quad \text{where } \rho \in \mathbb{R}_+ \text{ and } 0 < \rho < 1,$$

its *valuation ring* A, that is to say the set of elements with absolute value less than or equal to one, and its *maximal ideal* M in the valuation ring A, that is to say the set of elements with absolute value less than one. The *residue field of* K will be denoted $\bar{K} = A/M$, and if $y \in A$ the image of y in A/M under the canonical surjection will be denoted $\bar{y}$.

PROPOSITION 3: *Let L be an extension, of finite degree n, of a complete non-Archimedean field K. The relation:*

$$w(x) = \frac{1}{n} v(N_{L/K}(x))$$

defines the unique *valuation w of L extending the valuation v of K. ($N_{L/K}(x)$ denotes the norm of $x \in L$* over *K.*

$\mathbb{C}_p$ is defined to be the completion of the algebraic closure of $\mathbb{Q}_p$, the absolute value on $\mathbb{C}_p$ is normalised by $|p| = p^{-1}$. If $P(X) \in K[X]$ the degree of P is denoted deg$P$, and the image of P in $\bar{K}[X]$ by $\bar{P}$.

LEMMA: (Hensel): *Let K be a complete non-Archimedean field, and $P \in A[X]$ a non-zero polynomial of degree d. Let us assume that there exist two monic polynomials g, h in $A[X]$ such that $\bar{P} = \bar{g}\bar{h}$, $\bar{g}$ and $\bar{h}$ are relatively prime in $\bar{K}[X]$,* deg$g$ + deg$h \leqslant d$. *Then there exist G, H in $A[X]$ such that:*

$$\bar{G} = \bar{g}, \quad \bar{H} = \bar{h}, \quad \deg G = \deg g, \quad \text{and} \quad P = GH.$$

EISENSTEIN'S CRITERION: Let K be a non-Archimedean field with valuation ring A and maximal ideal M, and let P be a monic polynomial in $A[X]$:

$$P(X) = X^n + a_1 X^{n-1} + \cdots + a_n.$$

If $a_i \in M$ for $i = 1,\ldots,n$ and $a_n \notin M^2$, the polynomial P is irreducible in $K[X]$. If K is complete, P determines a totally ramified extension of K.

Let L be a finite extension, of degree n, of the complete non-Archimedean field K. Let w be the unique extension of the valuation v of K. By the proposition on extending valuations, *if the set of invertible elements of K is denoted K^*, then:*

$$w(L^*) \subseteq \frac{1}{n} v(K^*).$$

If $v(K^*)$ is a discrete subgroup of $\mathbb{R}$, then:

$$v(K^*) \subseteq w(L^*) \subseteq \frac{1}{n} v(K^*)$$

implies that $v(K^*)$ is a subgroup of $w(L^*)$ of finite index e which divides n and which is called the *RAMIFICATION INDEX of L over K*. Let f be the dimension of $\bar{L}$, since $\bar{K}$ is a vector space f is then the *RESIDUAL DEGREE of L over K*. We have the following result:

$$\boxed{ef = n.}$$

Let K be a complete non-Archimedean field. Let $a \in K$ and let $r \in \mathbb{R}_+$, then:

$$B(a,r^-) = \{x \in K: |x - a| < r\}, \quad B(a,r^+) = \{x \in K: |x - a| \leqslant r\}.$$

THEOREM: *Let* $F(X) = \sum_{n\geq 0} a_n(x - a)^n$ *be a Taylor series of* $K[[X]]$ *convergent in* $B(a,r^+)$, *then:*

$$\sup_{x\in B(a,r^+)} |F(x)| = \sup_{n\geq 0}|a_n|r^n.$$

If K is also algebraically closed, then:

$$\sup_{x\in B(a,r^+)} |F(X)| = \sup_{n\geq 0}|a_n|r^n.$$

Notice that if $B \subseteq K$ and if F is a function from B into K, then:

$$\|F\|_B = \sup_{X\in B}|F(X)|.$$

If B and C are two topological spaces we *denote* by $C(B,C)$ the *space of continuous functions from B to C with the norm given by uniform convergence on B.* If $m,n \in \mathbb{Z}$, (m,n) *denotes the greatest common divisor of m and n.* U_p *denotes* the *group of units of* $\mathbb{Q}_p$, $U_p = \{x \in \mathbb{Q}_p : |x| = 1\}$.

PROPOSITION 4: *In order that a complete valued field K be locally compact it is necessary and sufficient that the subgroup* $v(K)$ *of* $\mathbb{R}$ *be discrete and that the residue field,* $\bar{K}$ *be finite.*

PROPOSITION 5: $\mathbb{Z}_p$ *is a compact set of* $\mathbb{Q}_p$.

BIBLIOGRAPHY

[Ami], [Bac], [Bor], [Sam], [Ser1].

PROBLEMS

EXERCISE 10·1: VALUATIONS ON A FINITE FIELD

Let k be a finite field and let L be a finite (i.e., algebraic) extension of k.

Show that every absolute value on L is trivial.

EXERCISE 10·2: VALUATIONS ON THE RATIONAL FUNCTIONS

Let K be a field, $L = K(X)$ the field of rational functions in an indeterminate over K, $P(X)$ an irreducible polynomial of $K[X]$, and v_P the valuation on L associated with P.

Determine the valuation ring, the valuation ideal and the residue field of v_P. Describe the completion of L for the valuation v_P when $P = X$. From this deduce a description of the completion in general.

EXERCISE 10·3: The assumptions are the same as in Exercise 10·2.

When is the compltion of $K(X)$ locally compact for a $P(X)$-adic valuation?

EXERCISE 10·4: VALUATIONS OVER $K(X,Y)$

Let K be a field, and a,b two positve real numbers. Let $K(X,Y)$ be the *field of rational functions in two variables over* K. Let $P(X,Y) = \sum_{m,n} a_{m,n} X^m Y^n$ be a polynomial.

If $P \neq 0$, set $v(P) = \inf(am + bn)$ where m and n run through the set of pairs $(m,n) \in \mathbb{N}^2$ such that $a_{m,n} \neq 0$; if $P = 0$, set $v(0) = +\infty$.

If $R(X,Y) = P(X,Y)/Q(X,Y)$ where P and Q are polynomials; set $v(R) = v(P) - v(Q)$.

(a): Show that v is a valuation on $K(X,Y)$.

(b): Determine the valuation ring, the valuation ideal, as well as the residue field of $K(X,Y)$.

(c): Determine the completion $\hat{K}(X,Y)$ of $K(X,Y)$ for this valuation. Under what conditions on parts (a) and (b) and K is $\hat{K}(X,Y)$ locally compact?

EXERCISE 10·5: STUDY OF $\mathbb{Q}_p^*/(\mathbb{Q}_p^*)^p$

Let G be a multiplicative group; the subgroup of G formed by x^p, where $x \in G$, is denoted G^p. We propose to study $\mathbb{Q}_p^*/(\mathbb{Q}_p^*)^p$.

(a): If G is the direct product of the subgroups H_1 and H_2, G^p is the direct product $H_1^p \times H_2^p$.

(b): Let $a = 1 + p^2 b$, $b \in \mathbb{Z}_p$, show that the equation:

$$(1 + pX)^p = a$$

has a root in $\mathbb{Q}_p$.

(c): From this, deduce, for $p \neq 2$, that $\mathbb{Q}_p^*/(\mathbb{Q}_p^*)^p \simeq (\mathbb{Z}/p\mathbb{Z})^2$.

EXERCISE 10·6: NORMAL BASIS FOR A FINITE EXTENSION OF A COMPLETE NON-ARCHIMEDEAN FIELD

Let K be a complete non-Archimedean field with discrete valuation, and let L be a finite extension of K with the absolute value extending that of K.

Under what condition does L have a normal basis over K?

Let B be the valuation ring of L; what are the normal bases of L relative to the A-module B?

EXERCISE 10·7: STUDY OF THE MAPPING $n \to a^n$

Let $a \in A$ (a Banach algebra over $\mathbb{Q}_p$). Show that:

(a): The mapping $n \to a^n$ of $\mathbb{N}$ into A can be extended to a continuous mapping f_a of $\mathbb{Z}$ into A when $|a - 1| < 1$.

(b): The mapping $\varphi : a \to f_a$ of $L = \{a \in A : |a - 1| < 1\}$ to $C(\mathbb{Z}_p, A)$ is a continuous injective homomorphism of L into a multiplicative subgroup of $C(\mathbb{Z}_p, A)$.

EXERCISE 10·8: CAUCHY'S INEQUALITIES

Let K be a complete non-Archimedean field, and let $f(X) = \sum_{n \geq 0} a_n X^n$ be a Taylor series of $K[[X]]$ convergent for $0 \leq |X| < r_0$ with $r_0 \neq 0$. For $r < r_0$ let us set: $M(f,r) = \sup_{n \geq 0} |a_n| r$.

If the residue field k of K is finite and if $v(K^*)$ is a discrete subgroup of $\mathbb{R}$, show that we do not always have $M(f,r) = \sup_{|X| \leq r} |f(X)|$.

EXERCISE 10·9: AN ENTIRE FUNCTION BOUNDED ON $\mathbb{Q}_p$

We want to construct an entire function bounded on $\mathbb{Q}_p$.

(a): Let $P(X) = 1 - X^{p-1}$; determine:

$$M(P,K) = \inf_{\substack{X \in \mathbb{Q}_p \\ v(X)=k}} v(P(X)), \qquad k \in \mathbb{Z}.$$

(b): Determine a sequence of integers $h(k)$ such that if:

$$P_k(X) = P(X)(P(pX))^{h(1)} \cdots (P(p^k X)^{h(k)}),$$

we have:

$$\sup_{\substack{x \in \mathbb{Q}_p \\ v(x) \geq -k}} |P_k(x)| = 1.$$

(c): From this deduce an example of a bounded entire function on $\mathbb{Q}_p$.

EXERCISE 10·10: A CONGRUENCE IN $\mathbb{Z}_2$

Show that in $\mathbb{Z}_2$:

$$\alpha_n = 2 + \frac{2^2}{2} + \cdots + \frac{2^n}{n}$$

is divisible by 2^h ($h \in \mathbb{N}$) for n large enough.

EXERCISE 10·11: HENSEL'S EXPANSION FOR RATIONAL NUMBERS

We say that we have a *PERIODIC HENSEL EXPANSION* for a number $\alpha \in \mathbb{Q}$ if there exist two integers $n_0 \in \mathbb{Z}$ and $r > 0$ such that $\alpha_{n+r} = \alpha_n$ for $n > n_0$, where:

$$\alpha = \alpha_m p^m + \alpha_{m+1} p^{m+1} + \cdots + \alpha_n p^n + \cdots \qquad (m \in \mathbb{Z}).$$

Show that the Hensel expansion of $\alpha \in \mathbb{Q}_p$ is periodic if and only if $\alpha \in \mathbb{Q}$.

EXERCISE 10·12: p-th ROOTS OF UNITY IN $\mathbb{Q}_p$

Show that for $p \neq 2$ that 1 is the only p-th root of unity in $\mathbb{Q}_p$.

EXERCISE 10·13: A CALCULATION IN $\mathbb{Q}_3$

In $\mathbb{Q}_3$ determine the first five terms of the Hensel expansion of:

$$\log 4 = \log(1 + 3) = 3 - \frac{3^2}{2} + \frac{3^3}{3} - \frac{3^4}{4} + \frac{3^5}{5} \cdots .$$

EXERCISE 10·14: UPPER BOUND FOR THE NORM OF THE POLYNOMIALS $\binom{x}{k}$ IN $\mathbb{Z}_p$

Show that in $\mathbb{Q}_p$ $P_k(x) = \binom{x}{k}$, with $P_0(x) = 1$, satisfies:

$|P_k(x)| \leqslant 1$ if $|x| \leqslant 1$.

EXERCISE 10·15: $\mathbb{Q}_p^{*2}$ IS OPEN IN $\mathbb{Q}_p$

Show that $\mathbb{Q}_p^{*2} = \{x \in \mathbb{Q}_p - \{0\} : x = y^2, y \in \mathbb{Q}_p\}$ is open in $\mathbb{Q}_p$ $(p \neq 2)$.

EXERCISE 10·16: A FORM OF HENSEL'S LEMMA

Let F be a complete valuation field for a discrete valuation v of the valuation ring O. Let $f(x)$ be a monic polynomial in $O[x]$ such that there exist $\alpha_1 \in O$ with $|f(\alpha_1)| < 1$ and $|f'(\alpha_1)| = 1$. Then the sequence:

$$\alpha_1, \quad \alpha_2 = \alpha_1 - \frac{f(\alpha_1)}{f'(\alpha_1)}, \quad \ldots, \quad \alpha_n = \alpha_{n-1} - \frac{f(\alpha_{n-1})}{f'(\alpha_{n-1})}$$

converges in $\mathcal{O}$ to a root of f.

EXERCISE 10·17: Find a rational number in Q_5 that approximates a solution to $P(x) = x^3 - x^2 - 2x - 8$ within $\frac{1}{5^3}$.

EXERCISE 10·18: STUDY OF THE LOCUS OF THE p^h-th ROOTS OF UNITY IN $\mathbb{Q}_p$

Show that if μ_h is a p^h-th primitive root of unity in $\mathbb{C}_p$, with $h \in \mathbb{N}^*$, then we have:

$$|1 - \mu_h| = \left(\frac{1}{p}\right)^{1/p^{h-1}(p-1)}.$$

EXERCISE 10·19: SOLUTION OF CONGRUENCES mod p^k

Let $F(x_1, x_2, \ldots, x_n) \in \mathbb{Z}[x_1, \ldots, x_n]$.
Show that the congruences:

$$F(x_1, \ldots, x_n) \equiv 0 \bmod p^k \quad (\text{for all } k \in \mathbb{N}^*)$$

can be solved in $\mathbb{Z}$ if and only if the equation $F(x_1, \ldots, x_n) = 0$ has a solution in $\mathbb{Z}_p$.

EXERCISE 10·20: REDUCIBILITY OF POLYNOMIALS IN $\mathbb{Q}_2$

Show that the polynomial $X^2 + X + a$, where $a \in \mathbb{Z}$, has a solution in $\mathbb{Q}_2$ if and only if a is even.

EXERCISE 10·21: LIOUVILLE NUMBERS IN $\mathbb{Q}_p$

(1): Let $\alpha \in \mathbb{Z}_p$ be algebraic over $\mathbb{Q}$ of degree n, but not a rational integer.

Show that there exists a constant C depending only on α, such that:

$$|\alpha - A|_p \geqslant \frac{C}{|A|^n} \quad \text{for all } A \in \mathbb{Z}^*,$$

where $|A|$ is the ordinary absolute value on $\mathbb{Z}$.

(2): Let $\alpha \in \mathbb{Z}_p$. Show that if, for an infinite number of values of N, we can find a rational number A $(A \neq \mp 1)$ such that $0 < |\alpha - A|_p < \frac{1}{|A|^N}$, then the number α is transcendental over $\mathbb{Q}$. We say that α is a *LIOUVILLE NUMBER* if it satisfies the preceding property.

(3): Let $\alpha \in \mathbb{Z}_p$ and let $\alpha = \sum_{k=1}^{\infty} a_k p^{b_k}$ be its Hensel expansion. Assume that $1 \leqslant a_k \leqslant p - 1$, and that the sequence $(b_k)_{k \geqslant 0}$ is strictly increasing, and, moreover that:

$$\overline{\lim_{k \to +\infty}} \frac{b_{k+1}}{b_k} = +\infty.$$

Show that α is a Liouville number.

Example: $\alpha = \sum_{k=1}^{\infty} p^{k!}$.

EXERCISE 10·22: m-th ROOTS IN $\mathbb{Q}_p$

Let p be a prime number, and $\mathbb{Q}_p$ the field of p-adic numbers. Let m be an integer prime to p, and let ε be a p-adic unit such that $v_p(\varepsilon - 1) > 0$.

Show that ε is an m-th power in $\mathbb{Q}_p$.

EXERCISE 10·23: m-th ROOTS OF ELEMENTS OF $\mathbb{Q}_p$

(1): Recall that the series:

$$e^{\xi} = 1 + \sum_{n=1}^{\infty} \frac{\xi^n}{n!} \quad \text{and} \quad \log(1 + \xi) = \sum_{n \geqslant 1} (-1)^n \frac{\xi^n}{n}$$

converge for $|\xi|_p \leqslant \frac{1}{p}$, that:

$$|e^{\xi} - 1|_p = |\xi|_p, \quad |\log(1 + \xi)|_p = |\xi|_p,$$

and that:

$$e^{\log(1+\xi)} = 1 + \xi.$$

Let $m \geqslant 2$ be an integer; set $v_p(m) = f$ ($f \geqslant 0$).

Show that if $\gamma \in \mathbb{Q}_p$ satisfies $v_p(\gamma - 1) \geqslant f + 1$, then there exists $\rho \in \mathbb{Q}_p$ such that $\rho^m = \gamma$. We shall set $\rho = \sqrt[m]{\gamma} = \gamma^{1/m}$.

(2): From this show that there exists a *finite* algebraic extension K_p of $\mathbb{Q}_p$ containing the m-th roots of *all* the elements of $\mathbb{Q}_p$.

Give an upper bound for the number of m-th roots of elements of $\mathbb{Q}_p$ to be adjoined to $\mathbb{Q}_p$ in order to obtain K_p.

EXERCISE 10·24: m-th ROOTS OF ELEMENTS OF $\mathbb{Q}_p$

We use the same notation as in Exercise 10·23.

(1): Let $\beta \in \mathbb{Q}_p$ and satisfying $v_p(\beta) = 0$. Let us set $\beta = b + p\beta_1$, where $1 \leqslant b \leqslant p - 1$ and $v(\beta_1) \geqslant 0$.

Show that:

$$\beta^m = b^m + p^{f+1}\beta' \quad \text{with } v_p(\beta') \geqslant 0.$$

Note that, for $k \geqslant 1$:

$$\binom{m}{k} = \frac{m}{k}\binom{m-1}{k-1} .$$

(2): Let $\alpha \in \mathbb{Q}_p$ be such that $v_p(\alpha) = 0$, and set $\alpha = a + p^{f+1}\alpha_{f+1}$, where $v_p(\alpha_{f+1}) \geqslant 0$.

Show that a necessary and sufficient condition for $\alpha^{1/m} \in \mathbb{Q}_p$ is that there exists an integer b satisfying $1 \leqslant b \leqslant p - 1$ such that the congruence $a \equiv b^m \bmod p^{f+1}$ holds.

EXERCISE 10·25: p-th ROOTS OF ELEMENTS OF $\mathbb{Q}_p$

(1): Let $\gamma \in \mathbb{Q}_p$ satisfy $v(\gamma) \geqslant 0$.

Show that $v_p(\gamma^p - \gamma) \geqslant 1$, and prove that, for $k \geqslant 1$,

$$v_p(\gamma^{kp} - \gamma^k - k\gamma^{k-1}(\gamma^p - \gamma)) \geqslant 2.$$

(2): Let ρ, β, β' be elements of $\mathbb{Q}_p$ satisfying $v(\rho) = v(\beta) = v(\beta') = 0$.

Show that if there exist two *distinct* integers k, k' such that:

$$0 \leqslant k \leqslant p - 1, \quad 0 \leqslant k' \leqslant p - 1 \quad \text{and} \quad v(\rho^k\beta^p - \rho^{k'}\beta'^p) \geqslant 2,$$

then:

$$v(\rho^{p-1} - 1) \geqslant 2.$$

(3): Show that if there exist integers r such that:

$$v_p(r^{p-1} - 1) = 1,$$

then for $p \geqslant 3$ we can choose r satisfying $1 \leqslant r \leqslant p - 1$, and for $p = 2$ we have $r = 3$. Establish that the numbers $r^k b^p$, for $0 \leqslant k \leqslant p - 1$, $1 \leqslant b \leqslant p - 1$, form, modulo p^2, a system of residues.

From this deduce that the extension $\mathbb{Q}_p[r^{1/p}]$ of $\mathbb{Q}_p$ contains a p-th root of every element of $\vartheta \in U_p$.

EXERCISE 10·26: SQUARE ROOTS OF ELEMENTS IN $\mathbb{Q}_p$ ($p \neq 2$)

Let γ be a unit of $\mathbb{Q}_p$ such that the polynomial $X^2 - \gamma$ is irreducible in $\mathbb{Q}_p[X]$.

Show that all the polynomials $X^2 - \delta$, where $\delta \in U_p$, are reducible in $\mathbb{Q}_p[\sqrt{\gamma}]$.

(2): Show that the field $\mathbb{Q}_p([\sqrt{\gamma}])[\sqrt{p}]$ is a splitting field for every second degree polynomial of $\mathbb{Q}_p[X]$. Show that this field is identical with $\mathbb{Q}_p[\sqrt{\gamma} + \sqrt{p}]$ (Use Exercises 10·23, 10·24, 10·25).

EXERCISE 10·27: SQUARE ROOTS OF ELEMENTS OF $\mathbb{Q}_2$

(1): Show that the equation $X^2 = a$ (where $|a|_2 = 2^{-i}$, $i = 0$ or 1) has solutions in $\mathbb{Q}_2$ if and only if $|a - 1| < |2|_2^3$.

(2): Show that there are exactly seven different quadratic extensions of $\mathbb{Q}_2$ generated respectively by adjoining a root of the following polynomials:

$$X^2 - 3;\quad X^2 - 5;\quad X^2 - 7;\quad X^2 - 2;\quad X^2 - 6;\quad X^2 - 10;$$

$$X^2 - 14.$$

(3): Show, by a simple change of variables, that the extensions defined by $X^2 - 3$ and $X^2 - 7$ are ramified.

(4): Show that the extension defined by $X^2 - 5$ is not ramified.

EXERCISE 10·28: EXTENSIONS OF $\mathbb{Q}_p$ OF DEGREE TWO

Consider the polynomial $X^2 - X - 1$ and denote its two roots in an algebraic closure of $\mathbb{Q}_p$ by ϑ and $\bar{\vartheta}$.

(1): Show that if $p = 11$ or 19, then θ, $\bar{\theta}$ belong to $\mathbb{Q}_p$.

(2): Show that ϑ and $\bar{\vartheta}$ are not in $\mathbb{Q}_p$ if $p = 7$. Find the residual degree and the ramification index of the extension $\mathbb{Q}_7[\vartheta,\bar{\vartheta}]$.

(3): Calculate the 5-adic valuation of:

$N(\vartheta - c) = (\vartheta - c)(\bar{\vartheta} - c)$ for $c = 0,1,2,3,4$.

Find the residual degree and the ramification index of the extension of $\mathbb{Q}_5$ generated by ϑ and $\bar{\vartheta}$.

EXERCISE 10·29: EXTENSIONS OF $\mathbb{Q}_p$ BY A SEVENTH ROOT OF UNITY

Let ϑ be a primitive seventh root of unity.

(1): Show that the extension $\mathbb{Q}_7(\vartheta)$ of $\mathbb{Q}_7$ is completely ramified. Find a uniformizer for $\mathbb{Q}_7[\vartheta]$.

(2): Show that the equation $X^7 - 1 = 0$ splits completely into factors of the first degree in $\mathbb{Q}_p$ if and only if $p \equiv +1 \bmod 7$.

(3): Show that the extension $\mathbb{Q}_p[\vartheta]$ is of degree two over $\mathbb{Q}_p$ if $p = -1 \bmod 7$.

(4): Show that the extension $\mathbb{Q}_p[\vartheta]$ is of degree three over $\mathbb{Q}_p$ if $p \equiv 2$ or $p \equiv 4 \bmod 7$.

(5): Show that the extension $\mathbb{Q}_p[\vartheta]$ is of degree six over $\mathbb{Q}_p$ if $p \equiv 3$ or $p \equiv 5 \bmod 7$.

EXERCISE 10·30: CONTINUOUS FUNCTIONS AND GENERATING FUNCTIONS

(1): Set $P_0(x) = 1, \ldots, P_k(x) = \binom{x}{k}, \ldots$.

Show that if $x \in \mathbb{Z}_p$ we have:

$$|P_k(x)|_p \leqslant 1.$$

Let a_k be a sequence of numbers of $\mathbb{C}_p$ with:

$$\lim_{k\to\infty} |a_k|_p = 0.$$

Show that $\sum_{k=0}^{\infty} a_k P_k(x)$ defines a continuous function for $x \in \mathbb{Z}_p$.

(2): Let $f \in C(\mathbb{Z}_p, \mathbb{C}_p)$. Show that the analytic function:

$$F(X) = \sum_{n\geqslant 0} f(n) X^n, \quad \text{defined on } B(0,1^-) \subset \mathbb{C}_p,$$

is in fact an analytic element on $\mathbb{C}_p - B(1,1^-)$ (that is to say that $F(X)$ is the uniform limit on $\mathbb{C}_p - B(1,1^-)$ of a sequence of rational functions without poles in $\mathbb{C}_p - B(1,1^-)$, cf., [Ami]). Deduce from this that every function $f \in C(\mathbb{Z}_p, \mathbb{C}_p)$ can be put into the form:

$$f(x) = \sum_{k\geqslant 0} a_k(f) P_k(x) \quad \text{with} \quad \lim_{k\to\infty} |a_k(f)|_p = 0.$$

EXERCISE 10·31: THE p-ADIC LOGARITHM

Show that:

$$\lim_{n\to\infty} \frac{1}{p^n}\binom{p^n}{h}$$

exists for all $h \in \mathbb{N}^*$. Calculate this limit.

(2): Let $A_n(x) = \frac{1}{p^n}[(1 + x)^{p^n} - 1]$. Find the p-adic absolute values of the zeros of $A_n(x)$ in $\mathbb{C}_p$. Is $A_n(x)$ irreducible over $\mathbb{Q}_p$? Find the value of $|A_n(x)|_p$ for $x \in \mathbb{C}_p$ with $|x_p| = r$, where $0 < r < 1$ and $r \neq p^{-m_j}$ for every integer j, with:

$$m_j = \frac{1}{p^j(p - 1)} .$$

Deduce from this that $\lim_{n\to\infty} A_n(x) = f(x)$ exists in $\mathbb{C}_p$ for all x satisfying $|x|_p < 1$. Show that for all $x,y \in \mathbb{C}_p$ with $|x - 1|_p < 1$ and $|y - 1|_p < 1$ we have:

$$f(xy - 1) = f(x - 1) + f(y - 1).$$

Give the expansion of $f(x)$ as a Taylor series, and give its radius of convergence.

(3): For $n \geqslant 1$ set:

$$B_n(x) = \frac{A_n(x)}{A_{n-1}(x)} ;$$

Show that $B_n(x)$ is an irreducible polynomial in x on $\mathbb{Q}_p$. Show that the zeros of $B_n(x - 1)$ are the primitive p^n-th roots of unity In $\mathbb{C}_p$ what are the zeros of $f(x)$ inside $|x|_p < 1$?

EXERCISE 10·32: DIFFERENTIAL EQUATIONS

Let F be a continuous mapping of $\mathbb{Z}_p \times \mathbb{Z}_p$ into $\mathbb{Q}_p$. We propose to show that there exists a differentiable function f defined on $\mathbb{Z}_p$ to $\mathbb{Q}_p$ such that:

$$f'(x) = F(x,f(x)) \quad \text{with } f(0) = 0.$$

If $x \in \mathbb{Z}_p$ then x can be written:

$$x = a_0 + a_1 p + \cdots + a_n p^n + \cdots \quad \text{with } 0 \leqslant a_i \leqslant p - 1.$$

Set:

$$x_n = a_0 + a_1 p + \cdots + a_n p^n.$$

Set:

$$M = \sup_{\substack{x \in \mathbb{Z}_p \\ y \in \mathbb{Z}_p}} |F(x,y)|_p,$$

and let $k \in \mathbb{N}$ be such that:

$$Mp^{-(k+1)} \leqslant 1.$$

Set:

$$f_k(x) = (x - x_k)F(x_k, 0),$$

and for $n > k$ define f_n by the recurrence relation:

$$f_n(x) = f_{n-1}(x_n) + (x - x_n)F(x_n, f_{n-1}(x_n)).$$

(1): Show that $f_n(x)$ is defined for $x \in \mathbb{Z}_p$.

(2): Find an upper bound for:

$$|f_{n-1}(x) - f_n(x)|_p,$$

and deduce from it that $\lim_{n\to\infty} f_n(x)$ exists for all $x \in \mathbb{Z}_p$.

Set:

$$f(x) = \lim_{n} f_n(x).$$

Find an upper bound for:

$$|f(x) - f_n(x)_p|$$

and verify that $f(0) = 0$.

(3): Show that:

(a): For all $\varepsilon > 0$ there exists $N \in \mathbb{N}$ such that $n \geqslant N$ and:

$$|x - x'| \leqslant p^{-(n+1)}$$

implies:

$$\left|\frac{f_n(x) - f_n(x')}{x - x'} - F(x,f(x))\right| \leqslant \varepsilon.$$

(b): If $m > N \geqslant N$, then the inequality:

$$|x - x'| \leqslant p^{-(n+1)}$$

implies:

$$\left|\frac{f_m(x) - f_m(x')}{x - x'} - F(x,f(x))\right| \leqslant \varepsilon.$$

Deduce from this that f is differentiable on $\mathbb{Z}_p$ and that:

$$f'(x) = F(x,f(x)).$$

EXERCISE 10·33: CONGRUENCES FOR BELL NUMBERS

Let P_n be the n-th *BELL NUMBER*, that is to say the number of partitions of a set of n elements ($P_0 = 1$). We know (cf., [Com]) that:

$$\sum_{n\geqslant 0} P_n \frac{X^n}{n} = e^{e^X - 1}.$$

(1): Show that:

$$\sum_{n\geqslant 0} P_n X^n = \sum_{n\geqslant 0} \frac{X^n}{(1 - X)\cdots(1 - nX)} .$$

From this deduce that:

$$\sum_{n\geqslant 0} P_n X^n = \frac{\displaystyle\sum_{n=0}^{p^h - 1} X^n(1 - (n + 1)X)\cdots(1 - (p^h - 1)X)}{(1 - X)\cdots(1 - (p^h - 1)X) - X^{p^h}} \bmod p^h\mathbb{Z}[[X]].$$

Using this deduce that the $(P_n)_{n\geqslant 0}$ form, modulo p^h, a linear recurrent sequence, that is to say, that there exist integers $\lambda_{1,h},\ldots,\lambda_{k_h,h}$ such that:

$$\lambda_{1,h}P_{n-1} + \cdots + \lambda_{k_h,h}P_{n-k} \equiv 0 \bmod p^h.$$

(2): Show that if:

$$k(p) = \frac{p^p - 1}{p - 1} ,$$

then:

$$P_{n+k(p)} \equiv P_n \bmod p.$$

SOLUTIONS

SOLUTION 10·1: On k every absolute value is trivial. Let q = card$K$, let $x \in k^*$. We know that $x^{q-1} = 1$, therefore if the absolute value on k is denoted $|\cdot|$ we have $|x|^{q-1} = |1| = 1$, hence $|x| = 1$ for all $x \in k^*$. If L is an algebraic extension of k then L is itself a finite field, and the preceding argument can be applied.

SOLUTION 10·2: The valuation ring A of $K(X)$ for the P-adic valuation is the set of rational functions $\frac{R}{S}$ such that $(R,S) = 1$ and $(S,P) = 1$. The valuation ideal I is the set of rational functions $\frac{R}{S}$ such that $(S,R) = 1$, $(S,P) = 1$ and P/R. The residue field is A/I, and by Bezout's Identity:

$$A/I \simeq \frac{A \cap K[X]}{I \cap K[X]} \simeq \frac{K[X]}{P(X)K[X]} .$$

If $P(X) = X$, the completion of L is the set of Laurent series with coefficients in K having only finitely many negative indices. If P is an irreducible polynomial the completion of L is the set of Laurent series in $P(X)$ with coefficients in a complete system of representatives of $K[X]/P(X)K[X]$ (hence in a finite extension of K) and having only finitely many negative indices.

SOLUTION 10·3: For this to be so, it is necessary and sufficient (cf., [Ami]) that the group of values of v_p be discrete (which is the case here since it is $\mathbb{Z}$) and that the residue field be finite. Now, the residue field is a finite extension of K. Therefore it is necessary and sufficient that K be a finite field.

SOLUTION 10·4: (a): The valuation of $P(X,Y)$ is, in fact, the T-adic valuation of the polynomial $P(T^a,T^b)$. The value group of v is the additive subgroup of $\mathbb{R}$ generated by a and b on $\mathbb{Z}$, hence $v(K(X,Y)) = a\mathbb{Z} + b\mathbb{Z}$.

(b): The valuation ring is the set of rational functions $\frac{P(X,Y)}{Q(X,Y)}$ such that $\frac{P(T^a,T^b)}{Q(T^a,T^b)}$ has T-adic valuation $\geqslant 0$, hence, possibly after simplifying, such that $Q(0,0) \neq 0$. The valuation ideal is the set of rational functions $\frac{P(X,Y)}{Q(X,Y)}$ such that $\frac{P(T^a,T^b)}{Q(T^a,T^b)}$ has T-adic valuation greater than zero, hence, possibly after simplifying, such that $Q(0,0) \neq 0$ and $P(0,0) = 0$. The residue field is K.

(c): The completion $\hat{K}(X,Y)$ is isomorphic to the set of Laurent series:

$$\sum_{(m,n)\in\mathbb{Z}^2} a_{m,n} X^m Y^n$$

such that:

$$\operatorname*{Inf}_{a_{m,n}\neq 0} (am + bn) > -\infty.$$

$\hat{\mathcal{R}}(X,Y)$ is locally compact if K is finite and if the group generated by a and b on $\mathbb{Z}$ is discrete, hence if a and b are $\mathbb{Z}$-linearly dependent.

SOLUTION 10·5: (a): This is immediate.

(b): If $a = 1 + p^2 b$, $b \in \mathbb{Z}_p$, then:

$$(1 + p)^p - a \equiv 0 \bmod p^2,$$

and:

$$p(1 + p)^{p-1} \equiv 0 \bmod p,$$

and:

$$p(1 + p)^{p-1} \not\equiv 0 \bmod p^2.$$

Hensel's Lemma (cf., [Ami]) shows that there exists a solution of the equation:

$$(1 + pX)^p \equiv a$$

in $\mathbb{Q}_p$, and, furthermore,

$$X \equiv 1 \bmod p.$$

(c): We have:

$$\mathbb{Q}_p^* = \mathbb{Z} \times T \times (1 + p\mathbb{Z}_p) \text{ with } T \simeq (\mathbb{Z}/p\mathbb{Z})^*,$$

hence:

$$(\mathbb{Q}_p^*)^p \simeq p\mathbb{Z} \times T^p \times (1 + p\mathbb{Z}_p)^p;$$

now,

$$T^p = T \quad \text{and} \quad (1 + p\mathbb{Z}_p) = 1 + p^2\mathbb{Z}_p,$$

therefore:

$$\mathbb{Q}_p^*/(\mathbb{Q}_p^*)^p \simeq (\mathbb{Z}/p\mathbb{Z})\times(T/T)\times((1 + p\mathbb{Z}_p)/(1 + p^2\mathbb{Z}_p))$$

$$\simeq (\mathbb{Z}/p\mathbb{Z})^2.$$

SOLUTION 10·6: L has a normal basis over K if and only if L is not ramified (cf., [Sam]) over K (by Condition N, [Ami]).

If the valuation on L is normalised in such a way that its value group as well as that of K is $\mathbb{Z}$, then L has normal bases over K formed of elements of L integral over K.

SOLUTION 10·7: (a): We have:

$$f_a(x) = \sum_{n\geqslant 0} (a - 1)^n \binom{x}{n} \in C(\mathbb{Z}_p, a) \quad \text{if } a \in L.$$

In fact this formula is true for all $x \in \mathbb{N}$ and therefore, by continuity, for all $x \in \mathbb{Z}_p$ (since $|a - 1| < 1$).

(b): The mapping φ is injective, since $f_a(1) = a$; it is continuous, since:

$$\|f_a - f_{a'}\|_{\mathbb{Z}_p} \leqslant |a - a'|.$$

We have:

$$(f_a f_b)(x) = f_{ab}(x) \quad \text{for all } x \in \mathbb{N},$$

and therefore, by continuity, for all $x \in \mathbb{Z}_p$; in fact, if $a,b \in L$, then so does ab.

SOLUTION 10·8: Let us set card$k = q$. We have:

$$|x^q - x| < 1 \quad \text{if} \quad |x| \leqslant 1,$$

but:

$$M(f,1) = 1 \quad \text{if} \quad f(x) = x^q - x.$$

SOLUTION 10·9: (a): If $k > 0$, $M(P,k) > 0$; if $k = 0$, $M(P,k) = 1$; if $k < 0$, $M(P,k) = k(p - 1)$.

(b): Let us assume that $h(0) = 1, h(1), \ldots, h(k - 1)$ are determined; then:

$$\sup|P_k(x)| = 1 \quad \text{for } x \in \mathbb{Q}_p \text{ and } v(x) \geqslant -k.$$

Therefore:

$$k(p - 1) + (k - 1)(p - 1)h(1) + \cdots = h(k),$$

which determines $h(k)$ uniquely if $h(1),\ldots,h(k - 1)$ are known.

(c): Let us show that the sequence of polynomials $P_k(x)$ converges uniformly on every disc D of $\mathbb{Q}_p$. In fact,

$$P_{k+1}(X) - P_k(X) = P_k(X)(-1 + (1 - p^{k+1}X)^{p-1})^{h(k+1)}).$$

We have:

$$\sup_{x \in D}|P_{k+1}(x) - P_k(x)|$$

$$\leqslant \sup_{x \in D}|P_k(x)| \sup_{1 \leqslant i \leqslant h(k+1)} \sup_{x \in D}|p^{k+1}x|^{i(p-1)}.$$

For k large enough we have:

$$\sup_{x\in D}|P_k(x)| \leqslant 1 \quad \text{and} \quad \sup_{x\in D}|p^{k+1}x| \leqslant \varepsilon.$$

The sequence $k \to P_k$ is therefore a Cauchy sequence in the topology of uniform convergence on every bounded subset of $\mathbb{Q}_p$, and therefore has a limit f which is an entire function bounded by one on $\mathbb{Q}_p$.

SOLUTION 10·10: Let us consider the identity:

$$\frac{1-(1-2)^{2^k}}{2^k} = 0 \quad \text{for all } k > 0.$$

We have:

$$\frac{1-(1-2)^{2^k}}{2^k} = \sum_{n=1}^{2^k} -(-1)^n 2^n \frac{1}{2^k}\binom{2^k}{n}.$$

Now, if $n - k \geqslant h$, then:

$$\frac{2^n}{2^k} \geqslant 2^h,$$

therefore, for $n \geqslant k + h$,

$$(-1)^n \frac{2^n}{2^k}\binom{2^k}{n} \quad \text{is divisible by } 2^h$$

(since $\binom{2^k}{n} \in \mathbb{Z}$). On the other hand:

$$\frac{1}{2^k}\binom{2^k}{n} = \frac{(-1)^{n-1}}{n} + \frac{2}{n!}M(n), \quad \text{where } M(n) \in \mathbb{Z}.$$

Now, the largest power of 2 which divides $n!$ is:

$$v_2(n!) = \Omega(n) = \sum_{i\geqslant 1} \left[\frac{n}{2^i}\right].$$

And therefore if $n - k < h$ we have:

$$k - \sum_{i\geqslant 1} \left[\frac{n}{2^i}\right] > k - n > h,$$

hence:

$$\frac{2^n}{2^k}\binom{2^k}{n} = (-1)^{n-1}\frac{2^n}{n} + \frac{2^{k+n}}{n!}M(n) \equiv \frac{(-1)^{n-1}}{n}2^n \bmod 2^h.$$

Lastly, let us notice that:

$$\frac{1-(1-2)2^k}{2^k} \equiv \alpha_{2^k} \bmod 2^h \quad \text{if } k \geqslant h,$$

and that:

$$\alpha_n \equiv \alpha_m \bmod 2^{\lambda(n,m)},$$

where:

$$\lambda(n,m) = \mathrm{Inf}\left(n - \left[\frac{\log n}{\log_2}\right], m - \left[\frac{\log m}{\log 2}\right]\right),$$

hence the result.

SOLUTION 10·11: Let $\alpha \in \mathbb{Q}_p$, $\alpha = \alpha_m p^m + \cdots$, and let us assume that the Hensel expansion of α is periodic. Therefore there exist n_0 and r such that $\alpha_{n+r} = \alpha_n$ for $n \geqslant n_0$. Therefore:

$$\alpha = \alpha_m p^m + \cdots + \alpha_{n_0-1}p^{n_0-1} + \alpha_{n_0}(p^{n_0} + p^{n_0+r} + \cdots) +$$

$$+ \alpha_{n_0+1}(p^{n_0+1} + p^{n_0+r+1} + \cdots) + \cdots$$

$$+ \alpha_{n_0+r-1}(p^{n_0+r-1} + p^{n_0+2r-1} + \cdots)$$

$$= \alpha_m p^m + \cdots + \alpha n_{0-1} p^{n_0-1} + \frac{\alpha_{n_0} p^{n_0} + \cdots + \alpha_{n_0+r-1} p^{n_0+r-1}}{1 - p^r},$$

therefore $\alpha \in \mathbb{Q}$.

Conversely, if $\alpha \in \mathbb{Q}$, then $\alpha = \frac{m}{n}$ with $m, n \in \mathbb{Z}$. There exists n' such that if $n > 0$, then $n'n = (p^r - 1)p^s$ (this is an application of Lagrange's Theorem which says that $p^{\varphi(n)} - 1$ is divisible by n if $(n,p) = 1$, where $\varphi(n)$ is Euler's TOTIENT function). So let us assume that $m > 0$ and $n < 0$ (in order to fix our notation). Therefore there exists $n' > 0$ such that $\alpha = \frac{mn'}{p^s(1 - p^r)}$. By possibly modifying n' we may assume:

$$\alpha = \frac{mn'}{p^0(1 - p^s)} = \frac{mn'}{1 - p^s} + \frac{mn'}{p^s}.$$

Therefore:

$$\alpha = M + \frac{R}{1 - p^s} + \frac{mn'}{p^s} \quad \text{with } 0 \leqslant R < p^s - 1,\ M \in \mathbb{N}.$$

Hence, if we set:

$$M = m_0 + m_1 p + \cdots + m_i p^i, \qquad mn' = m'_0 + m'_1 p + \cdots + m'_j p^j,$$

$$R = r_0 + r_1 p + \cdots + r_k p^k,$$

with $k < s$, for $\ell + 1$ a multiple of s and $\ell \geqslant \sup(i,j)$, we have:

$$\alpha = \alpha_t p^t + \cdots + \alpha_\ell p^\ell + r_0 p^{\ell+1} + r_1 p^{\ell+2} + \cdots$$

$$\cdots + r_k p^{\ell+k+1} + \cdots + r_0 p^{s+\ell+1} + \cdots .$$

The Hensel expansion of α is therefore periodic with period s for $n \geqslant \ell + 1$.

SOLUTION 10·12: Let $\gamma \neq 1$ be a p-th root of unity in $\mathbb{Q}_p$. Set $\gamma - 1 = u$. Hence u is a root of the equation:

$$p + u\binom{p}{2} + u^2\binom{p}{3} + \cdots + u^{p-2}\binom{p}{p-1} + u^{p-1} = 0.$$

Now,

$$\left|\binom{p}{2}\right| = \left|\binom{p}{3}\right| = \cdots = \left|\binom{p}{p-1}\right| = \frac{1}{p},$$

therefore we must have:

$$|u^{p-1}| = \frac{1}{p}$$

by the ultrametric inequality, and therefore:

$$|u| = \left(\frac{1}{p}\right)^{1/(p-1)},$$

which is impossible if $u \in \mathbb{Q}_p$.

SOLUTION 10·13: We want to find $\alpha \in \mathbb{N}$ such that:

$$|\alpha - \log 4| \leqslant \left(\frac{1}{3}\right)^5 \quad \text{and} \quad \alpha < 3^5.$$

We have:

$$\left|\log 4 - \left(3 - \frac{3^2}{2} + \frac{3^3}{3} - \frac{3^4}{4}\right)\right| < \left(\frac{1}{3}\right)^5,$$

since:

$$\left|\frac{3^n}{n}\right| < \left(\frac{1}{3}\right)^5 \quad \text{whenever} \quad n > 5.$$

Now,

$$3 - \frac{3^2}{2} + \frac{3^3}{3} - \frac{3^4}{4} = 12 - \frac{99}{4} = 12 + \frac{243 - 99}{4} \bmod 3^5$$

$$= 48 \bmod 3^5$$

$$= 1.3 + 2.3^2 + 3^3 \bmod 3^5.$$

SOLUTION 10·14: We know that $\binom{n}{k} \in \mathbb{N}$ for all $n \in \mathbb{N}$ for all $n \in \mathbb{N}$.

Therefore:

$$\left|\binom{n}{k}\right| \leqslant 1 \quad \text{for } n \in \mathbb{N}$$

Now:

$$\mathbb{Z}_p = \{x \in \mathbb{Q}_p : |x| \leqslant 1\}$$

and $\mathbb{N}$ is dense in $\mathbb{Z}_p$, and finally $P_k(x)$ is a continuous function of $\mathbb{Z}_p$ into $\mathbb{Q}_p$, therefore:

$$|P_k(x)| \leqslant 1 \quad \text{if } x \in \mathbb{Z}_p.$$

SOLUTION 10·15: It suffices to show that if:

$$U_p = \{x \in \mathbb{Z}_p : |x| = 1\},$$

U_p^2 is open in $\mathbb{Z}_p$ and to end using a homothety. Therefore it suffices to show that if the equation $X^2 - x = 0$ has a root $(x \in U_p)$, then the equation $X^3 - (x + h) = 0$ has a root for every sufficiently small h.

Let y be a root of the equation $X^2 - x = 0$. Let us consider the sequence:

$$\alpha_0 = y, \quad \alpha_1 = \alpha_0 - \frac{\alpha_0^2 - (x + h)}{2\alpha_0}, \quad \ldots,$$

$$\alpha_n = \alpha_{n-1} - \frac{\alpha_{n-1}^2 - (x - h)}{2\alpha_{n-1}}, \quad \ldots .$$

Hence:

$$\alpha_1 = y + \frac{h}{2y} \quad \text{and} \quad \alpha_1^2 - (x + h) = \frac{h^2}{4y^2} .$$

Let us assume that:

$$\alpha_{n-1} = y + \frac{h}{2y} + \lambda_2 h^2 + \cdots + \lambda_{n-1} h^{n-1}$$

and:

$$\alpha_n - \alpha_{n-1} = - \frac{\alpha_{n-1}^2 - (x + h)}{2\alpha_{n-1}} = - \frac{\mu_n h^n}{2\alpha_{n-1}} ,$$

therefore:

$$\alpha_n = \alpha_{n-1} + \lambda_n h^n \quad \text{with} \quad -\lambda_n = \frac{\mu_n}{2\alpha_{n-1}} ,$$

and:

$$\alpha^2 - (x + h) = (\alpha_{n-1} - \frac{\mu_n}{2\alpha_{n-1}} h^n)^2 - (x + h) =$$

$$= \alpha_{n-1}^2 - \mu_n h^n + \frac{\mu_n^2 h^{2n}}{4\alpha_{n-1}^2} - (x + h) = \mu_{n+1} h^{n+1},$$

with $|\mu_{n+1}| \leqslant 1$. Therefore the sequence $(\alpha_n)_{n\in\mathbb{N}}$ is convergent if $|h| < 1$ and its limit satisfies:

$$\alpha = \alpha - \frac{\alpha^2 - (x + h)}{2\alpha},$$

or, again,

$$\alpha^2 - (x + h) = 0.$$

SOLUTION 10·16: We have:

$$(\alpha_2 - \alpha_1)f'(\alpha_1) = -f(\alpha_1).$$

By Taylor's formula, with n being the degree of f, we have:

$$f(\alpha_2) = (\alpha_2 - \alpha_1)^2\left[\frac{f''(\alpha_1)}{2!} + \cdots + \frac{(\alpha_2 - \alpha_1)^{n-2}}{n!} f^{(n)}(\alpha_1)\right].$$

Now,

$$\frac{f^{(k)}(X)}{k!} \in O[X],$$

therefore:

$$\left|\frac{f^{(k)}(\alpha_1)}{k!}\right| \leqslant 1.$$

Hence:

$$|f(\alpha_2)| \leqslant |\alpha_2 - \alpha_1|^2 = |f(\alpha_1)|^2,$$

and:

$$f'(\alpha_2) = f'(\alpha_1) + (\alpha_2 - \alpha_1)f''(\alpha_1) + \cdots .$$

Now, $|\alpha_2 - \alpha_1| < 1$, therefore:

$$|(f'(\alpha_2))| = |f'(\alpha_1)| = 1.$$

Then by induction we show that:

$$|f'(\alpha_i)| = 1, \qquad |f(\alpha_i)| < |f(\alpha_1)|^i,$$

and that:

$$|\alpha_i - \alpha_{i-1}| < |f(\alpha_1)|^{i-1}$$

Therefore:

$$\lim_{i\to\infty}|\alpha_i - \alpha_{i-1}| = 0.$$

The sequence $(\alpha_i)_{i\in\mathbb{N}}$ is therefore a Cauchy sequence in F, so it has a limit α. We have:

$$\alpha = \alpha - \frac{f(\alpha)}{f'(\alpha)},$$

therefore $f(\alpha) = 0$ because $|f'(\alpha)| = 1$.

SOLUTION 10·17: Use Hensel's Lemma in the form in Exercise 10·16 (Newton's Method). Thus we are looking for $\alpha_1 \in \mathbb{Z}_5$ such that:

$$|P(\alpha_1)|_5 < 1 \quad \text{and} \quad |P'(\alpha_1)|_5 = 1.$$

It suffices to look for α_1 in the set $\{0,1,2,3,4\}$ which forms a complete system of representatives for $\mathbb{Z}/5\mathbb{Z}$. We show that

$\alpha_1 = 1$, because $P(1) = -10$ and $P'(1) = -1$. Then (Exercise 10·16) there exists $\alpha \in \mathbb{Z}_5$ such that $P(\alpha) = 0$ and $|\alpha - \alpha_1|_5 \leq \frac{1}{5}$. In addition (Exercise 10·16), if:

$$\alpha_2 = \alpha_1 - \frac{P(\alpha_1)}{P'(\alpha_1)} = -9,$$

and if:

$$\alpha_3 = \alpha_2 - \frac{P(\alpha_2)}{P'(\alpha_2)} = -\frac{1,531}{259},$$

then:

$$|\alpha_3 - \alpha|_5 \leq \frac{1}{5^3},$$

and therefore:

$$|P(\alpha_3)|_5 < \left(\frac{1}{5}\right)^3.$$

Furthermore, we can replace α_3 by any number $\beta_3 \in \mathbb{Z}_5$ satisfying:

$$|\alpha_3 - \beta_3|_5 < \left(\frac{1}{5}\right)^3,$$

and by the ultrametric inequality we would also have:

$$|\beta_3 - \alpha| \leq \left(\frac{1}{5}\right)^3;$$

and so by any number of the form:

$$\beta_3 = -\frac{1,531}{259} + 125 \sum_{i \geq 0} a_i 5^i,$$

with $0 \leq a_i \leq 4$ (cf., Proposition 2 of the Introduction).

In particular, we can always choose $\beta_3 \in \mathbb{Z}$ and $0 \leqslant \beta_3 \leqslant 124$. Now, 1,531 = 31 + 1,500, and 259 = 9 + 250, therefore:

$$\frac{1{,}531}{259} = -\frac{31 + 1{,}500}{9 + 250} = -\frac{31}{9} + 125k', \quad \text{where } k' \in \mathbb{Z}_5$$

(for, $|31|_5 = 1$ and $|9|_5 = 1$); on the other hand:

$$-\frac{31}{9} = \frac{31}{1 - 10} = 31(1 + 10 + 10^2) + k''.125, \quad \text{where } k'' \in \mathbb{Z}_5.$$

What is left is to find a representation of 3,441 modulo 125 lying between 0 and 124. The remainder from dividing by 125 gives $\beta_3 = 66$. Therefore a solution in $\mathbb{Q}_5$ to $(1/5)^3$ near to

$$P(x) = x^3 - x^2 - 2x - 8$$

is $\beta_3 = 66$.

SOLUTION 10·18: Let us set $y_h = \mu_h - 1$. Let:

$$P_1(X) = \frac{X^p - 1}{X - 1},$$

then μ_h is a root of:

$$P_h(X) = P_1(X^{p^{h-1}}).$$

We show by induction on h that:

$$P_h(y + 1) = y^{p^{h-1}(p-1)} \bmod p\mathbb{Z}[X],$$

and that:

$$P_h(0) = p,$$

therefore we necessarily have:

$$|y_h|^{p^{h-1}(p-1)} = \frac{1}{p} \quad \text{and} \quad |y_h| = \frac{1}{p}^{1/p^{h-1}(p-1)}.$$

SOLUTION 10·19: Let $(x_1^{(k)},\ldots,x_h^{(k)})$ be a solution of the congruence $F(x_1,\ldots,x_n) \equiv 0 \bmod p^k$. As $(\mathbb{Z}_p)^n$ is compact we can extract from the sequence $k \to (x_1^{(k)},\ldots,x_n^{(k)})$ a subsequence converging towards $(x_1,\ldots,x_n) \in \mathbb{Z}_p^n$, and we have $F(x_1,\ldots,x_n) = 0$. Conversely, if there exists $(x_1,\ldots,x_n) \in \mathbb{Z}_p^n$ such that $F(x_1,\ldots,x_n) = 0$, then for all $k \in \mathbb{N}$ one can find $(x_1^{(k)},\ldots,x_n^{(k)}) \in \mathbb{Z}^n$ such that $F(x_1^{(k)},\ldots,x_n^{(k)}) \equiv 0 \bmod p^k$, for $\mathbb{Z}^n$ is dense in $\mathbb{Z}_p^n$ and F is a continuous mapping of $\mathbb{Z}_p^n$ into $\mathbb{Z}_p$, whose restriction to $\mathbb{Z}^n$ has image contained in $\mathbb{Z}$ (cf., [Bor] p. 44).

SOLUTION 10·20: If $X^2 + X + a$ has a root in $\mathbb{Q}_2$ this root must belong to $\mathbb{Z}_2$, for the polynomial is monic and has coefficients in $\mathbb{Z}$.

Let α be a root, $\alpha \in \mathbb{Z}_2$. Therefore $|\alpha| \leqslant 1$. If $|\alpha| < 1$, then the equality $a = -\alpha - \alpha^2$ implies $|a| < 1$, therefore a is even. If $|\alpha| = 1$, then:

$$\alpha = 1 + 2\alpha_1 \quad (|\alpha_1| \leqslant 1),$$

and therefore:

$$a = -1 - 2\alpha_1 - (1 + 2\alpha_1)^2 = -4\alpha_1^2 - 6\alpha_1 - 2 \equiv 0 \bmod 2.$$

Conversely, if a is even, the image of the polynomial $X^2 + X + a$ in $(\mathbb{Z}/2\mathbb{Z})[X] = \mathbb{F}_2[X]$ is the polynomial $X^2 + X$, which splits in $\mathbb{F}_2[X]$ into a product of two monic, relatively prime factors. Therefore by Hensel's Lemma the polynomial $X^2 + X + a$ is reducible in $\mathbb{Q}_2[X]$.

SOLUTION 10·21: (1): Let:

$$P(X) = X^n + a_1X^{n-1} + \cdots + a_n$$

be the minimal polynomial of α over $\mathbb{Q}$. Set $a_0 = 1$. For $A \in \mathbb{Z}^*$ we have:

$$P(A) = P(\alpha) + (\alpha - A)P'(\alpha) + \cdots + \frac{(\alpha - A)^n}{n!} P^n(\alpha)$$

$$= (\alpha - A)F(\alpha,A),$$

and:

$$F(\alpha,A) \neq 0,$$

since $P(A) \neq 0$, for P is irreducible over $\mathbb{Q}$. Therefore:

$$|\alpha - A|_p \geqslant |P(A)|_p \frac{1}{\max\limits_{A \in \mathbb{Z}} |F(\alpha,A)|_p} .$$

Now, by the Product Formula,

$$|P(A)|_p \geqslant \frac{1}{|P(A)|} ,$$

and:

$$|P(A)| \leqslant \left(\sum_{0 \leqslant i \leqslant n} |a_i|\right) |A|^n,$$

therefore:

$$|\alpha - A|_p \geqslant \frac{c}{|A|^n} , \quad \text{where } c = \frac{1}{\max\limits_{A \in \mathbb{Z}} |F(\alpha,A)|_p} \frac{1}{\sum |a_i|} ,$$

and $c \neq 0$, since:

$$\max_{A \in \mathbb{Z}} |F(\alpha,A)|_p < +\infty,$$

because $\mathbb{Z} \subset \mathbb{Z}_p$, which is compact. The result is now proved.

(2): Let us assume that for an infinite number of values of $N \in \mathbb{N}$ there exists $A \in \mathbb{Z}$ ($A \neq \pm 1$) such that:

$$0 < |\alpha - A|_p < \frac{1}{|A|^n}.$$

This implies that for every value of $N \in \mathbb{N}$ there exists $A \in \mathbb{Z}$ ($A \neq \pm 1$) such that:

$$0 < |\alpha - A|_p < \frac{1}{|A|^N},$$

in fact $|A| > 1$, and therefore if $m < N$ we have:

$$0 < |\alpha - A|_p < \frac{1}{|A|^m}.$$

Lastly, this implies that for every constant $c \in \mathbb{R}_+ - \{0\}$ and for every $n \in \mathbb{N}$ there exists $A \in \mathbb{Z}^*$ such that:

$$0 < |\alpha - A|_p < \frac{c}{|A|^n},$$

so there exists $m(n)$ such that:

$$\frac{c}{|A|^n} \leqslant \frac{1}{|A|^{m(n)}}.$$

The latter is the opposite of the result proved in Question (1) above, therefore α is transcendental.

(3): We have:

$$|\alpha - \sum_{k=1}^{N} a_k p^{b_k}|_p = \frac{1}{p^{b_{N+1}}} .$$

Now,

$$\frac{p^{b_{N+1}}}{\left(\sum_{k=1}^{N} a_k p^{b_k}\right)^n} \geqslant \frac{p^{b_{N+1}}}{p^{nb_N+n}} = p^{b_{N+1}-nb_N-n} ,$$

and therefore by hypothesis there exists N large enough such that:

$$1 - n \frac{b_N}{b_{N+1}} - \frac{n}{b_{N+1}} > 0,$$

which implies that:

$$\frac{1}{p^{b_{N+1}}} < \frac{1}{\left(\sum_{k=1}^{N} a_k p^{b_k}\right)^n} .$$

Hence we have found an infinite number of integers n, and for each n an $A \in \mathbb{Z}$ such that:

$$0 < |\alpha - A|_p < \frac{1}{|A|^n} .$$

EXAMPLE: Clearly:

$$\frac{k!}{(k + 1)!} = \frac{1}{k + 1} \to 0.$$

SOLUTION 10·22: This is a matter of showing that the polynomial:

$$P(X) = X^m - \varepsilon$$

has a root in $\mathbb{Q}_p$. To the polynomial P we associate its image $\bar{P}$ in $\mathbb{F}_p[X]$. We have

$$\bar{P}(X) = X^m - 1.$$

The identity:

$$X^m - 1 = (X - 1)(X^{m-1} + X^{m-2} + \cdots + 1)$$

shows that 1 is a root of the second polynomial only if $m \not\equiv 0 \bmod p$, the polynomials:

$$X - 1 \quad \text{and} \quad X^{m-1} + \cdots + 1$$

are relatively prime in $\mathbb{F}_p[X]$. Hensel's Lemma allows us to conclude that $X^m - \varepsilon$ is divisible by a polynomial of the first degree in $\mathbb{Q}_p[X]$.

SOLUTION 10·23: (1): The inequality:

$$v_p(\gamma - 1) \geqslant f + 1$$

implies that:

$$\gamma = 1 + p^{f+1}\gamma_1 \quad \text{with } |\gamma_1|_p \leqslant 1.$$

Therefore:

$$\log\gamma = \log(1 + p^{f+1}\gamma_1)$$

is meaningful, and hence:

$$\left|\frac{\log\gamma}{m}\right|_p \leqslant \left|\frac{p^{f+1}}{m}\right| = \frac{1}{p}.$$

Consequently $e^{\log\gamma/m}$ makes sense, and we clearly have:

$$(e^{\log\gamma/m})^m = \gamma.$$

Therefore $e^{\log\gamma/m}$ is an m-th root of γ.

(2): We are going to separate the proof into two steps:

(a): If an extension K_p of $\mathbb{Q}_p$ contains $p^{1/m}$ and $\alpha^{1/m}$ for all $\alpha \in U_p$, then K_p contains $\beta^{1/m}$ for all $\beta \in \mathbb{Q}_p$.

In fact, if $\beta \in \mathbb{Q}_p$, $\beta = p^h\alpha$ where $\alpha \in U_p$. Therefore, since:

$$\beta^{1/m} = p^{h/m}\alpha^{1/m},$$

and since:

$$p^{h/m} = (p^{1/m})^h \in K_p,$$

as is $\alpha^{1/m}$, we deduce that $\beta^{1/m} \in K_p$.

(b): Let us show that there exists a finite extension of $\mathbb{Q}_p$ containing $\alpha^{1/m}$ for all $\alpha \in U_p$.

Let A be the following set:

$$A = \{a \in \mathbb{N} : a = a_0 + a_1 p + \cdots + a_f p^f\},$$

$$1 \leqslant a_0 \leqslant p - 1, \quad 0 \leqslant a_i \leqslant p - 1, \quad 1 \leqslant i \leqslant f.$$

If an extension K_p of $\mathbb{Q}_p$ contains $\alpha^{1/m}$ for all $a \in A$, then for all $\alpha \in U_p$ we can find $a \in A$ such that:

$$v_p\left(\frac{\alpha}{a} - 1\right) \geqslant p^{f+1},$$

therefore:

$$\left(\frac{\alpha}{a}\right)^{1/m} \in \mathbb{Q}_p,$$

by Question (1) above, and therefore $\alpha^{1/m} \in K_p$.

Now A contains $p^f(p - 1)$ elements. Therefore by adjoining to $\mathbb{Q}_p$ the m-th roots of the $p^f(p - 1)$-elements of A and $p^{1/m}$, we have an extension of $\mathbb{Q}_p$ containing the m-th root of every element of $\mathbb{Q}_p$. Therefore $mp^f(p - 1) + 1$ is an upper bound for the number of m-th roots of elements of $\mathbb{Q}_p$ to be adjoined to $\mathbb{Q}_p$ so as to obtain an extension K_p of $\mathbb{Q}_p$ containing all the m-th roots of the elements of $\mathbb{Q}_p$.

SOLUTION 10·24: (1): Because of the equality:

$$\beta = b + p\beta_1 \qquad (v_p(\beta_1) \geqslant 0)$$

we have:

$$\beta^m = b^m + \binom{m}{1} p\beta_1 b^{m-1} + \cdots + \binom{m}{k} p^k \beta_1^k b^{m-k} + \cdots + p^m \beta_1^m.$$

Now the equality:

$$\binom{m}{k} = \binom{m-1}{k-1}\frac{m}{k}$$

implies:

$$v_p(\binom{m}{k}) \geqslant v_p(m) - v_p(k),$$

and therefore for $k \geqslant 1$:

$$v_p[\binom{m}{k}\beta_1^k p^k b^{m-k}] \geqslant v_p(m) + k - v_p(k).$$

Now, if $k = p^h a$, with $(a,p) = 1$, we have:

$$v_p(k) = h \leqslant \log_p k \leqslant k - 1,$$

because the curve $x \to \log_p x - x + 1$ has a maximum on $\mathbb{N}^*$ at $x = 1$.

Therefore:

$$v_p(\binom{m}{k}\beta_1^k p^k b^{m-k}) \geqslant v_p(m) + 1 = f + 1 \quad \text{for all } k,\ 1 \leqslant k \leqslant m.$$

Hence we have proved that:

$$\beta^m = b^m + p^{f+1}\beta' \quad \text{with } v(\beta') \geqslant 0.$$

(2): Let us assume that there exists $\beta \in \mathbb{Q}_p$ such that $\beta = \alpha^{1/m}$; the equality $v(\alpha) = 0$ implies $v(\beta) = 0$, and therefore β can be written:

$$\beta = b + p\beta_1 \quad \text{with } 1 \leqslant b \leqslant p - 1,$$

and $v(\beta_1) \geqslant 0$; the preceding question shows that:

$$\alpha = \beta^m = b^m + p^{f+1}\beta' \quad \text{with } v(\beta') \geqslant 0,$$

whence the congruence:

$$b^m \equiv a \bmod p^{f+1}.$$

Conversely, if the congruence:

$$b^m \equiv a \bmod p^{f+1}$$

holds then α can be written:

$$\alpha = b^m + p^{f+1}\alpha' \quad \text{with } v_p(\alpha') \geqslant 0,$$

and since $v_p(b) = 0$ we have:

$$\frac{\alpha}{b^m} = 1 + p^{f+1}\alpha'' \quad \text{with } v_p(\alpha'') \geqslant 0.$$

Then by Exercise 10·23(1):

$$\left(\frac{\alpha}{b^m}\right)^{1/m} \in \mathbb{Q}_p$$

and $\alpha^{1/m}$ therefore belongs to $\mathbb{Q}_p$.

SOLUTION 10·25: (1): By writing $\gamma = c + p\gamma_1$ with $0 \leqslant c \leqslant p - 1$ and $v_p(\gamma_1) \geqslant 0$, we obtain:

$$\gamma^p = c^p + p^2\gamma_2 \quad \text{with } v_p(\gamma_2) \geqslant 0$$

(since $v_p(\binom{p}{k}) \geqslant 1$ for $1 \leqslant k \leqslant p - 1$). Fermat's little Theorem shows that $c^p \equiv c \bmod p$, therefore:

$$v_p(\gamma^p - \gamma) \geqslant 1.$$

Let us set:

$$H = \gamma^{kp} - \gamma^k - k\gamma^{k-1}(\gamma^p - \gamma)$$

It follows that:

$$H = \gamma^k(\gamma^{p-1} - 1)[\gamma^{(p-1)(k-1)} - 1 + \cdots + \gamma^{p-1} - 1].$$

The term inside the square brackets contains $(\gamma^{p-1} - 1)$ as a factor, therefore:

$$H = \gamma^k(\gamma^{p-1} - 1)^2 H_1 \quad \text{with } v_p(H_1) \geqslant 0,$$

hence $v_p(H) \geqslant 2$ when $k \geqslant 2$. (For $k = 1$, $H = 0$).

(2): The preceding inequality shows that if, for an integer satisfying $1 \leqslant k \leqslant p - 1$, the inequality:

$$v_p(\rho^{kp} - \rho^k) \geqslant 2$$

is satisfied, then:

$$v_p(\rho^p - \rho) \geqslant 2.$$

Now,

$$v_p(\rho^k \beta^p - \rho^{k'} \beta'^p) \geqslant 2$$

implies the inequality:

$$v_p(\rho^h - \delta^p) > 2$$

if $k > k'$, and we set $k - k' = h$ and $\delta = \beta'/\beta$.

By writing δ, which is a unit of $\mathbb{Q}_p$, in the form:

$$\delta = d_0 + p\delta_1 \quad \text{with } 1 \leqslant d_0 \leqslant p - 1 \quad \text{and} \quad v_p(\delta_1) \geqslant 0,$$

we deduce from this:

$$\delta^p = d_0^p + p^2\delta_2 \quad (v_p(\delta_2) \geqslant 0).$$

From the inequality:

$$v_p(\rho^h - \delta^p) \geqslant 2$$

we deduce:

$$\rho^h = d_0^p + p^2\eta_1$$

with $v_p(\eta_1) \geqslant 0$ and $\rho^h = d_0 + p\eta'$, with $v(\eta') \geqslant 0$. Hence:

$$\rho^{hp} = d_0^p + p^2\eta_2 \quad \text{with} \quad v(\eta_2) \geqslant 0,$$

and therefore:

$$v_p(\rho^{hp} - \rho^h) \geqslant 2 \quad \text{with } 1 \leqslant h \leqslant p - 1.$$

(3): For $p = 2$, that $r = 3$ is obvious. If $p \neq 2$ let us consider $(p - 1)^{p-1}$:

$$(p - 1)^{p-1} = p^{(p-1)} + \sum_{j=1}^{p-2} \binom{p-1}{j} p^{p-1-j}(-1)^j + 1,$$

all the terms $\binom{p-1}{j} p^{p-1-j}$ are divisible by p^2 for $1 \leqslant j \leqslant p - 3$, and for $j = p - 2$ the corresponding term is divisible by p and not by p^2. Therefore $v_p((p - 1)^{p-1} - 1) = 1$. If there were k,b,k',b' satisfying the hypotheses of the Problem and such that $r^k b^p \equiv r^{k'} b'^p \bmod p^2$, we would then have:

$$v_p(r^{p-1} - 1) \geqslant 2$$

by Question (2) above).

Then letting ϑ be a unit of $\mathbb{Q}_p$, the numbers $r^k b^p$ forming, modulo p^2, a complete system of residues, there exist b and k with $1 \leqslant b \leqslant p - 1$ and $0 \leqslant k \leqslant p - 1$ such that:

$$v(r^k b^p - \vartheta) \geqslant 2,$$

or again:

$$v\left(b^p - \frac{\vartheta}{r^k}\right) \geqslant 2.$$

By Exercise 10·24:

$$\left(\frac{\vartheta}{r^k}\right)^{1/p} \in \mathbb{Q}_p,$$

therefore $\vartheta \in \mathbb{Q}_p(\sqrt[p]{r})$.

SOLUTION 10·26: (1): Let us set:

$$\gamma = c_0 + p\gamma_1 \quad \text{with } 1 \leqslant c_0 \leqslant p - 1 \text{ and } v_p(\gamma_1) \geqslant 0.$$

As $X^2 - \gamma$ is irreducible over $\mathbb{Q}_p$, c_0 is not a quadratic residue mod p (cf., Exercise 10·24(2)). Let $\delta \in U_p$ be such that:

$$\delta = d_0 + p\delta_1, \qquad 1 \leqslant d_0 \leqslant p - 1 \qquad \text{and} \qquad v_p(\delta_1) \geqslant 0.$$

First: d_0 may be a quadratic residue mod p, and then (cf., Exercise 10·24(2)) the polynomial $X^2 - \delta$ is reducible on $\mathbb{Q}_p$; otherwise d_0 is not a quadratic residue mod p, but then c_0d_0 is a quadratic residue mod p (this is a property of the Legendre symbol), and therefore $X^2 - \gamma\delta$ is reducible in $\mathbb{Q}_p[X]$, hence $\sqrt{\gamma\delta} \in \mathbb{Q}_p$, and therefore $\sqrt{\delta} \in \mathbb{Q}_p[\sqrt{\gamma}]$.

The polynomial $X^2 - \delta$ is therefore reducible over $\mathbb{Q}_p[\sqrt{\gamma}]$.

(2): Let:

$$P(X) = X^2 + 2\lambda X + \mu$$

be a polynomial of $\mathbb{Q}_p[X]$; put it into the canonical form:

$$P(X) = (X + \lambda)^2 + \mu - \lambda^2.$$

Let us set:

$$\mu - \lambda^2 = p^k\varepsilon \quad \text{where } k \in \mathbb{Z} \text{ and } \varepsilon \in U_p.$$

If $k = 0$, then $\mu - \lambda^2 \in U_p$, and therefore the polynomial P is reducible over $\mathbb{Q}_p[\sqrt{\gamma}]$.

If $k \neq 0$, and $k = 2h$, where $h \in \mathbb{Z}$, then:

$$\mu - \lambda^2 = p^{2h}\varepsilon,$$

therefore:

$$\sqrt{\mu - \lambda^2} = p^h\sqrt{\varepsilon},$$

and $P(X)$ is still reducible on $\mathbb{Q}_p[\sqrt{\gamma}]$.

If $k \neq 0$, and $k = 2h + 1$, then:

$$\sqrt{\mu - \lambda^2} = p^h\sqrt{\varepsilon}\sqrt{p},$$

and $P(X)$ is reducible over $\mathbb{Q}_p[\sqrt{\gamma}][\sqrt{p}]$. It is obvious that:

$$\mathbb{Q}_p[\sqrt{\gamma} + \sqrt{p}] \subset \mathbb{Q}_p[\sqrt{\gamma}][\sqrt{p}].$$

But conversely, since:

$$2\sqrt{\gamma p} = [\sqrt{\gamma} + \sqrt{p}]^2 - \gamma - p,$$

we see that:

$$\sqrt{\gamma p} \in \mathbb{Q}_p[\sqrt{\gamma} + \sqrt{p}],$$

and that:

$$\sqrt{\gamma p}(\sqrt{\gamma} + \sqrt{p}) \in \mathbb{Q}_p[\sqrt{\gamma} + \sqrt{p}],$$

and from this we deduce that:

$$\sqrt{\gamma} \in \mathbb{Q}_p[\sqrt{\gamma} + \sqrt{p}],$$

the same holds for $\sqrt{p}$.

$$\mathbb{Q}_p[\sqrt{\gamma}][\sqrt{p}] = \mathbb{Q}_p[\sqrt{\gamma} + \sqrt{p}].$$

The splitting field of all the second degree polynomials of $\mathbb{Q}_p[X]$ is a fourth degree extension of $\mathbb{Q}_p$ defined by the minimal polynomial of $\sqrt{\gamma} + \sqrt{p}$, which is:

$$X^4 - 1(\gamma + p)X^2 + (\gamma - p)^2 = 0.$$

SOLUTION 10·27: (1): We can either use Exercise 10·23(1), or we may use a slightly different method.

By hypothesis we have $a = 1 + 2^3 b$ where $|b|_2 \leqslant 1$. Let us set $x = 1 + 2y$. The equation becomes:

$$y^2 + y - 2b = 0.$$

Passing to the residue field the equation becomes:

$$\bar{y}(\bar{y} + \bar{1}) = 0.$$

Hensel's Lemma then implies that the polynomial $y^2 + y - 2b$ factors in $\mathbb{Q}_2$, and therefore that a is a perfect square. Conversely, if a is a perfect square, then:

$$|a|_2 = 1 \quad \text{and} \quad a = (1 + 2c)^2 \quad \text{where } |c|_2 \leqslant 1.$$

Therefore:

$$a - 1 = 4c(1 + c).$$

Now in $\mathbb{F}_2$ we have:

$$\bar{c}(1 + \bar{c}) = \bar{0} \qquad \text{for } \bar{c} = \bar{0} \text{ or } \bar{1},$$

therefore:

$$|c(1 + c)|_2 < \tfrac{1}{2},$$

and therefore:

$$|a - 1|_2 \leqslant (\tfrac{1}{2})^3.$$

(2): Let us assume that $|a|_2 = |b|_2 = 1$. The extensions determined by $X^2 - a$ and $X^2 - b$ are identical if and only if:

$$\left|\frac{a}{b} - 1\right|_2 \leqslant (\tfrac{1}{2})^3 \qquad \text{or} \qquad |a - b|_2 \leqslant (\tfrac{1}{2})^3,$$

or again:

$$b \equiv a \bmod 2^3\mathbb{Z}_2.$$

Now $\mathbb{Z}_2^*/2^3\mathbb{Z}_2$ has four elements and one can choose 1,3,5,7 as representatives in $\mathbb{Z}_2$. The extensions of $\mathbb{Q}_2$ by $X^2 - 1$, $X^2 - 3$, $X^2 - 5$, $X^2 - 7$ are therefore distinct. Let us note that the extension of $\mathbb{Q}_2$ by $X^2 - 1$ is $\mathbb{Q}_2$.

Assume that $|a|_2 = |b|_2 = \frac{1}{2}$. We then obtain four distinct extensions defined by $X^2 - 2$, $X^2 - 6$, $X^2 - 10$, $X^2 - 14$, and each of these extensions is ramified, for the polynomial which determines it is an Eisenstein polynomial. In fact $\mathbb{Q}_2[\sqrt{a}] = \mathbb{Q}_2[\sqrt{b}]$ if and only if $\mathbb{Q}_2[\sqrt{a}/2] = \mathbb{Q}_2[\sqrt{b}/2]$, and we finish using the result proved in the first part of this Exercise. Therefore altogether we have seven disinct quadratic extensions of $\mathbb{Q}_2$.

(3): In $X^2 - 3$ and $X^2 - 7$ let us make the change of variable $X = 1 + Y$, which gives the Eisenstein polynomials $Y^2 + 2Y - 2$ and $Y^2 + 2Y - 6$.

(4): Making the change of variable $X = 1 + 2Y$ in $X^2 - 5$ gives $Y^2 - Y - 1$. In $\mathbb{F}_2$ this polynomial becomes $\bar{y}^2 + \bar{y} + 1$, which does not have roots in $\mathbb{F}_2$; it therefore defines an extension of degree 2 of $\mathbb{F}_2$. The inertial degree f is therefore two, and the ramification index e is one, because ef is equal to the degree of the extension of $\mathbb{Q}_2$ by $X^2 - 5$.

In conclusion, we note that there are six ramified extensions and one non-ramified extension of $\mathbb{Q}_2$ of degree two.

SOLUTION 10·28: (1): Let us examine the polynomial $X^2 - X - \bar{1}$ in $\mathbb{F}_p$. If $p = 11$ we want to show that:

$$X^2 - X - \bar{1} = (X - \bar{a})(X - \bar{b}) \quad \text{with } \bar{a}\bar{b} = -1 \text{ and } \bar{a} + \bar{b} = 1.$$

Hence we must find two integers a,b such that:

$$ab \equiv 10 \bmod 11 \quad \text{and} \quad a + b \equiv 1 \bmod 11.$$

It suffices to take $a = 4$ and $b = 8$. By Hensel's Lemma the polynomial $X^2 - X - 1$ factors in $\mathbb{Q}_{11}$, and $\vartheta, \bar{\vartheta} \in \mathbb{Q}_{11}$. For $\mathbb{Q}_{19}$ we take $a = 5$ and $b = 15$, and the same result holds.

(2): If $p = 7$ let us show that $X^2 - X - \bar{1}$ has no roots in $\mathbb{Z}/7\mathbb{Z} = \mathbb{F}_7$. In fact it would be necessary to have two integers a,b such that $1 \leqslant a \leqslant 6$, $1 \leqslant b \leqslant 6$, $a + b = 1 \bmod 7$, and $ab \equiv 6 \bmod 7$, therefore:

$$b \equiv 8 - a \bmod 7 \quad \text{and} \quad a(1 - a) \equiv 6 \bmod 7,$$

which is impossible for $1 \leqslant a \leqslant 6$. The polynomial $X^2 - X - \bar{1}$ therefore determines an extension of degree 2 of $\mathbb{F}_7$, therefore the residual degree of $\mathbb{Q}_7[\vartheta]$ over $\mathbb{Q}_7$ is two, and its ramification index is one. Therefore the extension is not ramified.

(3): We have:

$$N(\vartheta - c) = \vartheta\bar{\vartheta} - c(\vartheta + \bar{\vartheta}) + c^2 = c^2 - c - 1.$$

Then:

$$v_5(c^2 - c - 1) = 1 \quad \text{if } c = 3 \text{ and } 0 \text{ otherwise.}$$

Making the change of variable $X = 3 + Y$, gives $Y^2 + 5Y + 5$. This polynomial is an Eisenstein polynomial. The extension $\mathbb{Q}_5[\vartheta]$ is therefore totally ramified over $\mathbb{Q}_5$.

SOLUTION 10·29: (1): Let us make the change of variable $X = 1 + Y$. For the equation $X^7 - 1$ satisfied by ϑ, this gives us:

$$Y(Y_1^6 + \binom{7}{1}Y^5 + \binom{7}{2}Y^4 + \binom{7}{3}Y^3 + \binom{7}{4}Y^2 + \binom{7}{5}Y + 7).$$

The polynomial:

$$Y^6 + \binom{7}{1}Y^5 + \cdots + 7$$

is an Eisenstein polynomial, it is therefore the irreducible polynomial for $(\vartheta - 1)$. Hence $\mathbb{Q}_7[\vartheta]$ is totally ramified and of degree six, and $\vartheta - 1$ is a uniformizer for $\mathbb{Q}_7[\vartheta]$, since:

$$v_7(\vartheta - 1) = \frac{1}{6}.$$

(2): In order that the equation $X^7 - 1$ be solvable in $\mathbb{Q}_p$ it is necessary that it be solvable in $\mathbb{F}_p$. Therefore $\mathbb{F}_p$ must contain the seventh roots of $\bar{1}$, and so $(\mathbb{F}_p)^*$ contains a subgroup of order seven, and hence that 7 divides $p - 1$. Conversely, if 7 divides $p - 1$ the polynomial $X^7 - \bar{1}$ factors in $\mathbb{F}_p$ as a product of seven factors of the first degree. In fact, the elements of $(\mathbb{F}_p)^*$ satisfy the equation:

$$X^{p-1} - \bar{1} = 0,$$

which therefore splits completely in $\mathbb{F}_p[X]$, and $X^7 - \bar{1}$ divides the polynomial $X^{p-1} - \bar{1}$. By Hensel's Lemma $X^7 - 1$ has seven distinct roots in $\mathbb{Q}_p$.

(3): If $\mathbb{Q}_p[\vartheta]$ is of degree two on $\mathbb{Q}_p$, the residue field of $\mathbb{Q}_p[\vartheta]$, $\mathbb{F}_{p^s}$ is of degree one or two over $\mathbb{F}_p$ (hence $s = 1$ or 2). In $\mathbb{F}_{p^s}$ we have:

$$\bar{\vartheta}^7 = \bar{1} \quad \text{and} \quad \bar{\vartheta}^{p^s-1} = \bar{1},$$

If $s = 1$, by the Question (2) preceding we have $\vartheta \in \mathbb{Q}_p$, which contradicts the hypotheses $[\mathbb{Q}_p[\vartheta]:\mathbb{Q}_p] = 2$. Therefore $s = 2$, and 7 divides $p^2 - 1$, so by Gauss's Lemma, 7 divides $p - 1$, or 7 divides $p + 1$; if 7 divides $p - 1$ then by part (2) above $\vartheta \in \mathbb{Q}_p$, hence 7 divides $p + 1$, or, again:

$$p \equiv -1 \bmod 7.$$

Conversely, if $p \equiv -1 \bmod 7$, then 7 divides $p + 1$, therefore 7 divides $p^2 - 1$, hence $X^7 - \bar{1}$ has seven *distinct* roots in $\mathbb{F}_{p^2}$, and therefore $X^7 - 1$ determines a non-ramified extension of degree two of $\mathbb{Q}_p$.

(4): Similarly to the preceding Question (3), in the residue field $\mathbb{F}_{p^2}$ of $\mathbb{Q}_p[\vartheta]$ we must have:

$$\bar{\vartheta}^7 = \bar{1} \quad \text{and} \quad \bar{\vartheta}^{p^s-1} = \bar{1},$$

therefore 7 divides $p^s - 1$. If the extension is of degree three, then $s = 1$, 2 or 3. Therefore 7 divides $p^3 - 1$, and so:

$$p^3 \equiv 1 \bmod 7,$$

which implies:

$$p \equiv 2 \text{ or } 4 \bmod 7.$$

Conversely, if $7|p^3 - 1$, the polynomial $X^7 - 1$, which determines the extension of $\mathbb{Q}_p$, defines an extension of the residue field of degree three. Therefore:

$$[\mathbb{Q}_p[\vartheta]:\mathbb{Q}_p] = 3,$$

since $X^7 - 1$ splits completely in $\mathbb{Q}_p[\vartheta]$, by Hensel's Lemma.

(5): If $[\mathbb{Q}_p[\vartheta]:\mathbb{Q}_p] = 6$, then necessarily:

$$p \equiv 3 \text{ or } 5 \mod 7,$$

for otherwise we are in the situation of parts (1),(2),(3) or (4). Conversely, if $p \equiv 3$ or $5 \mod 7$, the smallest s such that 7 divides $p^s - 1$ is $s = 6$. Therefore the polynomial $X^7 - 1$ determines an extension of degree six of $\mathbb{F}_p$, since we always have $[\mathbb{Q}_p[\vartheta]:\mathbb{Q}_p] \leqslant 6$. And so we have equality.

SOLUTION 10·30: (1): We notice that if $x \in \mathbb{N}$, then

$$\binom{x}{k} = P_k(x)$$

is a rational integer, therefore:

$$|P_k(x)|_p \leqslant 1 \quad \text{if } x \in \mathbb{N} \subset \mathbb{Z}_p$$

(cf., Exercise 10·14). Now $\mathbb{N}$ is dense in $\mathbb{Z}_p$, therefore:

$$\sup_{x \in \mathbb{Z}_p} |P_k(x)|_p = 1,$$

and it suffices to set $x = k$.

If $(a_k)_{k\geqslant 0}$ is a sequence of elements of $\mathbb{C}_p$ such that:

$$\lim_k |a_k|_p = 0,$$

then the series:

$$\sum_{k\geq 0} a_k P_k(x)$$

converges uniformly on $\mathbb{Z}_p$, and consequently it converges uniformly on $\mathbb{Z}_p$ towards $f \in C(\mathbb{Z}_p, \mathbb{C}_p)$ since, for all $n \in \mathbb{N}$:

$$\sum_{k=0}^{n} a_k P_k(x) \in C(\mathbb{Z}_p, \mathbb{C}_p).$$

(2): We have:

$$F(X) = \sum_{n\geq 0} f(n) X^n.$$

Now $f \in C(\mathbb{Z}_p, \mathbb{C}_p)$, therefore for all $\varepsilon > 0$ there exists $h \in \mathbb{N}$ such that:

$$|n - n'|_p \leqslant p^{-h}$$

implies:

$$|f(n) - f(n')|_p < \varepsilon.$$

Therefore if we set:

$$F_h(x) = \frac{\sum_{n=0}^{p^h-1} f(n) X^n}{1 - X^{p^h}},$$

we have:

$$\|F_h - F\|_{B(0,1^-)} \leqslant \varepsilon.$$

Notice that the sequence $h \to F_h$ converges uniformly on $\mathbb{C}_p - B(1,1^-)$. In fact:

$$F_{h+1}(X) - F_h(X) = \frac{\sum_{k=0}^{p-1} \sum_{n=0}^{p^h-1} (f(n + kp^h) - f(n)\ X^{n+kp^h}}{1 - X^{p^{h+1}}},$$

and on $\mathbb{C}_p - B(1,1^-)$ we have:

$$|1 - X^{p^{h+1}}|_p = 1 \quad \text{if } |X|_p \leqslant 1,$$

or:

$$|1 - X^{p^{h+1}}|_p = |X|_p^{p^{h+1}} \quad \text{if } |X|_p > 1$$

(cf., Exercise 10·18). Therefore:

$$\|F_{h'+1} - F_{h'}\|_{\mathbb{C}_p - B(1,1^-)} \leqslant \varepsilon$$

whenever $h' \geqslant h$. The sequence $(F_h)_{h\geqslant 0}$ of rational functions therefor converges on $\mathbb{C}_p - B(1,1^-)$ towards an analytic element that again is denoted by F. So there exist coefficients c_j such that for $|X - 1| \geqslant 1$ we have:

$$F(X) = \sum_{j\geqslant 0} \frac{c_j}{(1 - X)^{j+1}} \text{ with } \lim_{j\to\infty}|c_j| = 0.$$

Expanding the second member as a Taylor series in a neighbourhood of zero, and identifying it with:

$$F(X) = \sum_{n\geqslant 0} f(n)X^n,$$

we deduce that:

$$f(n) = \sum_{j\geqslant 0} c_j \binom{n + j - 1}{j} \quad \text{for all } n \in \mathbb{N},$$

and therefore, as $f \in C(\mathbb{Z}_p, \mathbb{C}_p)$,

$$f(X) = \sum_{j \geq 0} c_j P_j(x + j - 1).$$

From the identity:

$$P_j(X + j - 1) = \sum_{k=1}^{j} \binom{j-1}{j-k} P_k(X),$$

we deduce:

$$f(X) = \sum_{k \geq 0} a_k P_k(x)$$

with:

$$a_k = \sum_{j \geq k} c_j \binom{j-1}{j-k} \quad \text{and} \quad \lim_{k \to \infty} |a_k| = 0.$$

SOLUTION 10·31: (1): We have:

$$\frac{1}{p^n}\binom{p^n}{h} = \frac{1}{h}\binom{p^n - 1}{h - 1} .$$

Let us set:

$$\ell(n) = \left[\frac{\log n}{\log p}\right] .$$

Then:

$$\binom{p^n - 1}{h - 1} = (-1)^{h-1} + p^{n-\ell(h-1)}\lambda_n \quad \text{where } \lambda_n \in \mathbb{Z},$$

as is easily seen using the identity:

$$\binom{p^n - 1}{h - 1} = \sum_{k=0}^{h-1} \binom{p^n}{h - 1 - k}\binom{-1}{k} .$$

Therefore:

$$\lim_{n\to\infty} \frac{1}{p^n}\binom{p^n}{h} = \frac{(-1)^{h-1}}{h} .$$

(2): The p-adic absolute values of the zeros of A_n are, on the one hand,

$$\left(\frac{1}{p}\right)^{1/(p-1)p^h} \quad \text{with } 0 \leqslant h < n,$$

and 0 on the other hand (cf., Exercise 10·18). $A_n(x)$ is not irreducible on $\mathbb{Q}_p$, because:

$$A_n(x) = \frac{1}{p^n} [(1 + x)^{p^{n-1}(p-1)} + (1 + x)^{p^{n-1}(p-2)} + \cdots + 1]$$

$$\times [(1 + x)^{p^{n-2}(p-1)} + \cdots + 1]\cdots$$

$$\times [(1 + x)^{p-1} + \cdots + 1]x.$$

If:

$$r \neq p^{-m_j} \quad \text{with} \quad m_j = \frac{1}{p^j(p - 1)} ,$$

then:

$$|A_n(x)| = \left|\frac{1}{p^n}\binom{p^n}{p^{j+1}} r^{p^{j+1}}\right| = p^{-j-1} r^{p^{j+1}} ,$$

where j is defined by:

$$p^{-m_j} \geqslant |x| > p^{m_j+1};$$

this is seen either by considering the Newton polygon of A_n (cf., [Ami]), or by writing:

$$A_n(x) = \frac{1}{p^n} \prod_{\gamma \in \Gamma} (x + 1 - \gamma),$$

where Γ_h is the group of p-th roots of unity, and using Exercise 10·18.

$\lim_{n\to\infty} A_n(x)$ exists for all $x \in \mathbb{C}_p$ such that $|x|_p < 1$, as the Taylor series of the $A_n(x)$'s converge coefficientwise towards the series:

$$\sum_{h \geqslant 1} \frac{(-1)^{h-1}}{h} x^h$$

and the $A_n(x)$'s are bounded for fixed x independently of n, $|x|_p < 1$. We have:

$$f(x - 1) = \lim_{n\to\infty} A_n(x - 1),$$

$$f(y - 1) = \lim_{n\to\infty} A_n(y - 1),$$

$$f(xy - 1) = \lim_{n\to\infty} A_n(xy - 1),$$

Now,

$$A_n(x - 1) + A_n(y - 1) - A_n(xy - 1) = - \frac{(x^{p^n} - 1)(y^{p^n} - 1)}{p^n},$$

since:

$$\left|\frac{x^{p^n} - 1}{p^n}\right|_p \leqslant M(x),$$

where M does not depend upon n and:

$$\lim_{n\to\infty} |y^{p^n} - 1|_p = 0,$$

and we have:

$$\lim_{n\to\infty}\{A_n(x - 1) + A_n(y - 1) - A_n(xy - 1)\} = 0,$$

The radius of convergence of the series

$$\sum_{k\geqslant 0} \frac{(-1)^{h-1}}{h} x^h$$

is one, this is seen either because the sequence $A_n(x)$ converges for all $x \in \mathbb{C}_p$ such that $|x|_p < 1$, or because:

$$\overline{\lim} \sqrt[h]{\left|\frac{1}{h}\right|_p} = 1.$$

(3):

$$B_n(x) = \frac{A_n(x)}{A_{n-1}(x)} = (1 + x)^{p^{n-1}(p-1)} + \cdots + 1$$

is irreducible on $\mathbb{Q}_p$, since it is an Eisenstein polynomial. The roots of $B_n(x - 1)$ are the primitive p^n-th roots of unity, for these are the zeros of $A_n(x - 1)$ which are not zeros of $A_{n-1}(x - 1)$. If $x - 1$ is a p^n-th root of unity, then for $m \geqslant n$ we have $A_m(x) = 0$, and therefore:

$$f(x) = \lim_{n\to\infty} A_n(x) = 0.$$

The p^h-th roots of unity ($h \geqslant 0$) are therefore roots of:

$$f(x - 1) = 0.$$

Conversely, let $x \notin \bigcup_h \Gamma_h$, then $|A_n(x)|$ is constant for n large enough, and it is different from zero, therefore:

$$\lim_{n\to+\infty} A_n(x) \neq 0.$$

The zeros of $f(x - 1)$ are therefore exactly the p^h-th roots of unity for $h \geqslant 0$.

SOLUTION 10·32: (1): It suffices to prove that if f_n is defined for all $x \in \mathbb{Z}_p$ then so is f_{n+1}, and therefore it suffices to verify that $(x_{n+1}, f_n(x_{n+1}))$ belong to the domain of definition of F, that is to say that $f_n(x_{n+1}) \in \mathbb{Z}_p$. Now, we have:

$$f_n(x_{n+1}) = f_{n-1}(x_n) + (x_{n+1} - x_n)F(x_n, f_{n-1}(x_n)),$$

therefore:

$$f_n(x_{n+1})_p \leqslant \max(|f_{n-1}(x_n)|_p, |x_{n+1} - x_n|_p |F(x_n, f_{n-1}(x_n))|_p)$$

$$\leqslant \max(1, Mp^{-(n+1)}),$$

and as $n \geqslant k$ we have:

$$|f_n(x_{n+1})|_p \leqslant 1.$$

(2): By replacing $f_{n+1}(x)$ and $f_n(x)$ by their expressions, we obtain:

$$f_{n+1}(x) - f_n(x) = (x_{n+1} - x)$$

$$\times [F(x_n, f_{n-1}(x_n)) - F(x_{n+1}, f_n(x_{n+1}))],$$

whence the upper bound:

$$|f_{n+1}(x) - f_n(x)|_p \leqslant Mp^{-(n+2)}.$$

The series:

$$\sum_{n \geqslant k} (f_{n+1}(x) - f_n(x))$$

is therefore uniformly convergent for $x \in \mathbb{Z}_p$. Therefore $\lim_{n\to\infty} f_n(x)$ exists and is a continuous function on $\mathbb{Z}_p$. From the ultrametric inequality we deduce that if $m > n \geqslant k$ we have:

$$|f_m(x) - f_n(x)| \leqslant Mp^{-(n+2)},$$

and therefore that:

$$|f(x) - f_n(x)| \leqslant Mp^{-(n+2)},$$

Furthermore, as:

$$f_n(0) = f_{n-1}(0) \quad \text{for all } n \geqslant k,$$

we have:

$$f(0) = 0.$$

(3): (a): The inequality:

$$|x - x'|_p \leqslant p^{-(n+1)}$$

implies $x_n = x'_n$, therefore:

$$f_n(x) - f_n(x') = f_{n-1}(x_n) - f_{n-1}(x'_n) + (x - x_n)F(x_n, f_{n-1}(x_n))$$

$$- (x' - x'_n)F(x'_n, f_{n-1}(x'_n)) = \text{(Contd)}$$

(Contd) $$= (x - x')F(x_n, f_{n-1}(x_n)).$$

On $\mathbb{Z}_p \times \mathbb{Z}_p$, which is compact, $F(x,y)$ is uniformly continuous. Hence for all $\varepsilon > 0$ there exists $\eta > 0$ such that:

$$|x - x'|_p \leqslant \eta \quad \text{and} \quad |y - y'|_p \leqslant \eta$$

implies:

$$|F(x',y') - F(x,y)|_p \leqslant \varepsilon.$$

Let $N \in \mathbb{N}$ be such that

$$Mp^{-(N+1)} \leqslant \eta \quad \text{and} \quad p^{-(N+1)} \leqslant \eta,$$

then:

$$|f(x) - f_{n-1}(x_n)|_p \leqslant \max(|f(x) - f_n(x)|_p, |f_n(x) - f_{n-1}(x_n)|_p)$$

$$\leqslant \max(Mp^{-(N+2)}, Mp^{-(N+1)}) \leqslant \eta.$$

Therefore, for $n \geqslant N$ we have:

$$|x - x_n|_p \leqslant \eta \quad \text{and} \quad |f(x) - f_{n-1}(x_n)|_p \leqslant \eta,$$

hence:

$$|F(x,f(x)) - F(x_n, f_{n-1}(x_n))|_p \leqslant \varepsilon.$$

(b): The equality:

$$f_m(x) - f_m(x') = f_m(x) - f_n(x) + f_n(x)$$

$$- f_n(x') + f_n(x') - f_m(x')$$

implies:

$$\left|\frac{f_m(x) - f_m(x')}{x - x'} - F(x,f(x))\right|_p$$

$$\leqslant \max\left(\left|\frac{f_n(x) - f_n(x')}{x - x'} - F(x,f(x))\right|_p, \left|\frac{f_m(x) - f_n(x)}{x - x'}\right|_p,\right.$$

$$\left.\left|\frac{f_n(x') - f_m(x')}{x - x'}\right|_p\right) .$$

Now, the equality:

$$f_{n+1}(x) - f_n(x) = (x_n - x_{n+1})$$

$$\times [F(x_{n+1},f_n(x_{n+1})) - F(x_n,f_{n-1}(x_n))]$$

implies:

$$|f_{n+1}(x) - f_n(x)|_p < \varepsilon|x - x_{n+1}|_p < \varepsilon|x - x'|_p$$

and therefore for $m > n$:

$$|f_m(x) - f_n(x)|_p \leqslant \varepsilon|x - x'|_p$$

similarly:

$$|f_m(x') - f_n(x')|_p \leqslant \varepsilon|x - x'|_p$$

Hence for $m > n$ the inequality

$$|x - x'|_p \leqslant p^{-(n+1)}$$

implies:

$$\left|\frac{f_m(x) - f_m(x')}{x - x'} - F(x,f(x))\right|_p \leqslant \varepsilon.$$

By making $m \to \infty$ we thus obtain:

$$|x - x'| \leqslant p^{-(n+1)}$$

which implies:

$$\left|\frac{f(x) - f(x')}{x - x'} - F(x,f(x))\right| \leqslant \varepsilon,$$

hence:

$$\lim_{x' \to x} \frac{f(x) - f(x')}{x - x'} = F(x,f(x)).$$

Therefore f is differentiable on $\mathbb{Z}_p$, and:

$$f'(x) = F(x,f(x)).$$

SOLUTION 10·33: (1): A formal calculation shows that:

$$\sum_{n \geqslant 0} P_n \frac{X^n}{n!} = \sum_{n \geqslant 0} \frac{(e^X - 1)^n}{n!},$$

and therefore that:

$$\sum_{n \geqslant 0} P_n X^n = \sum_{n \geqslant 0} \frac{X^n}{(1 - X) \cdots (1 - nX)}.$$

(Decompose into partial fractions).

Modulo $p^h \mathbb{Z}[[X]]$ we have:

$$\sum_{n \geqslant 0} \frac{X^n}{(1 - X) \cdots (1 - nX)}$$

$$= \sum_{n=0}^{p^h - 1} \frac{X^{nn}}{(1 - X) \cdots (1 - nX)} \left[\sum_{\lambda \geqslant 0} \frac{X^{\lambda p^h}}{[(1 - X) \cdots (1 - (p^h - 1)X]^{\lambda}}\right] =$$

$$= \frac{\sum_{n=0}^{p^h-1} X^n(1-(n+1)X)\cdots(1-(p^h-1)X)}{(1-X)\cdots(1-(p^h-1)X) - X^{p^h}}.$$

Therefore $\sum_{n\geqslant 0} P_n X^n$ is a rational function modulo $p^h\mathbb{Z}[[X]]$, and so mod p^h the $(P_n)_{n\geqslant 0}$ form a linear recurrent sequence (cf., [Ami]).

(2): Modulo p we have:

$$\sum_{n\geqslant 0} P_n X^n = \frac{\sum_{n=0}^{p-1} X\ (1-(n+1)X)\cdots(1-(p-1)X)}{(1-X)\cdots(1-(p-1)X) - X^p}$$

$$= \frac{\sum_{n=0}^{p-1} X^n(1-(n+1)X)\cdots(1-(p-1)X)}{1 - X^{p-1} - X^p}.$$

The polynomial $1 - X^{p-1} - X^p$ is irreducible over $\mathbb{Q}_p$, for it is irreducible over $\mathbb{F}_p$. In the algebraic closure of $\mathbb{F}_p$ its roots satisfy:

$$\zeta^{1+p+\cdots+p^{p-1}} = 1.$$

(It suffices to calculate the norm of ζ to $\mathbb{F}_p$). Therefore:

$$\zeta^{k(p)} = 1,$$

and therefore in $\mathbb{C}_p$ its roots satisfy:

$$|\zeta^{k(p)} - 1|_p \leqslant \frac{1}{p}.$$

In addition, if ζ_i and ζ_j are two distinct roots of $1 - X^{p-1} - X^p$, then:

$$|\zeta_i - \zeta_j| = 1,$$

and the Mittag-Leffler Theorem (cf., [Ami]) then gives the result:

$$F_1(X) = \sum_{n=0}^{p-1} \frac{X^n(1 - (n + 1)X)\cdots(1 - (p - 1)X)}{1 - X^{p-1} - X^p}$$

$$= \sum_{i=1}^{p} \frac{\lambda_i}{1 - \zeta_i X}$$

where $|\lambda_i| \leqslant 1$. Consequently:

$$P_n = P_{n+k(p)} \bmod p.$$

Bibliography

[Ami] Y. AMICE: *Les nombres p-adiques*, Collection Sup. P.U.F., (Paris), (1975).

[Bac] G. BACHMANN: *Introduction to p-adic Numbers and Valuation Theory*, Academic Press (NY, London), 1964.

[Bak] A. BAKER: *Transcendental Number Theory*, (Cambridge University Press), (1975). Cambridge, NY.

[Bla] A. BLANCHARD: *Initiation à la théorie analytique des nombres premiers*, (Dunod, Paris), (1969).

[Bou] N. BOURBAKI: *Algèbre, chapitres IV et V*, (2nd edn.), (Hermann, Paris), (1959).

[Bor] Z.I. BOREVITCH and I.R. CHAFAREVITCH: *Théorie des Nombres*, (French Translation), (Gauthier-Villars, Paris), (1967). English trans., Number Theory, Academic Press, 1966.

[Cas] J.W.S. CASSELS: *An Introduction to Diophantine Approximation*, (Cambridge Tract No. 45), (Cambridge Univ. Press) (1957). Cambridge and NY

[CaF] J.W.S. CASSELS and A. FRÖHLICH: *Algebraic Number Theory* (Academic Press), (1967). New York and London

[Che] C. CHEVALLEY: "Sur la théorie du corps de classes', *J. Fac. Sci. Tokyo*, **II**, (1933), 365-476.

[Com] L. COMTET: *Analyse combinatoire, Vol.s 1 and 2*, Collection Sup. P.U.F., (Paris), (1970).

[Ell] W.J. ELLISON and M. MENDES-FRANCE: *Les nombres premiers*, (Hermann, Paris), (1975).

[Gun] R.C. GUNNING: *Lectures on Modular Forms*, Annals of Mathematics Studies No. 48, (Princeton University Press), (1962). Princeton.

[Hal] H. HALBERSTAM and K.F. ROTH: *Sequences, Vol. I*, (Oxford University Press), (1966). Reprint Springer 1983

[Har] G.H. HARDY and E.M. WRIGHT: *An Introduction to the Theory of Numbers*, (4th edn.), (Oxford University Press), (1960). Oxford, New York

[Has] H. HASSE: *Zahlentheorie*, (Akademie Verlag), (1963). English trans., Number Theory, Springer-Verlag 1981.

[Joly] J.R. JOLY: 'Equations et variétés algébriques sur un corps fini', *Enseignement Mathématique*, **XIX**, No.s 1-2, (1973).

[Kui] L. KUIPERS and H. NIEDERREITER: *Uniform Distribution of Sequences*, (J. Wiley & Sons), (1974).

[Lang1] S. LANG: *Algebra*, (Addison Wesley), (1965). Reading, Mass.

[Lang2] S. LANG: *Introduction to Transcendental Numbers*, (Addison Wesley), (1966). Reading, Mass.

[Lev] W.J. LEVEQUE: *Topics in Number Theory*, (Addison Wesley), (1965). Reading, Mass.

[Mac] P.J. MacCARTHY: *Algebraic Extensions of Fields*, (Blaisdell), (1966). Reprint Chelsea 1981 (NY).

[Mor] L.J. MORDELL: *Diophantine Equations*, (Academic Press, London and New York), (1969).

[Niv] I. NIVEN and H. ZUCKERMAN: *An Introduction to the Theory of Numbers*, (3rd edn.), (J. Wiley and Sons), (1972).

[Pis] C. PISOT: 'Familles compactes de fractions rationelles et ensembles dermés de nombres algébriques', *Ann. Sci. Ecole Normale Sup., Ser. 3*, **81**, (1964), 165-188.

[Rau] G. RAUZY: *Propriêtês statistiques de suites arithmêtiques*, Collection Sup. P.U.F., (Paris), (1976).

[Sam] P. SAMUEL: *Thêorie algêbrique des nombres*, Collect Méthodes, (Hermann, Paris), (1967). Trans. Algebraic Theory of Numbers, Hermon 1969

[Sal] R. SALEM: *Algebraic Numbers and Fourier Analysis*, (Heath Mathematical Monographs, Boston), (1973).

[Sch] T. SCHNEIDER: *Introduction aux nombres transcendants*, (Gauthier-Villars, Paris), (1957).

[Ser1] J.P. SERRE: *Cours d'arithmêtique*, Collection Sup. P.U.F., (Paris), (1970). Trans. A Course of Arithmetic, Springer N.Y. and Heidelberg.

[Ser2] J.P. SERRE: *Corps locaux*, (Hermann, Paris), (1962). Trans. Local Fields, Springer-Verlag

[Wal] M. WALDSCHMIDT: *Nombres Transcendants*, Lecture Notes Series No. 42, (Springer-Verlag), (1974).

[Wei] E. WEISS: *Algebraic Number Theory*, (McGraw-Hill), (1963). Reprint, Chelsea, 1980.

[Wid] D.V. WIDDER: *The Laplace Transform*, (Princeton University Press), (1946). Princeton.

Index of Terminology

Index of Symbols and Notations

$\mathbb{N}$, $\mathbb{Z}$, $\mathbb{Q}$, $\mathbb{R}$, $\mathbb{C}$: The entire, natural, rational, real, complex numbers.

$\mathbb{N}^*$, $\mathbb{Z}^*$, $\mathbb{Q}^*$, $\mathbb{R}^*$, $\mathbb{C}^*$, The same sets, without zero.

$\mathbb{F}_p = \mathbb{Z}/p\mathbb{Z}$; $(\mathbb{Z}/n\mathbb{Z})^*$ = The set of invertible elements of $\mathbb{Z}/n\mathbb{Z}$.

$\binom{m}{n}$ Binomial coefficient

i, j, k, ℓ, m, d Elements of N

p, q Prime numbers, except for Ch. 9, where p/z denotes a convergent.

(m,n), [m,n] g.c.d. and l.c.m. of m and n

$d|n$; $d\nmid n$; $p^r \| n$; d divides n; d does not divide n, p^r exactly divides n

$[x]$ = entire part of x; $\{x\} = x - [x]$; $\|x\| = \min |x-a|$.

$\overline{K}$: Algebraic closure of the field K, or residue class field.

$\pi(x)$; $\theta(x)$ Ch. 1; 1

$d(n)$, $\sigma(n)$, $\varphi(n)$, $\nu(n)$, $\Omega(n)$, $\lambda(n)$, $\mu(n)$, $\Lambda(n)$, $z(n)$, $i(n)$ Ch. 1; 4, 7, 17